Astronomy and Astrophysics: Principles and Practices

Astronomy and Astrophysics: Principles and Practices

Edited by **Audria Baldwin**

R CALLISTO REFERENCE

New York

Published by Callisto Reference,
106 Park Avenue, Suite 200,
New York, NY 10016, USA
www.callistoreference.com

Astronomy and Astrophysics: Principles and Practices
Edited by Audria Baldwin

International Standard Book Number: 978-1-63239-737-9 (Hardback)

Contents

Preface

Astronomy as a branch of science refers to the study of celestial bodies like planets, stars, galaxies, asteroids, etc. and the processes that take place in the universe. Astrophysics is a sub-field of astronomy which uses the elements of physics and chemistry to study the nature of heavenly bodies like extra solar planets, cosmic wave background, sun, stars, etc. This book will discuss in detail the principles of both of these branches. It will bring forth some of the most innovative concepts and elucidate the unexplored aspects of this field. Different approaches, evaluations, methodologies and advanced studies on these subjects have been included in this text. It also consists of contributions made by scientists from across the globe. Those in search of information to further their knowledge will be greatly assisted by this book. It will prove to be a beneficial text for astrophysicists, astronomists, climatologists, researchers and students.

The information contained in this book is the result of intensive hard work done by researchers in this field. All due efforts have been made to make this book serve as a complete guiding source for students and researchers. The topics in this book have been comprehensively explained to help readers understand the growing trends in the field.

I would like to thank the entire group of writers who made sincere efforts in this book and my family who supported me in my efforts of working on this book. I take this opportunity to thank all those who have been a guiding force throughout my life.

Editor

Approximate Metric for a Rotating Deformed Mass

Francisco Frutos-Alfaro[1], Paulo Montero-Camacho[1], Miguel Araya[1], Javier Bonatti-González[2]

[1]Space Research Center and School of Physics, University of Costa Rica, San José, Costa Rica
[2]Nuclear Research Center and School of Physics, University of Costa Rica, San José, Costa Rica
Email: frutos@fisica.ucr.ac.cr

Abstract

A new Kerr-like metric with quadrupole moment is obtained by means of perturbing the Kerr spacetime. The form of this new metric is simple as the Kerr metric. By comparison with the exterior Hartle-Thorne metric, it is shown that it could be matched to an interior solution. This approximate metric may represent the spacetime of a real astrophysical object with any Kerr rotation parameter a and slightly deformed.

Keywords

Approximate Solutions of Einstein Equations, Kerr Metric, Erez-Rosen Metric, Quadrupole

1. Introduction

In 1963, R. P. Kerr [1] proposed a metric that describes a massive rotating object. Since then, a huge amount of papers about the structure and astrophysical applications of this spacetime appeared. Now, it is widely believed that this metric does not represent the spacetime of a rotating astrophysical object. In 1967, Hernández [2] stated that *reasonable perfect fluid type solutions which might serve as source of the Kerr metric may not exist*. In 1971, Thorne [3] [4] added that *because of the relationship between multipole moments and angular momentum, the Kerr solution cannot represent correctly the external field of any realistic stars*. Moreover, the Kerr metric has difficulties when matching it to a realistic interior metric according to [5]. However, there has been a considerable amount of efforts trying to match the Kerr metric with a realistic interior metric that represents a physical source, see for example [6]-[12]. For a concise and comprehensive review of the different methods that have been used in order to try and obtain an interior solution for the Kerr metric, see [13].

In [14] and [15], the Newman-Janis algorithm was applied to look for interior solutions. Drake and Turolla

[14] also propose a general method for finding interior solutions with oblate spheroidal boundary surfaces and note that the boundary surfaces reduce to a sphere in the case with no rotation, however Vaggiu in [15] argues that it is more helpful to start with the Schwarzschild interior and then proceed to the Kerr interior. Vaggiu uses an anisotropic conformally flat static interior and is led to interior Kerr solutions with oblate spheroidal boundary surfaces, additionally he points out that his procedure can be applied to find interior solutions matching with a general asymptotically flat vacuum stationary spacetime.

Other exact rotating solutions to the Einstein field equations (EFE) containing mass multipoles and magnetic dipole were obtained by [16]-[23]. In the first four articles, they used the Ernst formalism [24], while in the four last ones, the solutions were obtained with the help of the Hoenselaers-Kinnersley-Xanthopoulos (HKX) transformations [25]. These authors obtain new metrics from a given seed metric. These formalisms allow to include other desirable characteristics (rotation, multipole moments, magnetic dipole, etc.) to a given seed metrics. Furthermore, Quevedo in [26] not only presents an exact electrovacuum solution that can be used to describe the exterior gravitational field of a rotating charged mass distribution, but also considers the matching using the derivatives of the curvature eigenvalues, this leads to matching conditions from which one can expect to obtain the minimum radius at which the matching can be made.

In Nature, it is expected that astrophysical objects are rotating and slightly deformed as is pointed out in [27] and in [28]. In addition, Andersson and Comer in [27] use a two-fluid model for a neutron star, one layer with neutrons that has a differential rotation and another layer consisting of a solid crust with constant rotation. The aim of this article is to derive an appropriate analytical tractable metric for calculations in which the quadrupole moment can be treated as a perturbation, but for arbitrary angular momentum. Moreover, this metric should be useful to tackle astrophysical problems, for instance, accretion disk in compact stellar objects [27]-[30], relativistic magnetohydrodynamic jet formation [31], astrometry [32] [33] and gravitational lensing [34]. Furthermore, software related with applications of the Kerr metric can be easily modified in order to include the quadrupole moment [35]-[37].

This paper is organized as follows. In Section 2, we give a succinct explanation of the Kerr metric, and the weak limit of the Erez-Rosen metric is presented. In Section 3, the Lewis metric is presented, and the perturbation method is discussed. The application of this method leads to a new solution to the EFE with quadrupole moment and rotation. It is checked by means of the REDUCE software [38] that the resulting metric is a solution of the EFE. In section 4, we compare our solution with the exterior Hartle-Thorne metric in order to assure that our metric has astrophysical meaning. Forthcoming works with this metric are discussed in Section 6.

2. The Kerr Metric and the Erez-Rosen Metric

2.1. The Kerr Metric

The Kerr metric represents the spacetime of a non-deformed massive rotating object. The Kerr metric is given by [1] [39]

$$ds^2 = \frac{\Delta}{\rho^2}\left[dt - a\sin^2\theta d\phi\right]^2 - \frac{\sin^2\theta}{\rho^2}\left[\left(r^2 + a^2\right)d\phi - adt\right]^2 - \frac{\rho^2}{\Delta}dr^2 - \rho^2 d\theta^2, \tag{1}$$

where $\Delta = r^2 - 2Mr + a^2$ and $\rho^2 = r^2 + a^2\cos^2\theta$. M and a represent the mass and the rotation parameter, respectively. The angular momentum of the object is $J = Ma$.

2.2. The Erez-Rosen Metric

The Erez-Rosen metric [39]-[42] represents the spacetime of a body with quadrupole moment. The principal axis of the quadrupole moment is chosen along the spin axis, so that gravitational radiation can be ignored. Here, we write down an approximate expression for this metric obtained by doing Taylor series [32]

$$ds^2 = \left(1 - \frac{2M}{r}\right)e^{-2\chi}dt^2 - \left(1 - \frac{2M}{r}\right)^{-1}e^{2\chi}dr^2 - r^2e^{2\chi}d\Sigma^2, \tag{2}$$

where $d\Sigma^2 = d\theta^2 + \sin^2\theta d\phi^2$, and

$$\chi = \frac{2}{15}q\frac{M^3}{r^3}P_2\left(\cos\theta\right), \tag{3}$$

where $P_2(\cos\theta)=(3\cos^2\theta-1)/2$. The quadrupole parameter is given by $q=15GQ/(2c^2M^3)$, with Q representing the quadrupole moment. This metric is valid up to the order $O(qM^4,q^2)$.

3. Perturbing the Kerr Metric

3.1. The Lewis Metrics

The Lewis metric is given by [39] [43]

$$ds^2 = Vdt^2 - 2Wdtd\phi - e^\mu d\rho^2 - e^\nu dz^2 - Zd\phi^2, \tag{4}$$

where we have chosen the canonical coordinates $x^1=\rho$ and $x^2=z$, V, W, Z, μ and ν are functions of ρ and z $(\rho^2=VZ+W^2)$. Choosing $\mu=\nu$ and performing the following changes of potentials

$$V=f, \quad W=\omega f, \quad Z=\frac{\rho^2}{f}-\omega^2 f \quad\text{and}\quad e^\mu=\frac{e^\gamma}{f},$$

we get the Papapetrou metric

$$ds^2 = f(dt-\omega d\phi)^2 - \frac{e^\gamma}{f}\left[d\rho^2+dz^2\right] - \frac{\rho^2}{f}d\phi^2. \tag{5}$$

3.2. The Perturbation Method

In order to include a small quadrupole moment into the Kerr metric we will modify the Lewis-Papapetrou metric (5). First of all, we choose expressions for the canonical coordinates ρ and z. For the Kerr metric [1], one particular choice is [39] [44]

$$\rho=\sqrt{\Delta}\sin\theta \quad\text{and}\quad z=(r-M)\cos\theta, \tag{6}$$

where $\Delta=r^2-2Mr+a^2$.

From (6) we get

$$d\rho^2+dz^2 = \left[(r-M)^2\sin^2\theta+\Delta\cos^2\theta\right]\left(\frac{dr^2}{\Delta}+d\theta^2\right). \tag{7}$$

If we choose

$$e^\mu = \tilde{\rho}^2\left[(r-M)^2\sin^2\theta+\Delta\cos^2\theta\right]^{-1},$$

the term (7) becomes

$$e^\mu\left[d\rho^2+dz^2\right] = \tilde{\rho}^2\left(\frac{dr^2}{\Delta}+d\theta^2\right),$$

where $\tilde{\rho}^2=r^2+a^2\cos^2\theta$.

From (5), we propose the following metric

$$ds^2 = \mathcal{V}dt^2 - 2\mathcal{W}dtd\phi - \mathcal{X}dr^2 - \mathcal{Y}d\theta^2 - \mathcal{Z}d\phi^2, \tag{8}$$

where

$$\mathcal{V}=Ve^{-2\psi}$$
$$\mathcal{W}=W$$
$$\mathcal{X}=Xe^{2\psi} \tag{9}$$
$$\mathcal{Y}=Ye^{2\psi}$$
$$\mathcal{Z}=Ze^{2\psi},$$

where the potentials V,W,X,Y,Z, and ψ depend on $x^1=r$ and $x^2=\theta$. The potential $\mathcal{W}=W$ is then chosen to maintain the same cross components of the Kerr metric.

Now, let us choose

$$V = f = \frac{1}{\tilde{\rho}^2}\left[\Delta - a^2 \sin^2\theta\right]$$

$$W = \frac{a}{\tilde{\rho}^2}\left[\Delta - \left(r^2 + a^2\right)\right]\sin^2\theta = -\frac{2Jr}{\tilde{\rho}^2}\sin^2\theta$$

$$X = \frac{\tilde{\rho}^2}{\Delta} \tag{10}$$

$$Y = \tilde{\rho}^2$$

$$Z = \frac{\sin^2\theta}{\tilde{\rho}^2}\left[\left(r^2 + a^2\right)^2 - a^2\Delta\sin^2\theta\right].$$

The only potential we have to find is ψ. In order to obtain this potential, the EFE must be solved

$$G_{ij} = R_{ij} - \frac{R}{2}g_{ij} = 0, \tag{11}$$

where R_{ij} $(i, j = 0,1,2,3)$ are the Ricci tensor components and R is the curvature scalar. The Ricci tensor components and the curvature scalar R for this metric can be found in the Appendix.

In our calculations, we consider the potential ψ as perturbation, *i.e.* one neglects terms of the form

$$\left(\frac{\partial\psi}{\partial r}\right)^2 = \left(\frac{\partial\psi}{\partial\theta}\right)^2 = \frac{\partial\psi}{\partial r}\frac{\partial\psi}{\partial\theta} \sim 0.$$

Terms containing factors of the form

$$a\frac{\partial\psi}{\partial x^i} = m\frac{\partial\psi}{\partial x^i} \sim 0 \qquad (i = 1,2)$$

are also neglected. Substituting the known potentials (V,W,X,Y,Z) into the expressions for the Ricci tensor and the curvature scalar (see **Appendix**) results only in one equation for ψ that we have to solve:

$$r^2\sin\theta\nabla^2\psi = \sin\theta\frac{\partial}{\partial r}\left(r^2\frac{\partial\psi}{\partial r}\right) + \frac{\partial}{\partial\theta}\left(\sin\theta\frac{\partial\psi}{\partial\theta}\right) = 0. \tag{12}$$

The solution for this equation is

$$\psi = \frac{\mathcal{K}}{r^3}P_2\left(\cos\theta\right), \tag{13}$$

where $\mathcal{K}$ is a constant. To determine this constant, we compare the weak limit of the metric (8) with the approximate Erez-Rosen metric (2). The result is $\mathcal{K} = 2qM^3/15$ $(\psi = \chi)$.

Then, the new modified Kerr metric containing quadrupole moment is

$$\begin{aligned}
ds^2 &= \frac{e^{-2\chi}}{\rho^2}\left[\Delta - a^2\sin^2\theta\right]dt^2 + \frac{4Jr}{\rho^2}\sin^2\theta dt d\phi - \frac{\rho^2 e^{2\chi}}{\Delta}dr^2 \\
&\quad - \rho^2 e^{2\chi}d\theta^2 - \frac{e^{2\chi}\sin^2\theta}{\rho^2}\left[\left(r^2 + a^2\right)^2 - a^2\Delta\sin^2\theta\right]d\phi^2 \\
&= \frac{\Delta}{\rho^2}\left[e^{-\chi}dt - ae^{\chi}\sin^2\theta d\phi\right]^2 - \frac{\sin^2\theta}{\rho^2}\left[\left(r^2 + a^2\right)e^{\chi}d\phi - ae^{-\chi}dt\right]^2 \\
&\quad - e^{2\chi}\left(\frac{\rho^2}{\Delta}dr^2 + \rho^2 d\theta^2\right),
\end{aligned} \tag{14}$$

where the tilde over ρ is now dropped.

We verified that the metric (14) is indeed a solution of the EFE using REDUCE [38] up to the order $O\left(qM^4, q^2\right)$.

Note that (14) has four important limiting cases. One obtains the Kerr metric (1) if $q = 0$, the weak metric of [32] if $a^2 = q^2 = qM^4 \simeq 0$, the Erez-Rosen-like metric (2) if $a = 0$, and the Schwarzschild metric if $q = a = 0$.

4. Comparison with the Exterior Hartle-Thorne Metric

In order to validate the metric (14) as representing the gravitational field of a real astrophysical object, one should show that it is possible to construct an interior solution, which can appropriately be matched with our exterior solution. To this aim, we employed the exterior Hartle-Thorne metric [5] [32] [45] [46]

$$
ds^2 = \left(1 - \frac{2M}{r} + \frac{2QM^3}{r^3} P_2(\cos\theta)\right) dt^2 - \left(1 + \frac{2M}{r} + \frac{4M^2}{r^2} - \frac{2QM^3}{r^3} P_2(\cos\theta)\right) dr^2
$$
$$
- r^2 \left(1 - \frac{2QM^3}{r^3} P_2(\cos\theta)\right) d\Sigma^2 + \frac{4J}{r}\sin^2\theta dt d\phi,
$$

(15)

where M, J, and Q are related with the total mass, angular momentum, and mass quadrupole moment of the rotating object, respectively. This approximation for the Hartle-Thorne metric (15) was obtained by Frutos-Alfaro $et\ al.$ using a REDUCE program [32].

The spacetime (14) has the same weak limit as the metric obtained by Frutos-Alfaro $et\ al.$ [32]. A comparison of the exterior Hartle-Thorne metric (15) with the weak limit of the metric (14) shows that upon defining

$$
M = M, \qquad J = J, \qquad 2QM^3 = -\frac{4}{15}qM^3,
$$

(16)

both metrics coincide up to the order $O\left(M^3, a^2, qM^4, q^2\right)$. Hence, the metric (14) may be used to represent a compact astrophysical object. Moreover, Berti $et\ al.$ [46] compared the Hartle-Thorne metric with Manko solution and a numerical solution of the EFE finding that this Hartle-Thorne approximation is very reliable for most astrophysical applications. Our metric has no approximation in the rotation terms as in this metric, this could be advantageous for astrophysical problems. The first approximate solution was obtained by Sato and Tomimatsu [47] and the second one by Hernández [48]. These solutions do not have a simple form as the metric we presented here.

5. Conclusions

The new Kerr metric with quadrupole moment was obtained by solving the EFE approximately. It may represent the spacetime of a rotating and slightly deformed astrophysical object, which is possible since it can match an interior solution. We showed this by comparison of our metric with the exterior Hartle-Thorne metric. The limiting cases for the new Kerr metric correspond to the Kerr metric, the Erez-Rosen-like metric, and the Schwarzschild metric as expected.

The inclusion of the quadrupole moment in the Kerr metric is more suitable for astrophysical calculations than the Kerr metric alone and there are a large variety of applications which can be tackled with this new metric. Amongst the applications for this metric are astrometry, gravitational lensing, relativistic magnetohydrodynamic jet formation, and accretion disks in compact stellar objects, additionally we would like to point out that previous works in superfluid neutron stars can be repeated using this new metric instead of the Hartle-Thorne metric as an exterior solution. Furthermore, existing software with applications of the Kerr metric can be easily modified to include the quadrupole moment.

Acknowledgements

We thank H. Quevedo for reading the manuscript.

References

[1]   Kerr, R.P. (1963) Gravitational Field of a Spinning Mass as an Example of Algebraically Special Metrics. $Physical\ Review\ Letters$, $\mathbf{11}$, 237-238. http://dx.doi.org/10.1103/PhysRevLett.11.237

[2]   Hernández, W. (1967) Material Sources for the Kerr Metric. $Physical\ Review$, $\mathbf{159}$, 1070-1072. http://dx.doi.org/10.1103/PhysRev.159.1070

[3]   Thorne, K.S. (1969) Relativistic Stars, Black Holes and Gravitational Waves. Sachs, B.K., Ed., *General Relativity and Cosmology, Proceedings of the International School of Physics Enrico Fermi, Course XLVII*, Academic Press, Waltham, 237-283. http://www.its.caltech.edu/kip/scripts/publications.html

[4]   Marsh, G.E. (2014) Rigid Rotation and the Kerr Metric. http://arxiv.org/abs/1404.5297

[5]   Boshkayev, K., Quevedo, H. and Ruffini, R. (2012) Gravitational Field of Compact Objects in General Relativity. *Physical Review D*, **86**, Article ID: 064043. http://dx.doi.org/10.1103/PhysRevD.86.064043

[6]   Cuchí, J.E., Molina, A. and Ruiz, E. (2011) Double Shell Stars as Source of the Kerr Metric in the CMMR Approximation. *Journal of Physics: Conference Series*, **314**, Article ID: 012070.

[7]   Krisch, J.P. and Glass, E.N. (2009) Counter-Rotating Kerr Manifolds Separated by a Fluid Shell. *Classical and Quantum Gravity*, **26**, Article ID: 175010. http://dx.doi.org/10.1088/0264-9381/26/17/175010

[8]   Haggag, S. and Marek, J. (1981) A Nearly-Perfect-Fluid Source for the Kerr metric. *Il Nuovo Cimento B*, **62**, 273-282. http://dx.doi.org/10.1007/BF02721277

[9]   Haggag, S. (1990) A Fluid Source for the Kerr Metric. *Il Nuovo Cimento B*, **105**, 365-370. http://dx.doi.org/10.1007/BF02728818

[10] Haggag, S. (1990) A Static Axisymmetric Anisotropic Fluid Solution in General Relativity. *Astrophysics and Space Science*, **173**, 47-51. http://dx.doi.org/10.1007/BF00642561

[11] Krasiński, A. (1980) A Newtonian Model of the Source of the Kerr Metric. *Physics Letters*, **80A**, 238-242. http://dx.doi.org/10.1016/0375-9601(80)90010-9

[12] Ramadan, A. (2004) Fluid Sources for the Kerr Metric. *Il Nuovo Cimento*, **119B**, 123-129.

[13] Krasiński, A. (1978) Ellipsoidal Space-Times, Sources for the Kerr Metric. *Annals of Physics*, **112**, 22-40. http://dx.doi.org/10.1016/0003-4916(78)90079-9

[14] Drake, S.P. and Turolla, R. (1997) The Application of the Newman-Janis Algorithm in Obtaining Interior Solutions of the Kerr Metric. *Classical and Quantum Gravity*, **14**, 1883-1897. http://dx.doi.org/10.1088/0264-9381/14/7/021

[15] Viaggiu, S. (2006) Interior Kerr Solutions with the Newman-Janis Algorithm Starting with Physically Reasonable Space-Times. *International Journal of Modern Physics D*, **15**, 1441-1453. http://dx.doi.org/10.1142/S0218271806009169

[16] Castejon-Amenedo, J. and Manko, V.S. (1990) Superposition of the Kerr Metric with the Generalized Erez-Rosen Solution. *Physical Review D*, **41**, 2018-2020. http://dx.doi.org/10.1103/PhysRevD.41.2018

[17] Manko, V.S. and Novikov, I.D. (1992) Generalizations of the Kerr and Kerr-Newman Metrics Possessing an Arbitrary Set of Mass-Multipole Moments. *Classical and Quantum Gravity*, **9**, 2477-2487. http://dx.doi.org/10.1088/0264-9381/9/11/013

[18] Manko, V.S., Mielke, E.W. and Sanabria-Gómez, J.D. (2000) Exact Solution for the Exterior Field of a Rotating Neutron Star. *Physical Review D*, **61**, Article ID: 081501. http://dx.doi.org/10.1103/PhysRevD.61.081501

[19] Pachón, L.A., Rueda, J.A. and Sanabria-Gómez, J.D. (2006) Realistic Exact Solution for the Exterior Field of a Rotating Neutron Star. *Physical Review D*, **73**, Article ID: 104038. http://dx.doi.org/10.1103/PhysRevD.73.104038

[20] Quevedo, H. (1986) Class of Stationary Axisymmetric Solutions of Einsteins Equations in Empty Space. *Physical Review D*, **33**, 324-327. http://dx.doi.org/10.1103/PhysRevD.33.324

[21] Quevedo, H. (1989) General Static Axisymmetric Solution of Einsteins Vacuum Field Equations in Prolate Spheroidal Coordinates. *Physical Review D*, **39**, 2904-2911. http://dx.doi.org/10.1103/PhysRevD.39.2904

[22] Quevedo, H. and Mashhoon, B. (1991) Generalization of Kerr Spacetime. *Physical Review D*, **43**, 3902-3906. http://dx.doi.org/10.1103/PhysRevD.43.3902

[23] Quevedo, H. (2011) Exterior and Interior Metrics with Quadrupole Moment. *General Relativity and Gravitation*, **43**, 1141-1152. http://dx.doi.org/10.1007/s10714-010-0940-5

[24] Ernst, F.J. (1968) New Formulation of the Axially Symmetric Gravitational Field Problem. *Physical Review*, **167**, 1175-1177. http://dx.doi.org/10.1103/PhysRev.167.1175

[25] Hoenselaers, C., Kinnersley, W. and Xanthopoulos, B.C. (1979) Symmetries of the Stationary Einstein-Maxwell Equations. VI. Transformations Which Generate Asymptotically Flat Spacetimes with Arbitrary Multipole Moments. *Journal of Mathematical Physics*, **20**, 2530-2536. http://dx.doi.org/10.1063/1.523580

[26] Quevedo, H. (2012) Matching Conditions in Relativistic Astrophysics. In: Damour, T., Jantzen, R.T. and Ruffini, R., Eds., *Proceedings of the Twelfth Marcel Grossmann Meeting on General Relativity*, World Scientific, Singapore. http://arxiv.org/abs/1205.0500

[27] Andersson, N. and Comer, G.L. (2001) Slowly Rotating General Relativistic Superfluid Neutron Stars. *Classical and Quantum Gravity*, **18**, 969-1002. http://dx.doi.org/10.1088/0264-9381/18/6/302

[28] Stergioulas, N. (2003) Rotating Stars in Relativity. *Living Reviews in Relativity*, **6**. http://www.livingreviews.org/lrr-2003-3

[29] Fragile, P.C., Blaes, O.M., Anninos, P. and Salmonson, J.D. (2007) Global General Relativistic Magnetohydrodynamic Simulation of a Tilted Black Hole Accretion Disk. *Astrophysical Journal*, **668**, 417-429. http://dx.doi.org/10.1086/521092

[30] Hawley, J.F. (2009) MHD Simulations of Accretion Disks and Jets: Strengths and Limitations. *Astrophysics and Space Science*, **320**, 107-114. http://dx.doi.org/10.1007/s10509-008-9799-2

[31] Fendt, C. and Memola, E. (2008) Formation of Relativistic MHD Jets: Stationary State Solutions and Numerical Simulations. *International Journal of Modern Physics*, **D17**, 1677-1686. http://dx.doi.org/10.1142/S0218271808013297

[32] Frutos-Alfaro, F., Retana-Montenegro, E., Cordero-Garca, I. and Bonatti-González, J. (2013) Metric of a Slow Rotating Body with Quadrupole Moment from the Erez-Rosen Metric. *International Journal of Astronomy and Astrophysics*, **3**, 431-437. http://dx.doi.org/10.4236/ijaa.2013.34051

[33] Soffel, M.H. (1989) Relativity in Astrometry, Celestial Mechanics and Geodesy (Astronomy and Astrophysics Library). Springer-Verlag, Berlin. http://www.springer.com/us/book/9783642734083

[34] Frutos-Alfaro, F. (2001) A Computer Program to Visualize Gravitational Lenses. *American Journal of Physics*, **69**, 218-222. http://dx.doi.org/10.1119/1.1290251

[35] Dexter, J. and Algol, E. (2009) A Fast New Public Code for Computing Photon Orbits in a Kerr Spacetime. *Astrophysical Journal*, **696**, 1616-1629. http://dx.doi.org/10.1088/0004-637X/696/2/1616

[36] Frutos-Alfaro, F., Grave, F., Müller, T. and Adis, D. (2012) Wavefronts and Light Cones for Kerr Spacetimes. *Journal of Modern Physics*, **3**, 1882-1890. http://dx.doi.org/10.4236/jmp.2012.312237

[37] Vincent, F.H., Paumard, T., Gourgoulhon, E. and Perrin, G. (2011) GYOTO: A New General Relativistic Ray-Tracing Code. *Classical and Quantum Gravity*, **28**, Article ID: 225011. http://dx.doi.org/10.1088/0264-9381/28/22/225011

[38] Hearn, A.C. (1999) REDUCE (User's and Contributed Packages Manual). Konrad-Zuse-Zentrum für Informationstechnik, Berlin. http://www.reduce-algebra.com/docs/reduce.pdf

[39] Carmeli, M. (2001) Classical Fields: General Relativity and Gauge Theory. World Scientific Publishing, Singapore. http://www.worldscientific.com/worldscibooks/10.1142/4843

[40] Winicour, J., Janis, A.I. and Newman, E.T. (1968) Static, Axially Symmetric Point Horizons. *Physical Review*, **176**, 1507-1513. http://dx.doi.org/10.1103/PhysRev.176.1507

[41] Young, J.H. and Coulter, C.A. (1969) Exact Metric for a Nonrotating Mass with a Quadrupole Moment. *Physical Review*, **184**, 1313-1315. http://dx.doi.org/10.1103/PhysRev.184.1313

[42] Zel'dovich, Y.B. and Novikov, I.D. (2011) Stars and Relativity. Dover Publications, New York. http://store.doverpublications.com/0486694240.html

[43] Lewis, T. (1932) Some Special Solutions of the Equations of Axially Symmetric Gravitational Fields. *Proceedings of the Royal Society of London A*, **136**, 176-192. http://dx.doi.org/10.1098/rspa.1932.0073

[44] Chandrasekhar, S. (2000) The Mathematical Theory of Black Holes. Oxford University Press, Oxford. http://www.oupcanada.com/catalog/9780198503705.html

[45] Hartle, J.B. and Thorne, K.S. (1968) Slowly Rotating Relativistic Stars. II. Models for Neutron Stars and Supermassive Stars. *Astrophysical Journal*, **153**, 807-834. http://dx.doi.org/10.1086/149707

[46] Berti, E., White, F., Maniopoulou, A. and Bruni, M. (2005) Rotating Neutron Stars: An Invariant Comparison of Approximate and Numerical Spacetime Models. *Monthly Notices of the Royal Astronomical Society*, **358**, 923-938. http://dx.doi.org/10.1111/j.1365-2966.2005.08812.x

[47] Sato, H. and Tomimatsu, A. (1973) Gravitational Field of Slowly Rotating Deformed Masses. *Progress of Theoretical Physics*, **49**, 790-799. http://ptp.oxfordjournals.org/content/49/3/790.full.pdf

[48] Hernández-Pastora, J.L. (2006) Approximate Gravitational Field of a Rotating Deformed Mass. *General Relativity and Gravitation*, **38**, 871-884. http://dx.doi.org/10.1007/s10714-006-0269-2

Appendix

The non-null Ricci tensor components for the metric (8) are given by (with the tilde over ρ dropped)

$$R_{00} = \frac{e^{-2\psi}}{4\rho^2 X^2 Y^2}\left(-4\rho^2 VX^2Y\frac{\partial^2\psi}{\partial\theta^2} + 8VW^2X^2Y\left(\frac{\partial\psi}{\partial\theta}\right)^2 - 2\rho^2 VXY\frac{\partial\psi}{\partial\theta}\frac{\partial X}{\partial\theta}\right.$$

$$+ 2VX^2Y\frac{\partial\psi}{\partial\theta}\frac{\partial\rho^2}{\partial\theta} - 4\rho^2 X^2Y\frac{\partial\psi}{\partial\theta}\frac{\partial V}{\partial\theta} - 4W^2X^2Y\frac{\partial\psi}{\partial\theta}\frac{\partial V}{\partial\theta} + 2\rho^2 VX^2\frac{\partial\psi}{\partial\theta}\frac{\partial Y}{\partial\theta}$$

$$- 4V^2X^2Y\frac{\partial\psi}{\partial\theta}\frac{\partial Z}{\partial\theta} - 4\rho^2 VXY^2\frac{\partial^2\psi}{\partial r^2} + 8VW^2XY^2\left(\frac{\partial\psi}{\partial r}\right)^2 + 2\rho^2 VY^2\frac{\partial\psi}{\partial r}\frac{\partial X}{\partial r}$$

$$+ 2VXY^2\frac{\partial\psi}{\partial r}\frac{\partial\rho^2}{\partial r} - 4\rho^2 XY^2\frac{\partial\psi}{\partial r}\frac{\partial V}{\partial r} - 4W^2XY^2\frac{\partial\psi}{\partial r}\frac{\partial V}{\partial r} - 2\rho^2 VXY\frac{\partial\psi}{\partial r}\frac{\partial Y}{\partial r}$$

$$- 4V^2XY^2\frac{\partial\psi}{\partial r}\frac{\partial Z}{\partial r} + \rho^2 XY\frac{\partial X}{\partial\theta}\frac{\partial V}{\partial\theta} - \rho^2 Y^2\frac{\partial X}{\partial r}\frac{\partial V}{\partial r} - X^2Y\frac{\partial\rho^2}{\partial\theta}\frac{\partial V}{\partial\theta}$$

$$- XY^2\frac{\partial\rho^2}{\partial r}\frac{\partial V}{\partial r} + 2\rho^2 X^2Y\frac{\partial^2 V}{\partial\theta^2} - \rho^2 X^2\frac{\partial V}{\partial\theta}\frac{\partial Y}{\partial\theta} + 2VX^2Y\frac{\partial V}{\partial\theta}\frac{\partial Z}{\partial\theta}$$

$$\left. + 2\rho^2 XY^2\frac{\partial^2 V}{\partial r^2} + \rho^2 XY\frac{\partial V}{\partial r}\frac{\partial Y}{\partial r} + 2VXY^2\frac{\partial V}{\partial r}\frac{\partial Z}{\partial r} + 2VX^2Y\left(\frac{\partial W}{\partial\theta}\right)^2 + 2VXY^2\left(\frac{\partial W}{\partial r}\right)^2\right)$$

$$R_{03} = \frac{e^{-2\psi}}{4\rho^2 X^2 Y^2}\left(8\rho^2 WX^2Y\left(\frac{\partial\psi}{\partial\theta}\right)^2 - 8W^3X^2Y\left(\frac{\partial\psi}{\partial\theta}\right)^2 - 4WX^2Y\frac{\partial\psi}{\partial\theta}\frac{\partial\rho^2}{\partial\theta}\right.$$

$$+ 8W^2X^2Y\frac{\partial\psi}{\partial\theta}\frac{\partial W}{\partial\theta} + 8VWX^2Y\frac{\partial\psi}{\partial\theta}\frac{\partial Z}{\partial\theta} + 8\rho^2 WXY^2\left(\frac{\partial\psi}{\partial r}\right)^2 - 8W^3XY^2\left(\frac{\partial\psi}{\partial r}\right)^2$$

$$- 4WXY^2\frac{\partial\psi}{\partial r}\frac{\partial\rho^2}{\partial r} + 8W^2XY^2\frac{\partial\psi}{\partial r}\frac{\partial W}{\partial r} + 8VWXY^2\frac{\partial\psi}{\partial r}\frac{\partial Z}{\partial r} - \rho^2 XY\frac{\partial X}{\partial\theta}\frac{\partial W}{\partial\theta}$$

$$+ \rho^2 Y^2\frac{\partial X}{\partial r}\frac{\partial W}{\partial r} + X^2Y\frac{\partial\rho^2}{\partial\theta}\frac{\partial W}{\partial\theta} + XY^2\frac{\partial\rho^2}{\partial r}\frac{\partial W}{\partial r} - 2WX^2Y\frac{\partial V}{\partial\theta}\frac{\partial Z}{\partial\theta}$$

$$- 2WXY^2\frac{\partial V}{\partial r}\frac{\partial Z}{\partial r} - 2\rho^2 X^2Y\frac{\partial^2 W}{\partial\theta^2} - 2WX^2Y\left(\frac{\partial W}{\partial\theta}\right)^2 + \rho^2 X^2\frac{\partial W}{\partial\theta}\frac{\partial Y}{\partial\theta}$$

$$\left. - 2\rho^2 XY^2\frac{\partial^2 W}{\partial r^2} - 2WXY^2\left(\frac{\partial W}{\partial r}\right)^2 - \rho^2 XY\frac{\partial W}{\partial r}\frac{\partial Y}{\partial r}\right) = R_{30}$$

$$R_{11} = \frac{1}{4\rho^4 XY^2}\left(-4\rho^4 X^2Y\frac{\partial^2\psi}{\partial\theta^2} - 2\rho^4 XY\frac{\partial\psi}{\partial\theta}\frac{\partial X}{\partial\theta} - 2\rho^2 X^2Y\frac{\partial\psi}{\partial\theta}\frac{\partial\rho^2}{\partial\theta} + 2\rho^4 X^2\frac{\partial\psi}{\partial\theta}\frac{\partial Y}{\partial\theta}\right.$$

$$- 4\rho^4 XY^2\frac{\partial^2\psi}{\partial r^2} - 8\rho^4 XY^2\left(\frac{\partial\psi}{\partial r}\right)^2 + 8\rho^2 W^2XY^2\left(\frac{\partial\psi}{\partial r}\right)^2 + 2\rho^4 Y^2\frac{\partial\psi}{\partial r}\frac{\partial X}{\partial r}$$

$$+ 6\rho^2 XY^2\frac{\partial\psi}{\partial r}\frac{\partial\rho^2}{\partial r} - 8\rho^2 WXY^2\frac{\partial\psi}{\partial r}\frac{\partial W}{\partial r} - 2\rho^4 XY\frac{\partial\psi}{\partial r}\frac{\partial Y}{\partial r} - 8\rho^2 VXY^2\frac{\partial\psi}{\partial r}\frac{\partial Z}{\partial r}$$

$$- 2\rho^4 XY\frac{\partial^2 X}{\partial\theta^2} + \rho^4 Y\left(\frac{\partial X}{\partial\theta}\right)^2 - \rho^2 XY\frac{\partial X}{\partial\theta}\frac{\partial\rho^2}{\partial\theta} + \rho^4 X\frac{\partial X}{\partial\theta}\frac{\partial Y}{\partial\theta} + \rho^2 Y^2\frac{\partial X}{\partial r}\frac{\partial\rho^2}{\partial r}$$

$$+ \rho^4 Y\frac{\partial X}{\partial r}\frac{\partial Y}{\partial r} - 2\rho^2 XY^2\frac{\partial^2\rho^2}{\partial r^2} + XY^2\left(\frac{\partial\rho^2}{\partial r}\right)^2 + 2VXY^2\frac{\partial\rho^2}{\partial r}\frac{\partial Z}{\partial r} + 2W^2XY^2\frac{\partial V}{\partial r}\frac{\partial Z}{\partial r}$$

$$\left. + 2\rho^2 XY^2\left(\frac{\partial W}{\partial r}\right)^2 - 4VWXY^2\frac{\partial W}{\partial r}\frac{\partial Z}{\partial r} - 2\rho^4 XY\frac{\partial^2 Y}{\partial r^2} + \rho^4 X\left(\frac{\partial Y}{\partial r}\right)^2 - 2V^2XY^2\left(\frac{\partial Z}{\partial r}\right)^2\right)$$

$$R_{12} = \frac{1}{4\rho^4 XY}\left(-8\rho^4 XY \frac{\partial \psi}{\partial \theta}\frac{\partial \psi}{\partial r} + 8\rho^2 W^2 XY \frac{\partial \psi}{\partial \theta}\frac{\partial \psi}{\partial r} + 4\rho^2 XY \frac{\partial \psi}{\partial \theta}\frac{\partial \rho^2}{\partial r} \right.$$

$$-4\rho^2 WXY \frac{\partial \psi}{\partial \theta}\frac{\partial W}{\partial r} - 4\rho^2 VXY \frac{\partial \psi}{\partial \theta}\frac{\partial Z}{\partial r} + 4\rho^2 XY \frac{\partial \psi}{\partial r}\frac{\partial \rho^2}{\partial \theta} - 4\rho^2 WXY \frac{\partial \psi}{\partial r}\frac{\partial W}{\partial \theta}$$

$$-4\rho^2 VXY \frac{\partial \psi}{\partial r}\frac{\partial Z}{\partial \theta} + \rho^2 Y \frac{\partial X}{\partial \theta}\frac{\partial \rho^2}{\partial r} - 2\rho^2 XY \frac{\partial^2 \rho^2}{\partial \theta \partial r} + W^2 XY \frac{\partial^2 \rho^2}{\partial \theta \partial r}$$

$$+ XY \frac{\partial \rho^2}{\partial \theta}\frac{\partial \rho^2}{\partial r} + \rho^2 X \frac{\partial \rho^2}{\partial \theta}\frac{\partial Y}{\partial r} + VXY \frac{\partial \rho^2}{\partial \theta}\frac{\partial Z}{\partial r} + VXY \frac{\partial \rho^2}{\partial r}\frac{\partial Z}{\partial \theta}$$

$$- W^2 XYZ \frac{\partial^2 V}{\partial \theta \partial r} - 2W^3 XY \frac{\partial^2 W}{\partial \theta \partial r} + 2\rho^2 XY \frac{\partial W}{\partial \theta}\frac{\partial W}{\partial r} - 2W^2 XY \frac{\partial W}{\partial \theta}\frac{\partial W}{\partial r}$$

$$\left. - 2VWXY \frac{\partial W}{\partial \theta}\frac{\partial Z}{\partial r} - 2VWXY \frac{\partial W}{\partial r}\frac{\partial Z}{\partial \theta} - VW^2 XY \frac{\partial^2 Z}{\partial \theta \partial r} - 2V^2 XY \frac{\partial Z}{\partial \theta}\frac{\partial Z}{\partial r} \right) = R_{21}$$

$$R_{22} = \frac{1}{4\rho^4 X^2 Y}\left(-4\rho^4 X^2 Y \frac{\partial^2 \psi}{\partial \theta^2} - 8\rho^4 X^2 Y \left(\frac{\partial \psi}{\partial \theta}\right)^2 + 8\rho^2 W^2 X^2 Y \left(\frac{\partial \psi}{\partial \theta}\right)^2 \right.$$

$$-2\rho^4 XY \frac{\partial \psi}{\partial \theta}\frac{\partial X}{\partial \theta} + 6\rho^2 X^2 Y \frac{\partial \psi}{\partial \theta}\frac{\partial \rho^2}{\partial \theta} - 8\rho^2 WX^2 Y \frac{\partial \psi}{\partial \theta}\frac{\partial W}{\partial \theta} + 2\rho^4 X^2 \frac{\partial \psi}{\partial \theta}\frac{\partial Y}{\partial \theta}$$

$$-8\rho^2 VX^2 Y \frac{\partial \psi}{\partial \theta}\frac{\partial Z}{\partial \theta} - 4\rho^4 XY^2 \frac{\partial^2 \psi}{\partial r^2} + 2\rho^4 Y^2 \frac{\partial \psi}{\partial r}\frac{\partial X}{\partial r} - 2\rho^2 XY^2 \frac{\partial \psi}{\partial r}\frac{\partial \rho^2}{\partial r}$$

$$-2\rho^4 XY \frac{\partial \psi}{\partial r}\frac{\partial Y}{\partial r} - 2\rho^4 XY \frac{\partial^2 X}{\partial \theta^2} + \rho^4 Y \left(\frac{\partial X}{\partial \theta}\right)^2 + \rho^4 X \frac{\partial X}{\partial \theta}\frac{\partial Y}{\partial \theta}$$

$$+ \rho^4 Y \frac{\partial X}{\partial r}\frac{\partial Y}{\partial r} - 2\rho^2 X^2 Y \frac{\partial^2 \rho^2}{\partial \theta^2} + X^2 Y \left(\frac{\partial \rho^2}{\partial \theta}\right)^2 + \rho^2 X^2 \frac{\partial \rho^2}{\partial \theta}\frac{\partial Y}{\partial \theta}$$

$$+ 2VX^2 Y \frac{\partial \rho^2}{\partial \theta}\frac{\partial Z}{\partial \theta} - \rho^2 XY \frac{\partial \rho^2}{\partial r}\frac{\partial Y}{\partial r} + 2W^2 X^2 Y \frac{\partial V}{\partial \theta}\frac{\partial Z}{\partial \theta} + 2\rho^2 X^2 Y \left(\frac{\partial W}{\partial \theta}\right)^2$$

$$\left. -4VWX^2 Y \frac{\partial W}{\partial \theta}\frac{\partial Z}{\partial \theta} - 2\rho^4 XY \frac{\partial^2 Y}{\partial r^2} + \rho^4 X \left(\frac{\partial Y}{\partial r}\right)^2 - 2V^2 X^2 Y \left(\frac{\partial Z}{\partial \theta}\right)^2 \right)$$

$$R_{33} = \frac{1}{4\rho^2 X^2 Y^2}\left(-4\rho^2 X^2 YZ \frac{\partial^2 \psi}{\partial \theta^2} - 8W^2 X^2 YZ \left(\frac{\partial \psi}{\partial \theta}\right)^2 - 2\rho^2 XYZ \frac{\partial \psi}{\partial \theta}\frac{\partial X}{\partial \theta} \right.$$

$$-2YX^2 Z \frac{\partial \psi}{\partial \theta}\frac{\partial \rho^2}{\partial \theta} + 8WX^2 YZ \frac{\partial \psi}{\partial \theta}\frac{\partial W}{\partial \theta} + 2\rho^2 X^2 Z \frac{\partial \psi}{\partial \theta}\frac{\partial Y}{\partial \theta} - 8W^2 X^2 Y \frac{\partial \psi}{\partial \theta}\frac{\partial Z}{\partial \theta}$$

$$-4\rho^2 XY^2 Z \frac{\partial^2 \psi}{\partial r^2} - 8W^2 XY^2 Z \left(\frac{\partial \psi}{\partial r}\right)^2 + 2\rho^2 Y^2 Z \frac{\partial \psi}{\partial r}\frac{\partial X}{\partial r} - 2XY^2 Z \frac{\partial \psi}{\partial r}\frac{\partial \rho^2}{\partial r}$$

$$+8WXY^2 Z \frac{\partial \psi}{\partial r}\frac{\partial W}{\partial r} - 2\rho^2 XYZ \frac{\partial \psi}{\partial r}\frac{\partial Y}{\partial r} - 8W^2 XY^2 \frac{\partial \psi}{\partial r}\frac{\partial Z}{\partial r} - \rho^2 XY \frac{\partial X}{\partial \theta}\frac{\partial Z}{\partial \theta}$$

$$+\rho^2 Y^2 \frac{\partial X}{\partial r}\frac{\partial Z}{\partial r} - X^2 Y \frac{\partial \rho^2}{\partial \theta}\frac{\partial Z}{\partial \theta} - XY^2 \frac{\partial \rho^2}{\partial r}\frac{\partial Z}{\partial r} - 2X^2 YZ \left(\frac{\partial W}{\partial \theta}\right)^2$$

$$\left. +4WX^2 Y \frac{\partial W}{\partial \theta}\frac{\partial Z}{\partial \theta} - 2XY^2 Z \left(\frac{\partial W}{\partial r}\right)^2 + 4WXY^2 \frac{\partial W}{\partial r}\frac{\partial Z}{\partial r} + \rho^2 X^2 \frac{\partial Y}{\partial \theta}\frac{\partial Z}{\partial \theta} \right.$$

$$\left. -\rho^2 XY \frac{\partial Y}{\partial r}\frac{\partial Z}{\partial r} - 2\rho^2 X^2 Y \frac{\partial^2 Z}{\partial \theta^2} + 2VX^2 Y \left(\frac{\partial Z}{\partial \theta}\right)^2 - 2\rho^2 XY^2 \frac{\partial^2 Z}{\partial r^2} + 2VXY^2 \left(\frac{\partial Z}{\partial r}\right)^2 \right)$$

Calculation of the scalar curvature (with the tilde over ρ dropped)

$$
\begin{aligned}
R = \frac{e^{-2\psi}}{2\rho^4 X^2 Y^2} \Bigg[& 4\rho^4 X^2 Y \frac{\partial^2 \psi}{\partial \theta^2} + 4\rho^4 X^2 Y \left(\frac{\partial \psi}{\partial \theta}\right)^2 - 4\rho^2 W^2 X^2 Y \left(\frac{\partial \psi}{\partial \theta}\right)^2 + 2\rho^4 XY \frac{\partial \psi}{\partial \theta}\frac{\partial X}{\partial \theta} \\
& -2\rho^2 X^2 Y \frac{\partial \psi}{\partial \theta}\frac{\partial \rho^2}{\partial \theta} + 4\rho^2 W X^2 Y \frac{\partial \psi}{\partial \theta}\frac{\partial W}{\partial \theta} - 2\rho^4 X^2 \frac{\partial \psi}{\partial \theta}\frac{\partial Y}{\partial \theta} + 4\rho^2 V X^2 Y \frac{\partial \psi}{\partial \theta}\frac{\partial Z}{\partial \theta} \\
& +4\rho^4 XY^2 \frac{\partial^2 \psi}{\partial r^2} + 4\rho^4 XY^2 \left(\frac{\partial \psi}{\partial r}\right)^2 - 4\rho^2 W^2 XY^2 \left(\frac{\partial \psi}{\partial r}\right)^2 - 2\rho^4 Y^2 \frac{\partial \psi}{\partial r}\frac{\partial X}{\partial r} \\
& -2\rho^2 XY^2 \frac{\partial \psi}{\partial r}\frac{\partial \rho^2}{\partial r} + 4\rho^2 W XY^2 \frac{\partial \psi}{\partial r}\frac{\partial W}{\partial r} + 2\rho^4 XY \frac{\partial \psi}{\partial r}\frac{\partial Y}{\partial r} + 4\rho^2 V XY^2 \frac{\partial \psi}{\partial r}\frac{\partial Z}{\partial r} \\
& +2\rho^4 XY \frac{\partial^2 X}{\partial \theta^2} - \rho^4 Y \left(\frac{\partial X}{\partial \theta}\right)^2 + \rho^2 XY \frac{\partial X}{\partial \theta}\frac{\partial \rho^2}{\partial \theta} - \rho^4 X \frac{\partial X}{\partial \theta}\frac{\partial Y}{\partial \theta} \\
& -\rho^2 Y^2 \frac{\partial X}{\partial r}\frac{\partial \rho^2}{\partial r} - \rho^4 Y \frac{\partial X}{\partial r}\frac{\partial Y}{\partial r} + 2\rho^2 X^2 Y \frac{\partial^2 \rho^2}{\partial \theta^2} - X^2 Y \left(\frac{\partial \rho^2}{\partial \theta}\right)^2 \\
& -\rho^2 X^2 \frac{\partial \rho^2}{\partial \theta}\frac{\partial Y}{\partial \theta} + 2\rho^2 XY^2 \frac{\partial^2 \rho^2}{\partial r^2} - XY^2 \left(\frac{\partial \rho^2}{\partial r}\right)^2 + \rho^2 XY \frac{\partial \rho^2}{\partial r}\frac{\partial Y}{\partial r} \\
& -\rho^2 X^2 Y \frac{\partial V}{\partial \theta}\frac{\partial Z}{\partial \theta} - \rho^2 XY^2 \frac{\partial V}{\partial r}\frac{\partial Z}{\partial r} - \rho^2 X^2 Y \left(\frac{\partial W}{\partial \theta}\right)^2 - \rho^2 XY^2 \left(\frac{\partial W}{\partial r}\right)^2 \\
& +2\rho^4 XY \frac{\partial^2 Y}{\partial r^2} - \rho^4 X \left(\frac{\partial Y}{\partial r}\right)^2 \Bigg]
\end{aligned}
$$

Distribution of Mass and Energy in Five General Cosmic Models

Fadel A. Bukhari

Department of Astronomy, Faculty of Science, King Abdulaziz University, Jeddah, KSA
Email: fdbukhari@gmail.com

Abstract

Distributions of the universe horizon distance and universe horizon volume were investigated in the light of five general cosmic models which were constructed in a previous study. Both distributions increase so slowly up to $t \approx 21.5444$ Myr, then they start raising very fast up to $t \approx 60$ Gyr. Afterwards, they increase again very slowly until $t \approx 124$ Gyr. Distributions of mass of radiation, matter and dark energy within the horizon volume of the universe were also studied in the five general cosmic models. The masses of both radiation and matter decrease gradually with time while the mass of dark energy increases. The mass of radiation prevailed in the early universe up to $t \approx 34627.5 - 55916.2$ yr, where it becomes equal to the mass of matter. Then the mass of matter dominated until $t \approx 9.4525 - 10.0632$ Gyr, where it becomes equal to the mass of dark energy. Thenceforward, the mass of dark energy prevails the universe. The cosmic space becomes approximately matter empty in the so far future of the universe.

Keywords

General Cosmic Models, Distribution of Mass and Energy

1. Introduction

In a previous study [1] the distribution of density parameters of radiation, matter and dark energy were investigated in details in five general cosmic models. Hence, it would be interesting to study the distributions of equivalent mass of radiation, mass of matter and equivalent mass of dark energy within the horizon volume of the universe in the general models.

Therefore, it is necessary to start this study by investigating the distributions of the horizon distance and horizon volume of the universe in the general models at different time intervals depending on the bases discussed in [2]. Description of methodology is given in Section 2 while algorithm would be illustrated in Section 3. Results and discussion are presented in Section 4. Conclusion is shown in Section 5.

2. Methodology

We have seen in [2] that the horizon distance and horizon volume of the universe at the present time are respectively

$$d_h(t_o) = \frac{c}{H_o} \int_0^1 \frac{1}{a} \left[1 - \Omega_{\Lambda,o}\left(1-a^2\right) + \Omega_{m,o}\left(\frac{1}{a}-1\right) + \Omega_{r,o}\left(\frac{1}{a^2}-1\right) \right]^{-\frac{1}{2}} da. \tag{1}$$

$$V_h(t_o) = \frac{8\pi}{3} d_h^3(t_o). \tag{2}$$

where $c, H_o, \Omega_{r,o}, \Omega_{m,o}, \Omega_{\Lambda,o}$ are all defined as in [1]. Thus the horizon distance of the universe at any given time is

$$d_h(t) = \frac{c}{H_o} \int_0^a \frac{1}{a} \left[1 - \Omega_{\Lambda,o}\left(1-a^2\right) + \Omega_{m,o}\left(\frac{1}{a}-1\right) + \Omega_{r,o}\left(\frac{1}{a^2}-1\right) \right]^{-\frac{1}{2}} da. \tag{3a}$$

Consequently the change in the horizon distance of the universe in the time interval between two instants of scale factors a_1, a_2 is written as

$$\Delta d_h(t) = \frac{c}{H_o} \int_{a_1}^{a_2} \frac{1}{a} \left[1 - \Omega_{\Lambda,o}\left(1-a^2\right) + \Omega_{m,o}\left(\frac{1}{a}-1\right) + \Omega_{r,o}\left(\frac{1}{a^2}-1\right) \right]^{-\frac{1}{2}} da. \tag{3b}$$

The horizon volume of the universe at any given time is

$$V_h(t) = \frac{8\pi}{3} d_h^3(t). \tag{4}$$

It is also obvious from [2] that the total density of the universe is given by

$$\rho(t) = \rho_{c,t} \Omega(t). \tag{5}$$

where

$$\Omega(t) = \rho_{c,t}\left(\Omega_{m,t} + \Omega_{r,t} + \Omega_{\Lambda,t}\right). \tag{6}$$

$$\rho_{c,t} = \frac{3H^2}{8\pi G}. \tag{7}$$

$$\Omega_{m,t} = \frac{\rho_{m,t}}{\rho_{c,t}} = \left(\frac{H_0}{H}\right)^2 \frac{\Omega_{m,o}}{a^3}. \tag{8}$$

$$\Omega_{r,t} = \frac{\rho_{r,t}}{\rho_{c,t}} = \left(\frac{H_0}{H}\right)^2 \frac{\Omega_{r,o}}{a^4}. \tag{9}$$

$$\Omega_{\Lambda,t} = \frac{\rho_{\Lambda,t}}{\rho_{c,t}} = \left(\frac{H_0}{H}\right)^2 \Omega_{\Lambda,o}. \tag{10}$$

$$H(t) = \frac{H_o}{a} \left[1 - \Omega_{\Lambda,o}\left(1-a^2\right) + \Omega_{m,o}\left(\frac{1}{a}-1\right) + \Omega_{r,o}\left(\frac{1}{a^2}-1\right) \right]^{\frac{1}{2}}. \tag{11}$$

Hence, the total mass within the horizon volume of the universe at any given time is expressed as

$$M_h(t) = V_h(t)\rho(t). \tag{12}$$

The mass of matter $M_{m,t}$, the equivalent mass of radiation $M_{r,t}$ and the equivalent mass of dark energy

$M_{\Lambda,t}$ within the horizon volume of the universe at any given time are given by

$$M_{m,t} = M_h(t)\frac{\rho_{m,t}}{\rho(t)}$$

$$M_{m,t} = M_h(t)\frac{\Omega_{m,t}}{\Omega(t)}. \tag{13}$$

$$M_{r,t} = M_h(t)\frac{\Omega_{r,t}}{\Omega(t)}. \tag{14}$$

$$M_{\Lambda,t} = M_h(t)\frac{\Omega_{\Lambda,t}}{\Omega(t)}. \tag{15}$$

The cosmic time is given by Equation (16) in [1] as

$$t = \frac{1}{H_o}\int_0^a \left[1 - \Omega_{\Lambda,o}\left(1-a^2\right) + \Omega_{m,o}\left(\frac{1}{a}-1\right) + \Omega_{r,o}\left(\frac{1}{a^2}-1\right)\right]^{-\frac{1}{2}} da. \tag{16a}$$

Thus the time interval between two instants with scale factors a_1, a_2 during the universe expansion is expressed as

$$\Delta t = \frac{1}{H_o}\int_{a_1}^{a_2} \left[1 - \Omega_{\Lambda,o}\left(1-a^2\right) + \Omega_{m,o}\left(\frac{1}{a}-1\right) + \Omega_{r,o}\left(\frac{1}{a^2}-1\right)\right]^{-\frac{1}{2}} da. \tag{16b}$$

3. Algorithm

In determination of the distributions of $d_h(t), V_h(t), M_h(t), M_{m,t}, M_{r,t}$ and $M_{\Lambda,t}$ we use the following steps:

i) Set $t = 0, d_h = 0, K_1 = 1, K_2 = 2000,$ then insert the value of $a_{max} = 0.098$ for $t \leq 0.5\,\mathrm{Gyr}$, $a_{max} = 10$ for $0.5 < t \leq 50\,\mathrm{Gyr}$ and $a_{max} = 1000$ for $50 < t \leq 124\,\mathrm{Gyr}$.

ii) Compute $DA = \dfrac{a_{max}}{DBLE(K_2)}.$

iii) Start general DO loop $I = K_1, K_2$ which includes the following sub steps:

iv) $a_1 = DA(I-1), a_2 = DAI.$

v) Calculate new value of cosmic time t numerically using(16-b), where $t = t + \Delta t$.

vi) Determinate new value of the universe horizon distance d_h numerically using (3-b) where, $d_h = d_h + \Delta d_h$.

vii) Obtain the corresponding values of $V_h, H, \rho_{c,t}, \Omega_{m,t}, \Omega_{r,t}, \Omega_{\Lambda,t}, \Omega(t), \rho(t), M_h(t), M_{m,t}, M_{r,t}, M_{\Lambda,t}$ using (4), (11), (7), (8), (9), (10), (6), (5), (12), (13), (14) and (15) respectively.

viii) Continue the general DO loop.

4. Results and Discussion

The distribution of the universe horizon distance in the general models up to $t = 0.5\,\mathrm{Gyr}$ is shown in **Figure 1(a)**. The distributions of all models coincide on each other until $t \approx 21.5444\,\mathrm{Myr}$. Then, the distributions of models A, B and C coincide on each other and get upper than the coincided distributions of models D and E. The universe horizon distance increases quite slowly with time in all general models up to $t \approx 21.5444\,\mathrm{Myr}$, hence it stats raising very fast. The distribution of the universe horizon distance in the general models in the range $t = 0.5 - 50\,\mathrm{Gyr}$ is illustrated in **Figure 1(b)**. The distributions of all models coincide on each other up to $t \approx 1.1111\,\mathrm{Gyr}$. Afterwards, the distributions of models A, B and C coincide on each other and become higher than the coincided distributions of models D and E. In all general models the universe horizon distance distributions increase very fast with time. The distribution of the universe horizon distance in the general models in the range $t = 50 - 124\,\mathrm{Gyr}$ is presented in **Figure 1(c)**. The distributions of models A, B and C are close to each other and lie upper than the distributions of models E and D. The increase in the universe horizon distance gets very small with increasing time in all general models.

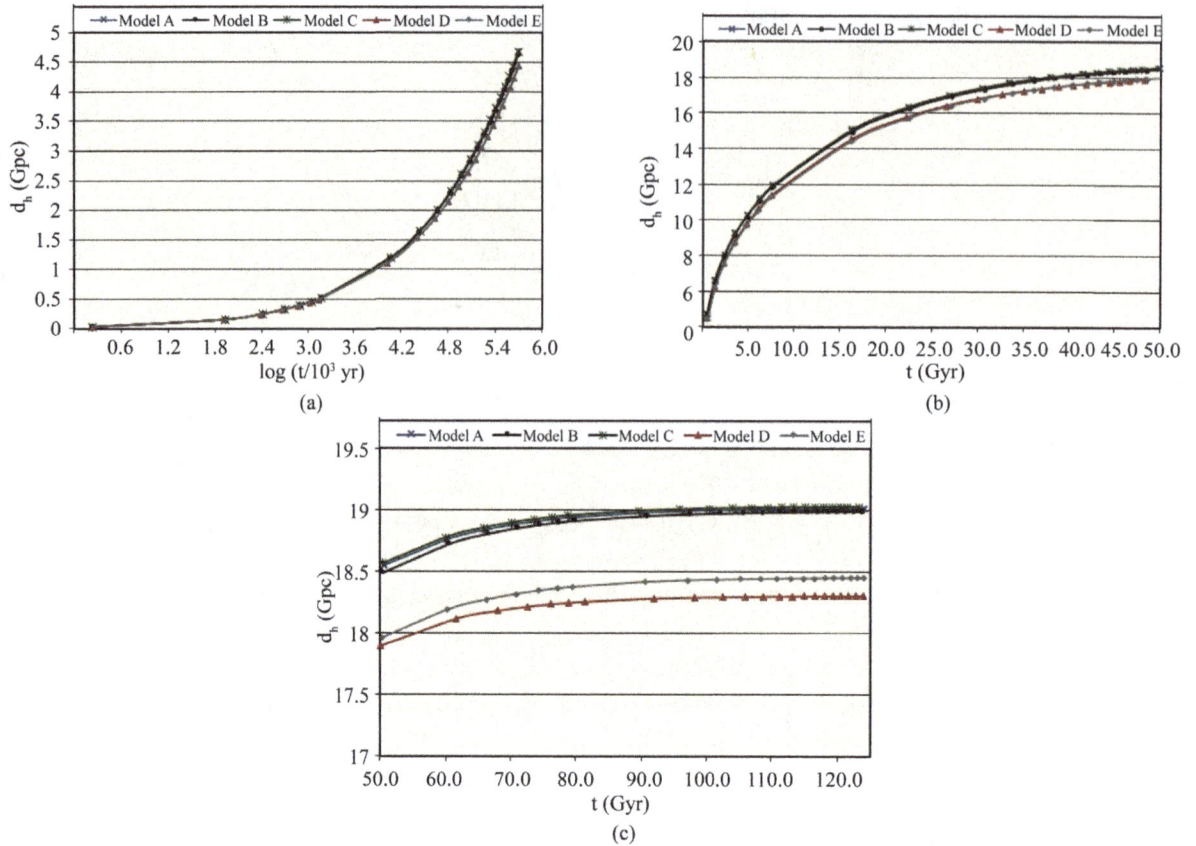

Figure 1. The distribution of the universe horizon distance in the general cosmic models (a) up to $t = 0.5$ Gyr; (b) in the range $t = 0.5$ - 50 Gyr; (c) in the range $t = 50$ - 124 Gyr.

Table 1 shows the universe horizon distances in the general models at special times. These times are the time of radiation-matter mass equivalence t_{rm}, the time of matter-dark energy mass equivalence $t_{m\Lambda}$, the present time $t_o = 13.7 \pm 0.2$ Gyr and the time $t_n = 124$ Gyr.

The results illustrated in **Figures 1(a)-(c)** are supported by those displayed in **Figures 2(a)-(c)** which show the distributions of the universe horizon volume in the general models in the ranges up to $t = 0.5$ Gyr, $t = 0.5$ - 50 Gyr and $t = 50$ - 124 Gyr respectively. **Table 2** presents the universe horizon volumes in the general models at the special times $t_{rm}, t_{m\Lambda}, t_o$ and t_n.

The distribution of mass and energy within the universe horizon volume of the universe in any general model up to $t = 0.5$ Gyr is exhibited in **Figure 3(a)**. The distributions of both radiation and matter decrease gradually with time and intersect at the time $t_{rm} = 34627.5$ - 55916.2 yr as shown in **Table 3**. On the other hand, the distribution of dark energy increases gradually until it intersects with the radiation distribution at the time $t_{r\Lambda} = 0.5166$ - 0.5839 Gyr as seen in **Table 4**. The distribution of total mass coincides with that of radiation up to $t \approx 5843.4141$ yr. Afterwards, the two distributions diverge from each other. However, the distribution of the total mass coincides on the distribution of matter from the time $t \approx 857695.9$ yr onwards.

The distribution of mass and energy within the universe horizon volume of the universe in any general model in the range $t = 0.5$ - 50 Gyr is displayed in **Figure 3(b)**. It is obvious that the distributions of matter and radiation decrease gradually with time and the former lies above the later. The distribution of dark energy increases with time and intersects with the distribution of matter at $t = 9.4525$ - 10.0632 Gyr as illustrated in **Table 5**. The distribution of the total mass coincides on the distribution of the matter up to $t = 4.5714$ Gyr, then they diverge from each other. Furthermore, the distribution of the total mass coincides on the distribution of the dark energy from $t = 18.2857$ Gyr onwards. Masses of radiation, matter and dark energy within the universe horizon volume in the general models at the present time are illustrated in **Table 6**.

The distribution of mass and energy within the universe horizon volume in any general model in the range

Table 1. Horizon distances of the universe in the general cosmic models at special times.

Model	$d_h(t_{rm})$ / Mpc	$d_h(t_{m\Lambda})$ / Gpc	$d_h(t_o)$ / Gpc	$d_h(t_n)$ / Gpc
A	118.1520	12.7667	14.2969	19.0103
B	112.1480	12.7586	14.1588	18.9849
C	118.0740	12.7812	14.2780	19.0274
D	98.0130	12.3367	13.8666	18.3062
E	94.3170	12.4269	13.8070	18.4510

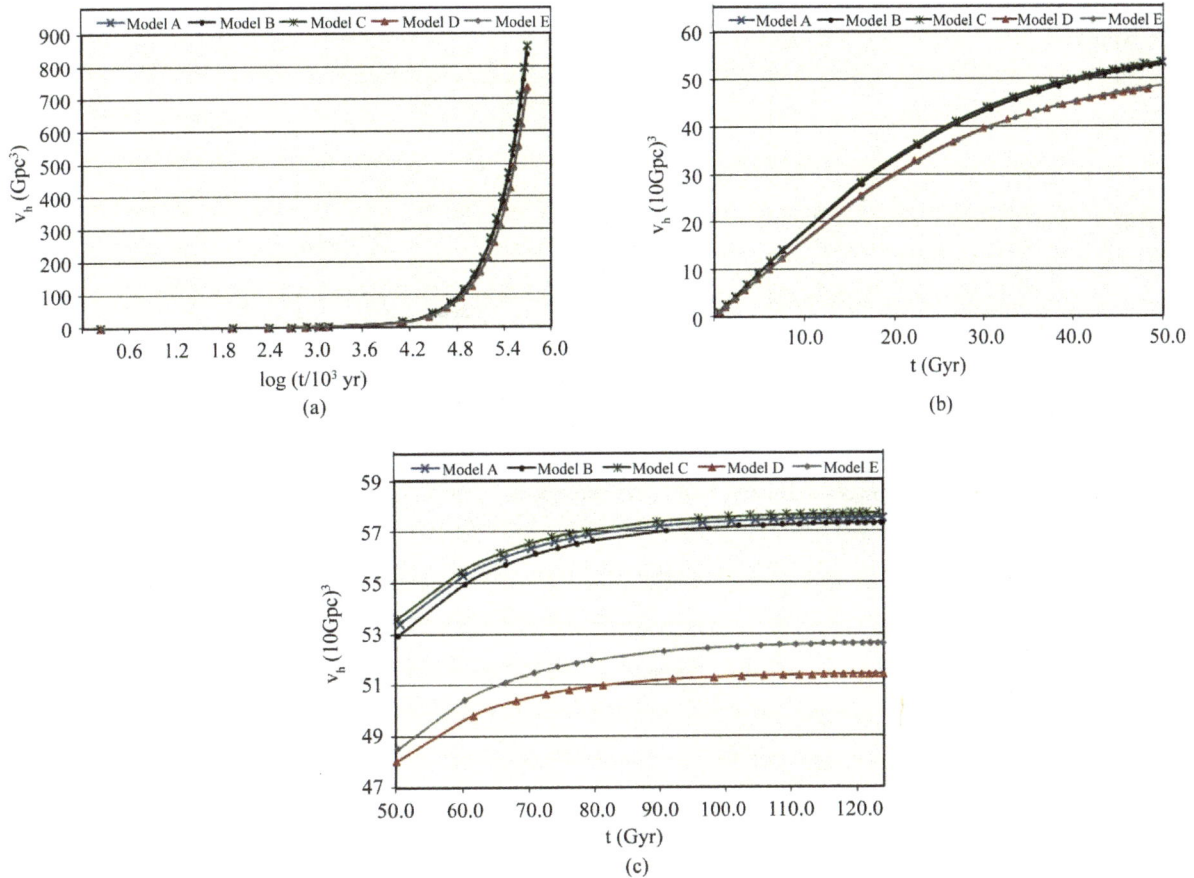

(a)

(b)

(c)

Figure 2. The distribution of the universe horizon volume in the general cosmic models (a) up to $t = 0.5$ Gyr; (b) in the range $t = 0.5 - 50$ Gyr; (c) in the range $t = 50 - 124$ Gyr.

Table 2. Horizon volumes of the universe in the general cosmic models at special times.

Model	$V_h(t_{rm})$ / Mpc^3	$V_h(t_{m\Lambda})$ / $(10\ \text{Gpc})^3$	$V_h(t_o)$ / $(10\ \text{Gpc})^3$	$V_h(t_n)$ / $(10\ \text{Gpc})^3$
A	13820.0	17.4322	24.4819	57.5555
B	11820.0	17.3990	23.7792	57.3248
C	13790.0	17.4916	24.3849	57.7111
D	7890.0	15.7295	22.3372	51.3939
E	7030.0	16.0772	22.0504	52.6234

(a)

(b)

(c)

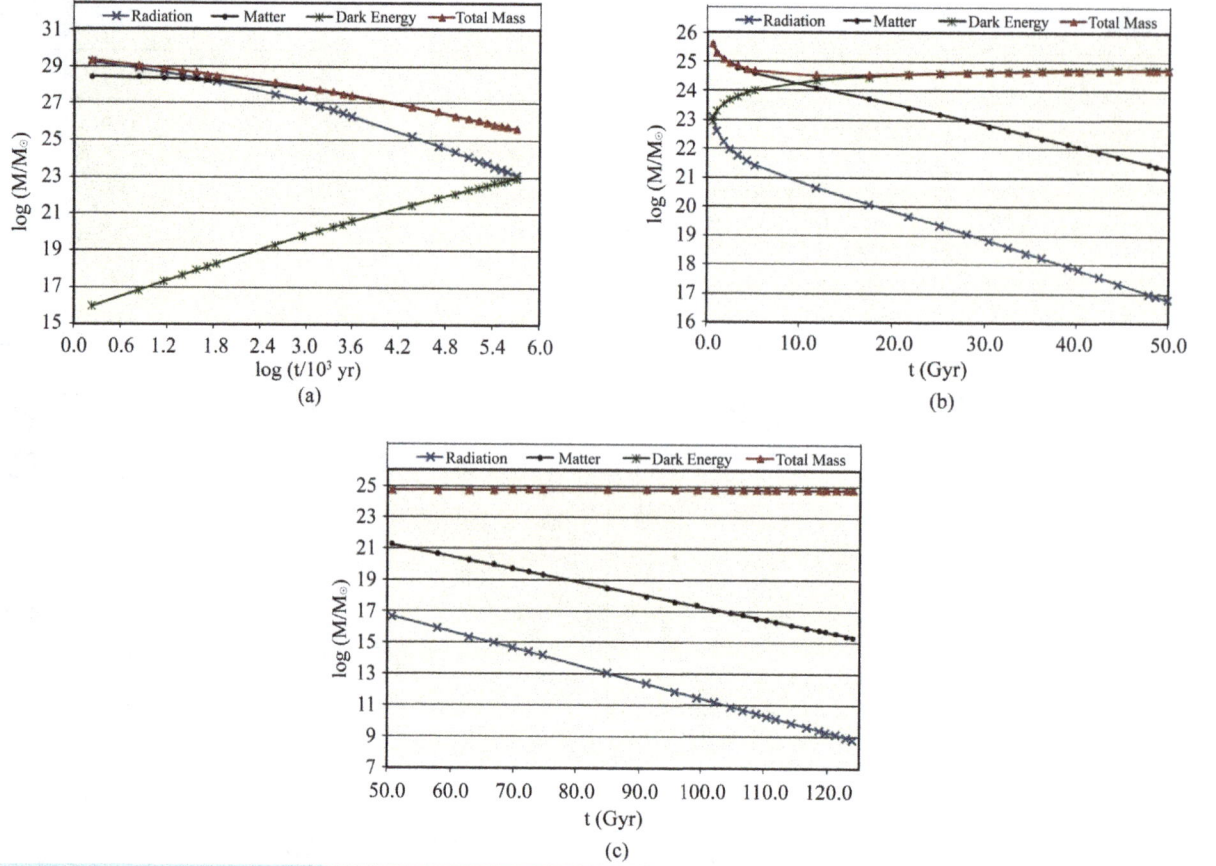

Figure 3. The distribution of mass and energy within the universe horizon volume in any general cosmic model (a) up to $t = 0.5$ Gyr; (b) in the range $t = 0.5$ - 50 Gyr; (c) in the range $t = 50$ - 124 Gyr.

Table 3. Cosmic times at which $M_r(t) = M_m(t)$ within the universe horizon volume in the general cosmic models.

Model	$t_{rm}/10^3$ yr	$\text{Log}(M_{rm}/M_\odot)$	$\text{Log}(M_\Lambda/M_\odot)$
A	55.9162	28.2526	18.1498
B	49.6665	28.2876	18.0712
C	55.8428	28.2529	18.1562
D	38.4783	28.3338	17.9566
E	34.6275	28.3753	17.8515

Table 4. Cosmic times at which $M_r(t) = M_\Lambda(t)$ within the universe horizon volume in the general cosmic models.

Model	$t_{r\Lambda}/$ Gyr	$\text{Log}(M_{r\Lambda}/M_\odot)$	$\text{Log}(M_m/M_\odot)$
A	0.5839	23.0153	25.5410
B	0.5736	22.9834	25.5375
C	0.5808	23.0202	25.5445
D	0.5166	22.9407	25.5350
E	0.5259	22.8906	25.5216

$t = 50 - 124 \, \text{Gyr}$ is exhibited in **Figure 3(c)**. Again the distribution of both matter mass and radiation mass decrease with time and the former is higher than the later. The distributions of dark energy mass and total mass coincide on each other. Masses of radiation, matter and dark energy within the universe horizon volume in the general models at t_n are given in **Table 7**.

Table 8 shows the equivalent number of the Coma-like clusters to the mass of matter within the universe horizon volume $N_{COMA}(t)$ in the general models at the special times $t_{rm}, t_{m\Lambda}, t_o$ and t_n. It is obvious that this content of matter strongly decreases with time such that the cosmic space becomes almost matter empty in the far future of the universe.

5. Conclusion

In this article distributions of the universe horizon distance and universe horizon volume were determined in the five general cosmic models which were established previously. The two distributions were found increasing slowly up to $t \approx 21.5444 \, \text{Myr}$, hence they raise appreciably fast up to $t = 60 \, \text{Gyr}$, then they increase again so slowly until $t = 124 \, \text{Gyr}$. Distributions of mass of radiation, matter and dark energy within the universe horizon volume were also investigated in the five general models. The masses of radiation and matter are decreasing with time although the mass of dark energy is increasing. The mass of radiation was dominant in the early

Table 5. Cosmic times at which $M_m(t) = M_\Lambda(t)$ within the universe horizon volume in the general cosmic models.

Model	$t_{m\Lambda} / \text{Gyr}$	$\text{Log}(M_{m\Lambda}/M_\odot)$	$\text{Log}(M_r/M_\odot)$
A	9.4650	24.2507	20.8831
B	9.6930	24.2393	20.8339
C	9.4525	24.2594	20.8937
D	9.6289	24.2564	20.7974
E	10.0632	24.2109	20.7030

Table 6. Masses of radiation, matter and dark energy within the universe horizon volume in the general cosmic models at $t = t_o$.

Model	$\text{Log}(M_r/M_\odot)$	$\text{Log}(M_m/M_\odot)$	$\text{Log}(M_\Lambda/M_\odot)$
A	20.4531	23.9625	24.3973
B	20.4324	23.9683	24.3785
C	20.4570	23.9663	24.4086
D	20.4035	23.9954	24.3981
E	20.3684	23.9921	24.3328

Table 7. Masses of radiation, matter and dark energy within the universe horizon volume in the general cosmic models at $t = t_n$.

Model	$\text{Log}(M_r/M_\odot)$	$\text{Log}(M_m/M_\odot)$	$\text{Log}(M_\Lambda/M_\odot)$
A	8.8400	15.3480	24.7694
B	8.9797	15.4781	24.7571
C	8.7546	15.2847	24.7779
D	8.1326	14.8863	24.7706
E	8.8615	15.4585	24.7258

Table 8. Equivalent number of the Coma-like clusters to the mass of matter within the universe horizon volume in the general cosmic models at special times.

Model	t_{rm}	$t_{m\Lambda}$	t_o	t_n
A	8.9448×10^{12}	8.9057×10^8	4.5864×10^8	1.1142
B	9.6955×10^{12}	8.6750×10^8	4.6480×10^8	1.5034
C	8.9510×10^{12}	9.0859×10^8	4.6267×10^8	0.9631
D	10.7838×10^{12}	9.0234×10^8	4.9473×10^8	0.3848
E	11.8651×10^{12}	8.1259×10^8	4.9099×10^8	1.4370

universe up to $t = 34627.5 - 55916.2$ yr, where it becomes equivalent to the mass of matter. Afterwards, the mass of matter prevailed until $t = 9.4525 - 10.0632$ Gyr, where it becomes equal to the mass of dark energy. From this time onwards the mass of dark energy dominates the universe. The cosmic space gets approximately matter empty in the very remote future of the universe.

References

[1] Bukhari, F.A. (2013) Five General Cosmic Models. *Journal of King Abdulaziz University: Science*, **25**.

[2] Bukhari, F.A. (2013) Cosmological Distances in Five General Cosmic Models. *International Journal of Astronomy and Astrophysics*, **3**, 183-188.

3

Two Stream Instability as a Source of Coronal Heating

Antony Soosaleon*, Blesson Jose

SPAP, Mahatma Gandhi University, Kottayam, India
Email: *antonysoosaleon@yahoo.com

Abstract

Recent observation of oscillating the two stream instability (TSI) in a solar type III radio bursts and spatial damping of Langmuir oscillations has made this instability as an important candidate to understand the coronal heating problem. This instability has been studied by several authors for cold plasma found to be stable for high frequencies (greater than plasma frequency ω_p). In this paper, we prove that this instability is unstable for warm plasma for higher frequencies (greater than plasma frequency ω_p) and much suitable to study the solar coronal heating problem. We have derived a general dispersion relation for warm plasma and discussed the various methods analyzing the instability conditions. Also, we derived an expression for the growth rate of TSI and analyzed the growth rate for photospheric and coronal plasmas. A very promising result is that the ion temperature is the source of this instability and shifts the growth rate to high frequency region, while the electron temperature does the reverse. TSI shows a high growth rate for a wide frequency range for photosphere plasma, suggesting that the electron precipitation by magnetic reconnection current, acceleration by flares, may be source of TSI in the photosphere. But for corona, these waves are damped to accelerate the ions and further growing of such instability is prohibited due to the high conductivity in coronal plasma. The TSI is a common instability; the theory can be easily modifiable for multi-ion plasmas and will be a useful tool to analyze all the astrophysical problems and industrial devices, too.

Keywords

Coronal Heating, Two Stream Instability, Langmuir Waves, Ion Temperature, Drift Velocity, Photosphere, Fusion Plasma

---

*Corresponding author.

1. Introduction

Solar corona is hot to the level of million degree kelvin; several mechanisms have been put forward for the last 60 years which is continued as the problem exist. In this paper, we try to add one more to the list of possible mechanisms as a source of coronal heating, which is the streaming instability. Streaming instabilities arise when there is relative velocity between ions and electrons in a plasma. The simplest type of streaming instability is the two stream instability (TSI). This arises in an electron-proton plasma with electrons in relative motion with ions. This type of streaming instability arises in stellar atmospheres, since the stellar plasmas are predominantly electron-proton type. A specific example for this instability is an electron-precipitation related phenomenon in solar chromosphere [1], and also the presence of oscillating TSI in a solar type III radio bursts [2] is a strong experimental support for our theoretical calculation which shows a greater growth rate of TSI in the photosphere. The very important point presented by them is the spatial damping of Langmuir waves induced by the TSI, which is also our conclusion that these waves are damped out in coronal plasma which heats the ions. The streaming instability can also be ignited when a high energy beam of electrons created during the process of reconnection taking place at the site of a solar flare comes down towards the chromosphere. Since the density of chromosphere plasma is high, the energetic electrons from the flare suffer collisions and transfers their energy to electrons and protons and to the small population of heavier ions too. These secondary electrons and ions are accelerated with different speeds [3] [4]. This creates a plasma in which electrons and protons are at different velocities and always electrons drive fast to ions which create the situation for the TSI. The TSI for hot plasmas also arises at chromospheric foot points heating by energetic streams from magnetic reconnection, which is discussed in [5] and [6]. Similar situation happens at helmet streamers, cometary atmospheres and earth magneto sphere, etc.

The electromagnetic wave propagation and instabilities for the counter streaming astrophysical situations for cold plasma have been discussed by several authors [7]-[12], who find them unimportant, because it is unstable for very low frequency (less than plasma frequency ω_p), but TSI is a high frequency wave and also the growth rate is negligible for cold plasma. But for warm plasma, we find it is unstable for a very wide range of frequencies and even for the frequencies much greater than ω_p and that the growth rate is very large and sensitively depends on the ion (electron) temperatures which is ideal for solar atmosphere.

An interesting result is that the growth rate shifts to the higher frequency region as the ion temperature increases, while the electron temperature shifts it to the lower frequency region. It is the ion temperature that becomes the source of TSI as the high frequency oscillations, for it is a necessary condition presumed for this instability and hence the damping of this instability will heat the ions. Also, this instability depends on the electron temperature, but stabilises the plasma. It is found that the growth rate region shifts towards the low frequency regions as the electron temperature increases in reverse of the shift due to ion temperature. As the kinetic energy of electrons shifts the growth rate to the low frequency region which is nothing, damping of high frequency oscillation results in the heating of ions. The high frequency oscillations are induced by the electric field due to the electron drift which accelerates the ions at the expense of kinetic energy of electrons, and this is true because the increase in drift velocity decreases the instability. This instability is a common one and hence this theory can be applied to any hot electron-proton plasma and can be better for fusion studies too.

2. Theory

For the theoretical study of streaming instability in a hot plasma, we consider an electron-proton plasma, with ions assumed stationary and electrons moving with a velocity v_0 relative to ions. This is same as we assume the observer moving with a stream of ions. We consider hot plasma $KT \neq 0$. For simple analysis we consider the case of zero ambient magnetic field $\left(B_0 = 0 \right)$. It can easily shown that the same results can be applied for electrostatic waves along magnetic field.

The linearised equation of motion for protons and electrons are respectively:

$$Mn_0 \frac{\partial \boldsymbol{v}_{i1}}{\partial t} = en_0 \boldsymbol{E}_1 - \gamma_i KT_i \nabla n_{i1} \tag{1}$$

$$mn_0 \frac{\partial \boldsymbol{v}_{e1}}{\partial t} = -en_0 \boldsymbol{E}_1 - \gamma_e KT_e \nabla n_{e1} \tag{2}$$

We consider electrostatic waves of the form,

$$E_1 = E \exp\left[i(kx - \omega t)\right]\hat{x} \tag{3}$$

where $\hat{x}$ is in the direction of v_0 and k.

Under these conditions Equation (1) gives

$$-i\omega M n_0 v_{i1}\hat{x} = e n_0 E\hat{x} - \gamma_i K T_i i k n_{i1}\hat{x}$$

i.e.,

$$v_{i1} = \frac{ie}{M\omega}E\hat{x} + \frac{\gamma_i K T_i}{M\omega}k\left(\frac{n_{i1}}{n_0}\right)\hat{x} \tag{4}$$

Similarly, Equation (2) gives

$$v_{e1} = \frac{-ie}{m}\frac{E}{(\omega - kv_0)}\hat{x} + \frac{\gamma_e K T_e}{m(\omega - kv_0)}k\left(\frac{n_{e1}}{n_0}\right)\hat{x} \tag{5}$$

The same results hold good for parallel electrostatic plasma oscillations (i.e., wave propagation parallel to B_0.) The ion continuity equation for our case is

$$\frac{\partial n_{i1}}{\partial t} + n_0\left(\nabla \cdot v_{i1}\right) = 0 \tag{6}$$

Linearising this and noting that $\nabla \cdot v_0$ and ∇n_0 vanishes, and using the value of v_{i1}, we get for protons

$$n_{i1} = \frac{k}{\omega}n_0 v_{i1} = \left(\frac{ie n_0 k}{M\omega^2}E + \frac{k^2 \gamma_i K T_i}{M\omega^2}n_{i1}\right) \tag{7}$$

Similarly the electron continuity equation is,

$$\frac{\partial n_{e1}}{\partial t} + n_0\left(\nabla \cdot v_{e1}\right) + \left(v_0 \cdot \nabla\right)n_{e1} = 0 \tag{8}$$

Linearising this we get

$$n_{e1} = \frac{k n_0}{\omega - kv_0}v_{e1} = \left(-\frac{ie k n_0}{m(\omega - kv_0)^2}E + \frac{k^2 \gamma_e K T_e}{m(\omega - kv_0)^2}n_{e1}\right) \tag{9}$$

Simplifying Equation (7) and Equation (9) we get,

$$n_{i1} = \frac{ie n_0 kE}{M\omega^2 - k^2 \gamma_i K T_i} \tag{10}$$

and

$$n_{e1} = \frac{-ie n_0 kE}{m(\omega - kv_0)^2 - k^2 \gamma_e K T_e} \tag{11}$$

The plasma waves leading to TSI are high frequency plasma oscillations. To deal with these type of waves it is well known that we should use Poisson's equations

$$\epsilon_0 \nabla \cdot E = \rho \tag{12}$$

i.e.,

$$\epsilon_0 \nabla \cdot E_1 = e\left(n_{i1} - n_{e1}\right) \tag{13}$$

Here the electric field induced is due to the perturbation in the density. Assuming the perturbation as $e^{i(kx - \omega t)}\hat{x}$ and substituting the value of n_{i1} and n_{e1} in Equation (13) and simplifying we obtain the following dispersion relation

$$1 = \omega_p^2\left[\frac{m/M}{\omega^2 - k^2\dfrac{\gamma_i K T_i}{M}} + \frac{1}{(\omega - kv_0)^2 - k^2\dfrac{\gamma_e K T_e}{m}}\right] \tag{14}$$

3. Results and Discussions

Equation (14) is a fourth-order equation in ω. and there will be four roots for this equation and each root represents a possible oscillation $E_1 = E e^{i(kx - \omega_j t)} \hat{x}$, if all roots are real or if some roots are complex, they occur in complex conjugate pairs, and real part represents the propagation modes and imaginary part represents the instability. This could be analyzed by putting $\omega_i = \alpha_i + i\gamma_i$. Here α and γ are the real part and imaginary part of frequency respectively. Now the oscillation is re expressed as

$$E_1 = E e^{i(kx - \alpha_i t)} e^{\gamma_i t} \hat{x}$$

Positive $\operatorname{Im}(\omega)$ denotes a growing wave and negative $\operatorname{Im}(\omega)$ indicates a damping wave.

As a check for knowing whether there is any unstable modes in the plasma, we follow the procedure in [9] and re-write the equation 14 as

$$1 = \frac{m/M}{x^2 - a^2 y^2} + \frac{1}{(x-y)^2 - b^2 y^2} \tag{15}$$

where $a^2 = \dfrac{v_i^2}{v_0^2}$, $b^2 = \dfrac{v_e^2}{v_0^2}$, $x = \dfrac{\omega}{\omega_p}$ and $y = \dfrac{kv_0}{\omega_p}$. $v_i^2 = \dfrac{\gamma_i K T_i}{M}$ and $v_e^2 = \dfrac{\gamma_e K T_e}{m}$ are the thermal velocities of ion and electron respectively. The dispersion relation shown in Equation (15) is a function of four variables $F(x, y, a, b)$. To make more clarity, it is better to discuss the cold plasma first and then extend to warm plasma. The dispersion relation for cold plasma is obtained by putting $a = b$ in Equation (15) and hence the dispersion relation becomes

$$1 = \frac{m/M}{x^2} + \frac{1}{(x-y)^2} = F(x, y) \tag{16}$$

For any given y, we can plot $F(x, y)$ as a function of x. This function will have singularities at certain particular values of x which can be found graphically. The singularities of cold plasma case is at $x = 0$ and $x = y$. We can plot $F(x, y)$ as a function of x. The intersection of the curve with the line $F(x, y) = 1$, gives the values of x satisfying the dispersion relation. Since it is a fourth power of equation, there must be four intersection for four roots if all roots are real. Anything less than four intersection signifies the instability. The first graph is plotted to the value of $y = 4$ and second graph is for $y = 1$ shown in **Figure 1**.

We could see four intersection in the first graph, which means the plasma is stable. But for in the second graph, there is only two intersection means the plasma is unstable, for $y = 1$. As we know that $y = \dfrac{kv_0}{\omega_p}$, i.e,

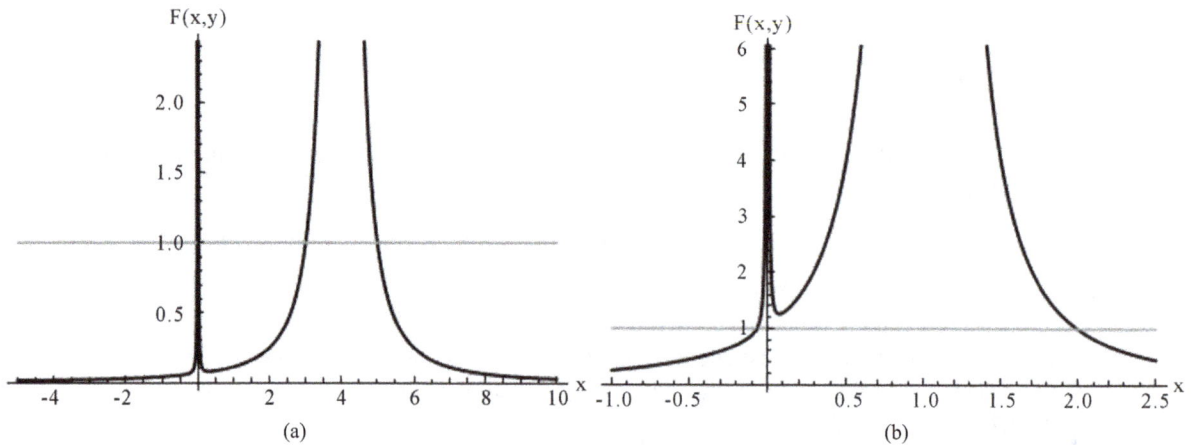

Figure 1. Figure shows the streaming instability for cold plasmas. The first graph (a) shows the intersection of $F(x, y)$ with the line $F(x, y) = 1$, with four intersections and hence no imaginary roots for a higher value of $y = 4$. The second graph shows the case of $y = 1$ with two intersections and hence shows instability.

when kv_0 is less than or equal to ω_p (the plasma frequency), the plasma becomes unstable. This could understood as that, when the frequency of induced oscillations is less than the plasma oscillation, the plasma derives the energy which could be probably at the expense of the energy of plasma waves. When the frequency is higher than the plasma frequency $(y > 1)$ the plasma becomes stable because the the oscillation does not grow due to the lack of internal energy to support it.This shows that the cold plasma is stable for large values of y, but for sufficiently small values of drift velocity the plasma becomes unstable.

Now we can look into the dispersion relation of warm plasma, as it is to compare with the cold plasma, the plot is done for $y = 4$ in which the cold plasma is stable for this value. Since we have additional two more parameters, we want to see the effect of a and b separately, so we neglect the thermal effect of electron by $(T_e = 0)$ by setting $b = 0$ and a is not equal to zero, shown in **Figure 2**.

We see the plasma is unstable with finite ion thermal velocity even without any thermal velocity of electron. For any higher values of b the plot is same and hence the plasma shows instability. This is quite interesting that the ion temperature becomes a source of instability without the factor of considering the electron temperature. The additional mobility given to the ions by the temperature, negate the deficit of the density due to the high drift speed of electrons or the moving ions could easily find the sufficient electrons in the new spatial situation which sustains the field.

Also this could be verified by looking in to the Poissons equation that the electric field induced is depends on the perturbation density of ions and electrons. If we look into Equation (10)

$$n_{i1} = \frac{i e n_0 k E}{M \omega^2 - k^2 \gamma_i K T_i}$$

we see the second term in the denominator is with negative sign, which shows that the density of ions is increased by the temperature ions and hence the ion temperature becomes the source of instability. This could also viewed as the electric field induced by the drift of steaming electrons accelerate the ions to maintain the instability or the damping waves heats the ions.

Then we wish to analyze the effect of electron temperature, we set the ion temperature zero $(a = 0)$ and plot for different values of b, which is shown in **Figure 3**.

Figures 3(a)-(c) show the plot for the value of $y = 4$ and with zero ion temperature with different values of $b = 1, 2, 3$. The first two graph shows that the plasma is stable for the ratio of the thermal velocity of electron to drift velocity till the value 2, but when the thermal velocity becomes more than twice the value of drift, the plasma becomes unstable (shown in the third graph **Figure 3(c)**). This suggests that the minimum thermal velocity to ignite the instability must be greater than twice of the drift velocity. This result could be analyzed by little more detailed way by looking in to the result of cold plasma, we see from the figure the cold plasma is unstable for lower values of drift speed ($y = \dfrac{kv_0}{\omega_p}$, $y = 1$, shown in **Figure 2**). The plasma oscillations

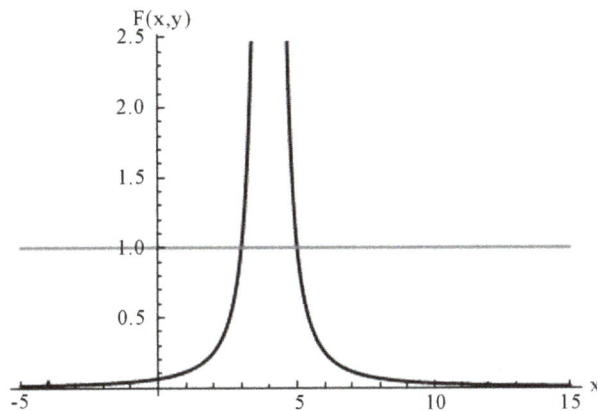

Figure 2. The streaming instability of hot plasmas is shown here. The figure is for $y = 4$, $a = 1$ respectively and $b \left(\dfrac{v_{the}}{v_0} \right) = 0$.

(a)

(b)

(c)

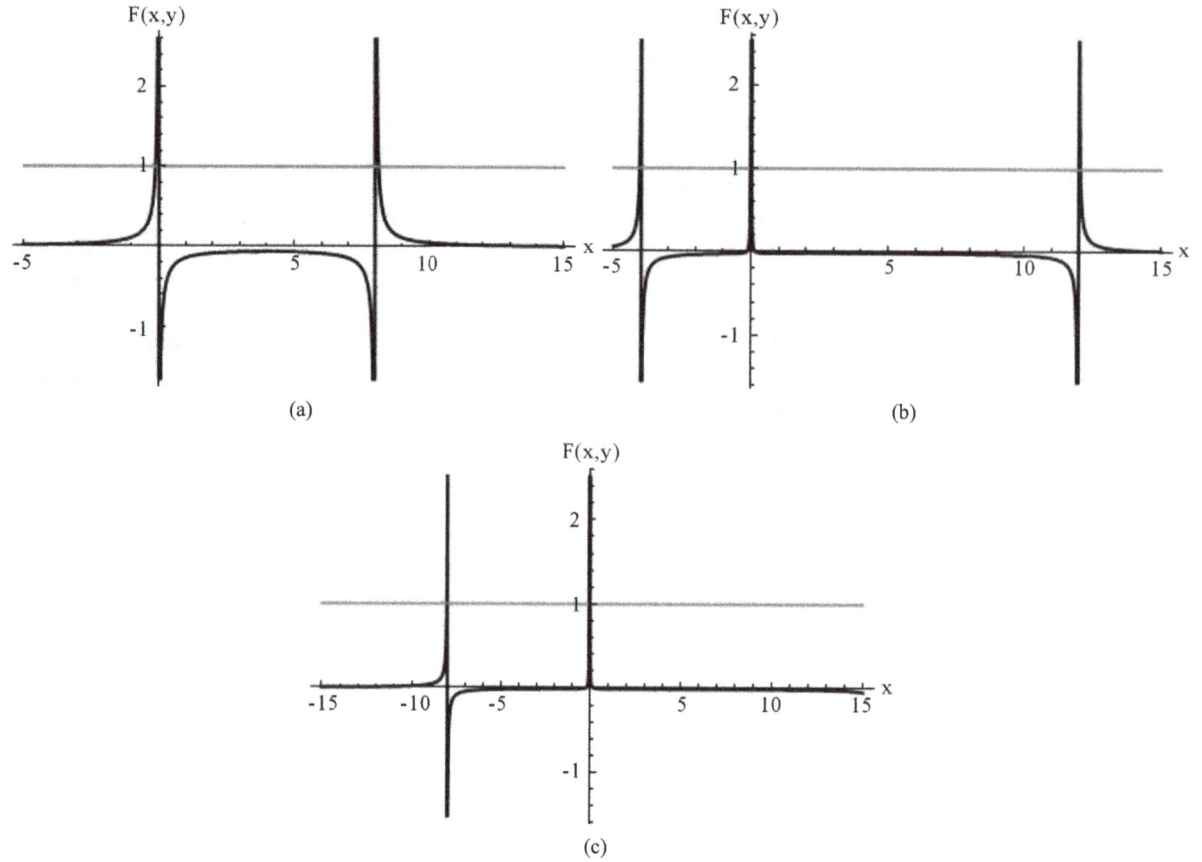

Figure 3. The streaming instability of hot plasmas is shown here. First three graphs (a), (b), and (c) are for $y = 4$ at normalized electron thermal speeds $\dfrac{v_{the}}{v_0} = 1, 2, 3$ respectively with the ion temperature zero.

depends on the sustainability of the electric field, which requires the charge neutrality conditions should be maintained for the oscillations. When the drift velocity of electrons increases, the plasma loses the charge neutrality conditions or local energy is reduced and hence the oscillations easily die out for larger drift speed. When the thermal electrons are added, it gives sufficient background density to support the electric field or it gives an additional local energy to maintain the oscillations. This could be verified from the equation, *i.e.*, density of perturbation of electrons given as

$$ n_{el} = \frac{-ien_0 kE}{m\left(\omega - kv_0\right)^2 - k^2\gamma_e KT_e} $$

If we look into the equation, we see the second term in the denominator contains the square of drift speed which decreases the electron density, but the last term in the denominator is the kinetic pressure term it also reduces the density of electrons. So, it justifies the result that drift velocity of electrons and electron temperature reduces the instability.

As we have seen that the instability is sensitive to the drift speeds and thermal speeds, we have to analyze the growth rate for various temperatures. For deducing the growth rate, expand Equation 16, we arrive at the following fourth power equation. We know that for sufficiently small y, that is the plasma is unstable for $kv_0 \approx y\omega_{pe}$ [5]. The dispersion relation can be written in the following form,

$$ 1 - \frac{\omega_{pe}^2\left(m/M\right)}{\left(\omega^2 - \omega_{pe}^2 a^2 y^2\right)} - \frac{\omega_{pe}^2}{\left(\omega - y\omega_{pe}\right)^2 - \omega_{pe}^2 b^2 y^2} = 0 \qquad (17) $$

i.e.,

$$1 - \frac{\omega_{pi}^2}{\left(\omega^2 - \omega_{pe}^2 a^2 y^2\right)} - \frac{1}{\left(y - \dfrac{\omega}{\omega_{pe}}\right)^2 - b^2 y^2} = 0 \tag{18}$$

Since $\dfrac{1}{\left(y - \dfrac{\omega}{\omega_{pe}}\right)^2 - b^2 y^2} \approx \dfrac{1}{y^2}\left(1 + \dfrac{2\omega}{y\omega_{pe}}\right) + \dfrac{1}{y^2}\dfrac{b^2 y^2\left(1 + \dfrac{2\omega}{y\omega_{pe}}\right)}{-b^2 y^2 + \left(y^2 - \dfrac{2y\omega}{\omega_{pe}}\right)}$, the dispersion relation can be brought

to the form,

$$-\frac{\omega_{pi}^2 y^2}{\omega^2 - \omega_{pe}^2 a^2 y^2} + \left(y^2 - 1\right) - \frac{2\omega}{y\omega_{pe}} - \frac{b^2 y^2\left(1 + \dfrac{2\omega}{y\omega_{pe}}\right)}{-b^2 y^2 + \left(y^2 - \dfrac{2y\omega}{\omega_{pe}}\right)} = 0 \tag{19}$$

After some simplifications and, the dispersion relation can be brought to the form

$$4x^4 - 2y^3 x^3 + \left[y^4 - y^2\left(1 + b^2 y^2 + 4a^2\right)\right]x^2 + 2y^3\left[\frac{m}{M} + a^2 y^2\right]x$$
$$+ y^2\left[y^2\left\{\frac{m}{M}\left(b^2 - 1\right) + a^2\left[1 + \left(b^2 - 1\right)y^2\right]\right\}\right] = 0, \tag{20}$$

This is the general dispersion relation for warm plasma for two stream instability. This is a forth power equation in x $\left(\omega/\omega_p\right)$, has four roots and one of these roots shows high growth rate growth. The growth rate is plotted for coronal plasmas. Here also we want to check the influence of the electron and ion temperatures separately.

Figure 4(a) gives the growth rate for the different values of a by keeping b as zero. As our earlier discussion on the influence of ion temperature on the instability, this graph clearly shows that the growth rate of the instability region shifted to the higher values of y $\left(kv_0\right)$, which is high frequency oscillations when ion temperature increases. As this is a high frequency oscillations which is driven by the ion temperature will be an effective way to heat the ions by the inducing these oscillations. This result could be more confirmed by analyzing the graph **Figure 4(b)** is a plot for different electron temperature with ion temperature zero. Here, in this graph, the growth rate shifts to the low frequency region for higher and higher electron temperature. Also it is to be noted that the growth rate is found only for high value of b which is greater than 2, as it is discussed earlier. The instability arises only when the kinetic energy is twice the drift energy as we know that for higher drift the plasma is stabilized, so any increase in the kinetic energy will be used for increasing the kinetic energy of ions to maintain the oscillations.

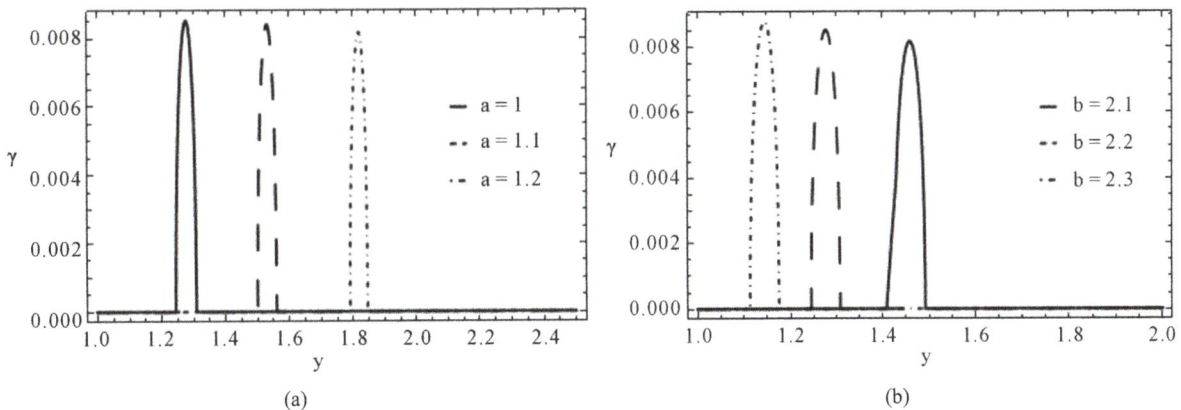

(a)   (b)

Figure 4. (a) for different a values with $a = 0$; (b) for different b values with $a = 0$.

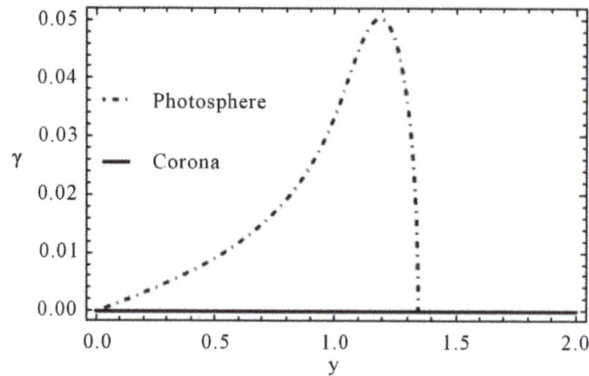

Figure 5. Plotted for drift velocity $v_0 = 10^6$.

To check the effect these waves over the solar atmosphere, we plot the growth rate for photosphere and corona and found the results are as expected shown in **Figure 5**. The high growth rate of photosphere plasma assures the possibility of such oscillations [1] [2] and these waves could drive the ions and these waves are damped out in the coronal plasma, which confirms the heating ions by these waves in the corona.

4. Conclusions

Streaming instabilities arise when there is relative velocity between ions and electrons in a plasma. The simplest type of streaming instability is the two stream instability which arises in an electron-proton plasma with electrons moving faster relative to ions. This instability has been studied by several authors for cold plasma, but for warm plasma, it is the first time that a study has been done by deriving a general dispersion relation. The dispersion relation for TSI is a fourth power equation in the angular frequency and wave vector, which has been studied for different conditions. For cold plasma, this instability arises only for lower values kv_0, which is less than the plasma frequency ω_p, and hence the growth rate is very small, which is proportional to $(m/M)^{1/3}$ where m is the mass of electron and M is the mass of proton. But for warm plasma, it is unstable for higher frequencies greater than the plasma frequency, which shows a good growth rate.

This instability sensitively depends on the ion temperature, the growth rate shifts to the higher, and higher frequency region as the ion temperature increases. The high frequency oscillations are necessary conditions or pre-assumptions for this instability, that is the ion temperature become the source of this instability or the reverse can be more sensible that the damping of this instability will heat the ions.

Also, this instability depends on the electron temperature; the thermal velocity of electrons must be greater than a critical value which is double of the drift velocity, which means that the kinetic energy of electrons must be double times to negate the loss of the energy due to the drift velocity. It is found that the growth rate shifts towards the low frequency regions when the electron temperature increases. This is also quite exciting as it is a high frequency oscillation, if the kinetic energy of electrons shifts the growth rate to the low frequency region, to conserve the energy; the ions must be heated at the expense of the kinetic drift of electrons. This is quite true from the study when the drift velocity increases the instability decreases.

We have studied this instability for the coronal plasma and photospheric plasma, and found the result as predicted that there is no growth rate for coronal plasmas; the oscillations are damped out to heat the ions but for photosphere there is a good growth rate. This instability is a common one and hence this result can be applied to any hot electron-proton plasma and can be more useful to fusion devices.

Acknowledgements

The authors acknowledge the financial help made by the UGC for this work as a minor project on "Role of Macro Instabilities in Solar Coronal Heating".

References

[1]   Aschwanden, M.J. (2005) Physics of Solar Corona. Springer, Berlin.

[2] Thejappa, G., MacDowall, R.J., Bergamo, M. and Papadopoulos, K. (2012) Evidence for the Oscillating Two Stream Instability and Spatial Collapse of Langmuir Waves in a Solar Type III Radio Burst. *Astrophysical Journal*, **747**, L1. http://dx.doi.org/10.1088/2[041-8205/747/1/L1

[3] Winglee, R.M. (1989) Heating and Acceleration of Heavy Ions during Solar Flares. *Astrophysical Journal*, **343**, 511-525. http://dx.doi.org/10.1086/167726

[4] Brown, J.C., Karlicky, M., MacKinnon, A.L., *et al.* (1990) Beam Heating in Solar Flares—Electrons or Protons? *Astrophysical Journal*, **73**, 343-348. http://dx.doi.org/10.1086/191470

[5] Dwivedi, B.N. and Narain, U. (2006) Physics of Sun and Its Atmosphere. *Proceedings of the National Workshop (India) on Recent Advances in Solar Physics*, Meerut, November 2008.

[6] Sturrock, P.A., Holzer, T.E., Mihalas, D.M. and Ulrich, R.K. (1986) Physics of the Sun. Vol. 2, D. Reidel Publishing Company, Dordrecht.

[7] Tautz, R.C. and Schlickeiser, R. (2005) Covariant Kinetic Dispersion Theory of Linear Waves in Anisotropic Plasmas. III. Counterstreaming Plasmas. *Physics of Plasmas*, **12**, Article ID: 072101. http://dx.doi.org/10.1063/1.1939967

[8] Tautz, R.C. and Schlickeiser, R. (2005) Counterstreaming Magnetized Plasmas. I. Parallel Wave Propagation. *Physics of Plasmas*, **12**, Article ID: 122901. http://dx.doi.org/10.1063/1.2139505

[9] Tautz, R.C., Schlickeiser, R., (2006) Counterstreaming Magnetized Plasmas. II. Perpendicular Wave Propagation. *Physics of Plasmas*, **13**, Article ID: 062901. http://dx.doi.org/10.1063/1.2207588

[10] Chen, F.F. (1981) Introduction to Plasma Physics and Controlled Fusion. Plenum Press, New York.

[11] Nicholson, D.R. (1983) Introduction to Plasma Theory. Wiley, Hoboken.

[12] Treumann, R.A. and Baumjohann, W. (2001) Advanced Space Plasma Physics. Imperial College Press, London.

4

Continued Fraction Evaluation of the Universal Y's Functions

Mohammed Adel Sharaf[1], Abdel-naby Saad Saad[2,3], Nihad Saad Abd El Motelp[2,4]

[1]Department of Astronomy, Faculty of Science, King Abdulaziz University, Jeddah, KSA
[2]Department of Astronomy, National Research Institute of Astronomy and Geophysics, Cairo, Egypt
[3]Department of Mathematics, Preparatory Year, Qassim University, Buraidah, KSA
[4]Department of Mathematics, Preparatory Year for Girls Branch, Hail University, Hail, KSA
Email: Sharaf_adel@hotmail.com, Saad6511@gmail.com, nihad_planet@hotmail.com

Abstract

In the present paper, an efficient algorithm based on the continued fractions theory was established for the universal Y's functions of space dynamics. The algorithm is valid for any conic motion (elliptic, parabolic or hyperbolic).

Keywords

Universal Kepler Equation, Continued Fraction Technique, Y-Universal Function

1. Introduction

Today, one of the well-known facts of space dynamics is the desperate needs of the universal formulations of orbital motion. This is because, in complete interplanetary transfer, all types of the two body motion (elliptic, parabolic, or hyperbolic) appear, moreover, the given type of an orbit is occasionally changed by perturbing forces acting during finite interval of time. Thus far, we have been obliged to use different functional representations for motion depending upon the energy state (elliptic, parabolic, or hyperbolic) and a simulation code must then contain branching to handle a switch from one state to another. In cases where this switching is not smooth, branching can occur many times during a single integration time-step causing some numerical "chatter". Consequently, through the use of the universal formulations, orbit predictions will be free of the troubles, since a single functional representation suffices to describe all possible states.

Recently Sharaf and Saad [1] (hereafter will be referred to as Paper I) established new set of the universal functions (Y-functions) for the two-body initial value problem. Due to the importance of accurate universal orbital predications using the Y-functions, an efficient algorithm based on the continued fractions theory was established for these functions.

2. The Universal Y's Functions

The universal Y's functions are given by:

$$Y_n(\chi;\alpha) = \left(\chi\sqrt{\mu}\right)^n \sum_{k=0}^{\infty} (-1)^k \frac{\left(\alpha\mu\chi^2\right)^k}{(2k+n)!}, \tag{1}$$

where χ is to be considered, as a new independent variable—a kind of *generalized anomaly*, α is just the inverse of the semi-major axis a given as:

$$\alpha = \frac{1}{a} = \frac{2}{r} - \frac{v^2}{\mu}, \tag{2.1}$$

$$\alpha = \begin{cases} = 0 \ (\text{or } e = 1) & \text{Parabolic orbits,} \\ > 0 \ (\text{or } e < 1) & \text{Elliptic orbits,} \\ < 0 \ (\text{or } e > 1) & \text{Hyperbolic orbits,} \end{cases} \tag{2.2}$$

μ is the gravitational parameter, finally, r and v are the magnitudes of the position and velocity vectors respectively.

What concerns us among the properties of the Y's functions given in Paper I are:

$$Y_n(\chi;0) = \frac{\left(\chi\sqrt{\mu}\right)^n}{n!}, \tag{3.1}$$

$$\alpha Y_{n+2}(\chi;\alpha) = \frac{1}{n}\left\{\alpha\chi\sqrt{\mu}Y_{n+1}(\chi;\alpha) - nY_n(\chi;\alpha) + \chi\sqrt{\mu}Y_{n-1}(\chi;\alpha)\right\}. \tag{3.2}$$

Figure 1 and **Figure 2** show the three dimension visualizations of Y_1 & Y_2 with $\mu = 1, -2\pi \le \chi \le 1.5\pi$ and $-3 \le \alpha \le 3$.

3. Continued Fraction Method

In fact, continued fraction expansions are generally far more efficient tools for evaluating the classical functions

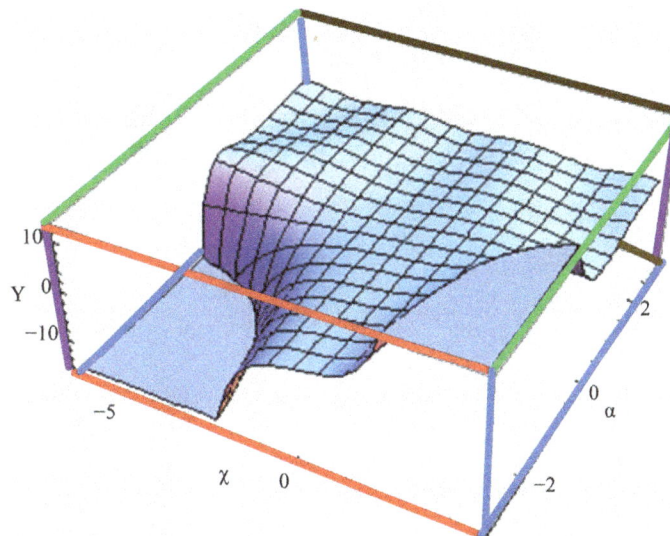

Figure 1. Visualization of Y_1 function in three-dimensional space.

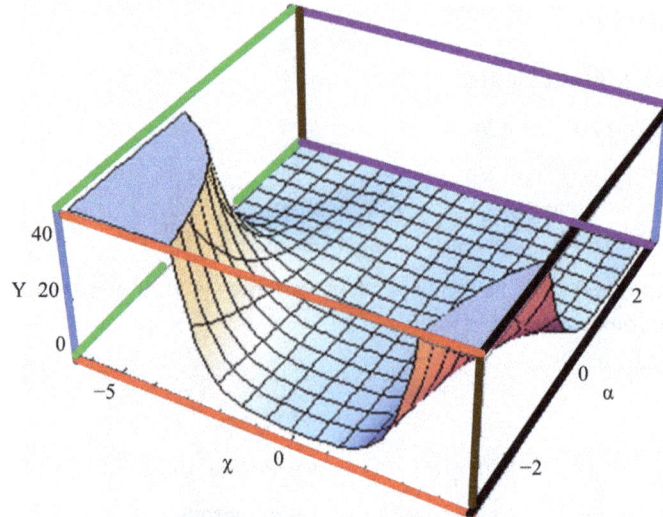

Figure 2. Visualization of Y_2 function in three-dimensional space.

than the more familiar infinite power series. Their convergence is typically faster and more extensive than the series.

Top-Down Continued Fraction Evaluation

There are several methods available for the evaluation of continued fraction. Traditionally, either the fraction was computed from the bottom up, or the numerator and denominator of the nth convergent were accumulated separately with three-term recurrence formulae. The drawback of the first method is obviously, having to decide far down the fraction to being in order to ensure convergence. The drawback to the second method is that the numerator and denominator rapidly overflow numerically even though their ratio tends to a well-defined limit. Thus, it is clear that an algorithm that works from top down while avoiding numerical difficulties would be ideal from a programming standpoint.

Gautschi [2] proposed very concise algorithm to evaluate continued fraction from the top down and may be summarized as follows. If the continued fraction is written as

$$c = \cfrac{n_1}{d_1 + \cfrac{n_2}{d_2 + \cfrac{n_3}{d_3 + \cdots}}},$$

then initialize the following parameters

$$a_1 = 1, \quad b_1 = n_1/d_1, \quad c_1 = n_1/d_1$$

and iterate $\left(k = 1, 2, \cdots\right)$ according to:

$$a_{k+1} = \cfrac{1}{1 + \left[\cfrac{n_{k+1}}{d_k d_{k+1}}\right] a_k}$$

$$b_{k+1} = \left[a_{k+1} - 1\right] b_k,$$

$$c_{k+1} = c_k + b_{k+1}.$$

In the limit, the c sequence converges to the value of the continued fraction. Continued fraction method was used in many problems in astrophysics [3] [4] as well as in special functions of astrodynamics [5] [6].

4. Evaluation of the Y's Functions

In the following, we shall consider the evaluations of the four functions $Y_j\left(\chi; \alpha\right); j = 0, 1, 2, 3$ only, because these four functions appear in the orbital motion when treated by the Y's functions (see Paper I) , on the other hand, the functions $Y_i(\chi;\alpha); i \geq 4$ could be obtained from $Y_j(\chi;\alpha); j = 0,1,2,3$ by using the recurrence relation (3.2) for $\alpha \neq 0$ and directly from Equation (3.1) if $\alpha = 0$.

4.1. Expression of $u = Y_1\left(\frac{1}{2}\chi;\alpha\right)\Big/Y_0\left(\frac{1}{2}\chi;\alpha\right)$ as Continued Fractions

From the expressions of $\tan x$ and $\tanh x$ as continued fractions [7] for any α we can show that,

$$u = \frac{a_0}{1+}\frac{a_1}{1+}\frac{a_2}{1+}\cdots$$

where

$$a_j = -\frac{\alpha\mu\chi^2}{4(4j^2-1)}; j = 1,2,\cdots; \quad a_0 = \frac{1}{2}\chi\sqrt{\mu}$$

4.2. Computational Algorithm

Input: α, μ, χ
Output $Y_j(\chi;\alpha); j = 0,1,2,3$
Computational sequence
1-Compute a's from

$$a_j = -\frac{\alpha\mu\chi^2}{4(4j^2-1)}; \quad j = 1,2,\cdots; \quad a_0 = \frac{1}{2}\chi\sqrt{\mu}$$

2-Compute u from the continued fraction

$$u = \frac{a_0}{1+}\frac{a_1}{1+}\frac{a_2}{1+}\cdots$$

by using Gautschi's algorithm of Subsection 3.1
3- $A = 1 + \alpha u^2$
4- $Y_0(\chi;\alpha) = (1-\alpha u^2)\big/A$
5- $Y_1(\chi;\alpha) = 2u/A$
6- $Y_2(\chi;\alpha) = uY_1(\chi;\alpha)$

$$7\text{-} Y_3(\chi;\alpha) = \begin{cases} \left\{\alpha\chi\sqrt{\mu}Y_2(\chi;\alpha) - Y_1(\chi;\alpha) + \chi\sqrt{\mu}Y_0(\chi;\alpha)\right\}\big/\alpha; & \alpha \neq 0, \\ \dfrac{\left(\chi\sqrt{\mu}\right)^3}{6}; & \alpha = 0. \end{cases}$$

8-The algorithm is completed.

4.3. Numerical Applications

The applications of the above algorithm for the numerical values of Y_0, Y_1, Y_2 and Y_3, $\mu = 1$ and for some values of α and χ, are listed in **Table 1**.

Table 1. Numerical values of $Y_{0,1,2,3}$, $\mu = 1$ for some values of α and χ.

No	α	χ	Y_0	Y_1	Y_2	Y_3
1	−3	−3.14159	115.384	−66.6147	38.1282	−21.1577
2	−2	−2.14159	10.359	−7.29074	4.67952	−2.57457
3	−1	−1.14159	1.72553	−1.40622	0.725531	−0.264628
4	0	−0.141593	1.00000	−0.141593	0.0100242	−0.0004731
5	1	0.858407	0.653644	0.756802	0.346356	0.1016050
6	2	1.85841	−0.871076	0.347294	0.935538	0.7555560
7	3	2.85841	0.236263	−0.561005	0.254579	1.198000

The more accurate calculation of $Y_j(\chi;\alpha); j = 0,1,2,3$, the more accurate orbit determination. That is because the universal Kepler's equation is expressed in terms of Y's functions [1]. Thus efficient tools used for evaluating Y's functions have contributions in well describing the two-body initial value problem.

5. Conclusion

In concluding the present paper, an efficient algorithm based on the continued fractions theory was established for the recent universal Y's functions of space dynamics. The algorithm is valid for any conic motion (elliptic, parabolic or hyperbolic).

References

[1]   Sharaf, M.A. and Saad, A.S. (2014) New Set of Universal Functions for the Two Body-Initial Value Problem. *Astrophysics and Space Science*, **349**, 71-81. (Paper I)

[2]   Gautschi, W. (1967) Computational Aspects of Three-Term Recurrence Relations. *SIAM Review*, **9**, 24-82. http://dx.doi.org/10.1137/1009002

[3]   Sharaf, M.A. (2006) Computations of the Cosmic Distance Equation. *Applied Mathematics and Computation*, **174**, 1269-1278. http://dx.doi.org/10.1016/j.amc.2005.05.054

[4]   Sharaf, M.A., Almleaky, Y.M., Malawi, A.A., Goharji, A.A. and Basurah, H.M. (2004) Symbolic Analytical Expressions of the Physical Characteristics for N-Dimensional Radially Symmetrical Isothermal Models. *AJSE, King Fahd, Univ.*, **29**, 67-82.

[5]   Sharaf, M.A. and Banajh, M.A. (2001) Continued Fraction Evaluation of the Stumpff Functions of Space Dynamics. *Scientific Journal of Faculty of Science, Minufiya University*, **XV**, 267-282.

[6]   Sharaf, M.A. and Najmuldeen, S.A. (2001) Continued Fraction Evaluation of the Normal Distribution Function. *Scientific Journal of Faculty of Science, Minufiya University*, **XV**, 311-324.

[7]   Battin, R.H. (1999) An Introduction to the Mathematics and Methods of Astrodynamics. Revised Edition, AIAA, Education Series, Reston, Virginia.

5

Expected Radial Column Density of Methylcyanopolyynes in CW Leonis (IRC+10216)

Raul G. E. Morales[1], Carlos Hernández[2]

[1]Centre for Environmental Sciences and Department of Chemistry Faculty of Sciences, Universidad de Chile, Santiago, Chile
[2]Department of Chemistry, Faculty of Basic Sciences Universidad Metropolitana de Ciencias de la Educación, Santiago, Chile
Email: correo@raulmorales.cl, carloshernandez.carlos@gmail.com

Abstract

Methylcyanopolyynes ($CH_3-[C\equiv C]_n-CN$) are a particular kind of linear molecular wires, where the first three oligomers have been detected in the interstellar medium, particularly in CW Leonis (IRC + 10216), as well as in the envelopes of carbon-rich stars in a similar way to the unsubstituted cyanopolyynes. Based on the projected natural distribution in cold clouds under LTE, we have determined the radial column density of new expected methylcyanopolyynes to be present in CW Leonis (IRC + 10216). By following, we have made use of the inner molecular resistances of the internal charge transfer process presenting in these oligomeric species in order to determine the reactivity trends between them. Therefore, geometrical parameters and dipole moments determinations for these methylcyanopolyynes involving the $n = 1$ to 14 molecular species were obtained from *Ab initio* molecular orbital calculations by means of a GAUSSIAN Program, using a restricted Hartree-Fock approach and 6-311G* basis set. Our results present a similar behavior observed in cyanopolyynes, where this series reaches a saturation level at the 14th oligomer with a maximum dipole moment of 8.21 ± 0.01 (Debyes). Thus, this molecular wire model permits us to comprehend how these methylcyanopolyynes reach a maximum length in such chemical environment, in agreement to the astronomical observations and cosmological chemical models. The following CH_3C_9N and $CH_3C_{11}N$ oligomers in CW Leonis should be expected near to 3.52×10^{10} [cm^2] and 1.82×10^{10} [cm^2], respectively.

Keywords

Methylcyanopolyynes, Dipole Moments, Inner Molecular Resistances, Organic Molecular Wires, ISM Radial Column Density

1. Introduction

New observational evidences in some interstellar mediums (ISM) such as circumstellar shells and cold nebulae, it has been recently reported in Sagittarius B2, HL Tauri, Rho Ophiuchi cloud complex and Orion KL, between others, show the presence and abundance of different molecular species [1]-[8]. Large molecules in heterogeneous astronomical environments is one of the main goals of the frontier Astrochemistry, where project as ALMA (Atacama Large Millimeter Array) is being implemented in the south hemisphere, in order to reach new molecular observations and therefore, new explanations of complex molecular systems such as fullerenes in the ISM originated from long linear molecular compounds such as cyanopolyynes.

Several smaller members of the polyynic series have been detected in the ISM, particularly in cold circumstellar envelopes surrounding carbon-rich stars. These unsaturated carbon chains comprise at one end by hydrogen or methyl group and a CN group in the other extreme, a classical electron acceptor group [9]. The rotational spectra of smaller members of the simplest cyanopolyynes up to $HC_{17}N$ are well known at the laboratory [10], but under hard experimental conditions and extreme detection sensitivity. However, $HC_{11}N$ has been the last-discovered large linear interstellar molecule [11] [12]. Whenever spectral measurements of these types of molecular wires involving the CN group [13] [14] from various interstellar regions have been reported since 1978 up to date, the astronomical observations have systematically proved the difficult to detect molecular wires longer than five units.

These linear oligomeric molecular structures are constituted by an electron-donor group (D), in this case CH_3 group, a molecular wire bridge (W), represented by the unsaturated carbon chain ($-[C≡C]_n-$), and the CN electron-acceptor group (A) [15]-[17]. By following, when these molecules are formed an electronic charge transfer process occurring from D to A through W, *i.e.*, materialized the dipole moment. In several previous studies [18]-[21], we have developed an one-dimensional molecular model for these systems based on the electronic conduction properties of linear conjugated oligomeric compounds of the D-W-A type. These studies have allowed us to focus our particular attention on these new interstellar cyanopolyynes.

In the present work, we have extended our cyanopolyynes study [19] to the methylcyanopolyynes, in order to find out the expected radial column density of new species to be present in CW Leonis (IRC +10216). Geometrical and dipole moment parameters calculated from *Ab initio* molecular orbital calculations, permitted us to determine the molecular resistances, necessary data to correlate the radial column densities in a similar trend previously observed in cyanopolyynes species [19].

2. The One-Dimensional Conduction Model

Based on our one-dimensional conduction model for D-W-A molecular systems developed some years ago [9] [18], we have determined the molecular resistance of the first fourteen methylcyanopolyynes and the linear molecular resistivity. Therefore, the dipole moment of every linear n^{th} oligomer (μ_n) can be represented as [9]:

$$\mu_n = \mu_o + \mu_\infty \left\{ 1 - e^{-\gamma L(n)} \right\}$$ (1)

where μ_o is the dipole moment of the first compound of the oligomeric series without a bridge unit ($n = 0$), μ_∞ is a molecular constant of the oligomeric series at the limit value for $L \to \infty$, $L(n)$ is the molecular wire length of the nth oligomer ($-[C≡C]_n-$), and γ is the molecular wire conduction constant. Thus, the inner resistance of the molecular wire can be depicted as follow [9]:

$$R(L) = \left(h/2e^2 \right) \left[e^{gL} - 1 \right]$$ (2)

where $(h/2e^2) = 12.91$ kΩ. However, in order to separate the linear and non linear contributions of the molecular wire resistance, we can expand $R(L)$ in Equation (2) as a Maclaurin series:

$$R(L) = 12.91 \left[\gamma L + 1/2! \gamma^2 L^2 + \cdots + 1/m! \gamma^m L^m \right]$$ (3)

where the first term defines the linear contribution given by $R_l = 12.91 [\gamma L]$ (kΩ) and the remainder terms of the series define the non linear contribution of the molecular wire resistance. Thus, the linear resistivity (ρ), an intrinsic property of the conductor wires, can be calculated as $\rho = R_l \cdot (S/L)$, where S is the molecular wire cross-section estimated to be 4.5 Å^2 [18].

3. Results and Discussion

The best molecular geometry optimization and the electronic charge distribution as a function of the length of the molecular wire for the methylcyanopolyynes under study were obtained by means of the GAUSSIAN Quantum Mechanics Program software [22]. From this calculation program we have used the restricted Hartree-Fock (RHF) approach and the 6-311G* basis set, up to reach the best linear equilibrium geometry. All these oligomeric compounds, from $n = 1$ to 14, have been well described by means of these molecular-orbital calculations.

However, it is well known fact by experimentalist the inherent difficulties that these cyanopolyynes present for laboratory synthesis according to increase the length of the molecular series, due to high molecular instability, as well as, hard experimental conditions in order to reach dipole moment measurements in vacuum. Therefore, it is not possible to compare systematic dipole moment measurements to our or other calculations, however, the fresh results providing by the recent Astrochemistry research can open new expectative in a near future.

In spite of the scarce studies on this subject, our results permit to project the interstellar chemistry of these particular methylcyanopolyynes [9] [19]. Hence, $L(n)$ as the molecular wire lengths and μ_n as the dipole moments of these methylcyanopolyynes were calculated using this theoretical approach (**Table 1**). **Figure 1** shows the expected interdependence between μ_n and $L(n)$ in agreement to the molecular model developed for these molecular wires [9]. In this Figure, the observed trend is better than $r^2 > 0.999$ and the final parameters described by Equation (1) in **Figure 1** are presented in **Table 2**.

Furthermore, from **Figure 1** we can observe how this molecular series converge to a maximum dipole moment as function of the polyynic wire in a similar behavior than the cyanopolyynes [19]. Thus, according to Equation (1), the observed limit is determined by $(\mu_o + \mu_\infty)$, where the dipole moment in this oligomeric series converge to 8.21 ± 0.01 (Debyes).

Therefore, the oligomeric series under study have been extended to the first fourteen species, where after the $CH_3C_{29}N$ compound, the dipole moment does not change significantly.

On the other hand, the γ molecular wire conduction constant of methylcyanopolyynes exhibits a similar behavior to the cyanopolyynes [19]. Certainly, the methyl group induces a slight perturbation to the molecular wire nevertheless the observed change is lower than 3%.

Thus, by means of this γ conduction constant, we have determined the molecular resistances as a function of the wire length according to Equation (2). **Table 3** depicts the molecular resistance of these methylcyanopolyynes and we have determined a linear resistivity of 74.9 $\mu\Omega$/cm, similar to cyanopolyynes (72.6 kΩ/cm) and other molecular wires previously reported [19].

Based on these results, we have used the molecular resistance as a criterion for chemical reactions feasibility of these linear molecular species [19]. Thus, the molecular resistance to the internal charge transfer emerges as

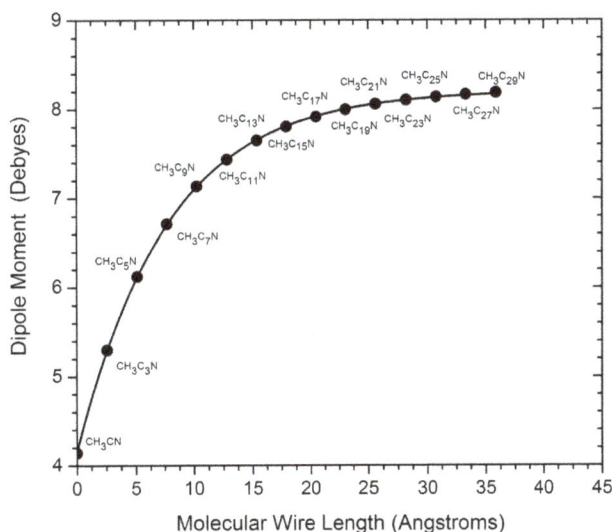

Figure 1. Dipole moments of the methylcyanopolyynes ($n = 0$ to 14) versus the molecular wire lengths in the RHF approach and 6-311G* basis set.

Table 1. Molecular wire lengths and dipole moments of the Methylcyanopolyynes according to Gaussian calculations in the RHF approach and the 6-311G* basis set.

Molecular Series	Formula	Molecular Wire Length (Å)	Dipole Moment (Debye)
$CH_3-C\equiv N$	CH_3CN	0.00	4.14
$CH_3-(C\equiv C)_1-C\equiv N$	CH_3C_3N	2.56	5.30
$CH_3-(C\equiv C)_2-C\equiv N$	CH_3C_5N	5.13	6.12
$CH_3-(C\equiv C)_3-C\equiv N$	CH_3C_7N	7.69	6.71
$CH_3-(C\equiv C)_4-C\equiv N$	CH_3C_9N	10.25	7.14
$CH_3-(C\equiv C)_5-C\equiv N$	$CH_3C_{11}N$	12.82	7.43
$CH_3-(C\equiv C)_6-C\equiv N$	$CH_3C_{13}N$	15.38	7.65
$CH_3-(C\equiv C)_7-C\equiv N$	$CH_3C_{15}N$	17.94	7.80
$CH_3-(C\equiv C)_8-C\equiv N$	$CH_3C_{17}N$	20.50	7.91
$CH_3-(C\equiv C)_9-C\equiv N$	$CH_3C_{19}N$	23.07	8.00
$CH_3-(C\equiv C)_{10}-C\equiv N$	$CH_3C_{21}N$	25.63	8.06
$CH_3-(C\equiv C)_{11}-C\equiv N$	$CH_3C_{23}N$	28.19	8.10
$CH_3-(C\equiv C)_{12}-C\equiv N$	$CH_3C_{25}N$	30.75	8.14
$CH_3-(C\equiv C)_{13}-C\equiv N$	$CH_3C_{27}N$	33.31	8.17
$CH_3-(C\equiv C)_{14}-C\equiv N$	$CH_3C_{29}N$	35.88	8.19

Table 2. Dipole moment parameters of the Methylcyanopolyynes.

Dipole Parameters	
μ_o (Debyes)	4.147 ± 0.007
μ_∞ (Debyes)	4.065 ± 0.007
γ (Å^{-1})	0.1292 ± 0.0005

Table 3. Inner molecular resistance of the Methylcyanopolyynes.

$CH_3-(C\equiv C)_n-C\equiv N$	R (kΩ)
1	5.06
2	12.14
3	21.96
4	35.63
5	54.74
6	81.26
7	118.2
8	169.6
9	241.4
10	341.1
11	479.9
12	673.1
13	942.0
14	1318

an indicator of the polarity strength of the ground state during the molecular formation process for every one molecule of the polyynic series. Consequently, every new polyynic unit extension of the molecular wire of the methylcyanopolyynes gradually presents an additional resistance to the internal charge transfer process and, subsequently, their reaction feasibility necessarily decreases, weakening the attraction force of the CN electron-acceptor group trough the molecular wire.

Therefore it is not possible to expect long wire cyanopolyynes when its internal charge transfer activation energy is higher than the thermal energy of the reaction bulk, in other words, when the dipole moment reaches infinitesimal changes between two molecules of the series. This observation is particularly interesting to the interstellar cold dense-clouds. In this case the molecular reactivity for every molecule of the series must be understood how a dependent parameter of the sequential reaction scheme and under the same LTE conditions in a bulk delimited by low-temperature dense-cloud regions.

Therefore, the role of the CN group in the chemical reactions associated to the cyanopolyynes synthesis is determined by the electronic feasibility of the charge transfer between the molecular wire and the electron acceptor group. Thus, we can expect a relationship between the oligomeric species distribution and the molecular resistance to the internal charge transfer of the molecular wires that determines the final probability of the molecular array density under LTE.

IRC +10216 or CW Leonis, a carbon-rich star embedded in a thick dust envelope, has been a specific natural reactor of molecular reactions. This stellar source has been well studied both observationally and theoretically and Millar *et al.* [7] have developed chemical models in order to reproduce detailed radial distributions of different molecules. In particular, they report the cyanopolyynes chemistry based on the positive ion-molecule and neutral-neutral reactions leading to the production of these oligomers. Reactions including the radical CN and hydrocarbons were involved in the formation of cyanopolyynes, as they are in dense clouds, but reactions between the radical C_2H and smaller cyanopolyynes were far more important in the IRC +10216 envelope chemistry.

Radial column densities for the cyanopolyynes in IRC +10216 are well represented by the Millar *et al.* model [7], which includes several cyanopolyynes and the only three observed methylcyanopolyynes. Therefore, we have made use of the radial column densities of these last reported oligomers [7]. In **Figure 2**, we can appreciate the methylcyanopolyynes expected behavior between radial column density and the molecular resistance. A very good linear correlation can be observed in the logarithmic scale in a similar behavior than the cyanopolyynes previously reported [19]. Therefore, this result in a new molecular series as the methylcyanopolyynes gives a new basement to our hypothesis respect to the chemical reaction feasibility in terms of the molecular wire length. Furthermore, the present methodology determines a new approximation to the radial column density estimation for astronomical observations in order to project the presence of new molecular species of these oligomeric varieties.

Thus, based on this linear correlation we have estimated the expected molecular distributions through the radial column density of this series from the relationship established by **Figure 2**. By following the radial column density (RCD) of these methylcyanopolyynes can be determined as follow:

$$\text{Log}(\text{RCD}) = 12.95 - 1.548 \text{Log}(R(L))$$

where $R(L)$ is in kΩ and RCD in cm^2. By following, the expected new radial column densities for the next two members of this oligomers in CW Leonis, CH_3C_9N and $CH_3C_{11}N$, should be expected near to 3.52×10^{10} [cm^2] and 1.82×10^{10} [cm^2], respectively.

4. Conclusions

Our results show a new route of linear oligomeric species analysis presenting in the ISM. Furthermore, the methodology developed from inner resistance to the dipole moments in these molecular wire systems opens new lines of research in the astronomical distribution and detection limits of these specific chemicals.

Particularly, in the present work, we show how dipole moments of these oligomeric series increase with chain length up to reach a certain saturation point, as can be seen in cyanopolyynes [19] and methylcyanopolyynes (this work), where R–[C≡C]$_{14}$–CN establishes a molecular limit in every series. Therefore, we have developed a simple approach extendable to other similar molecular series.

Furthermore, our results and methodology on these molecular wires determine new tools of analysis to be

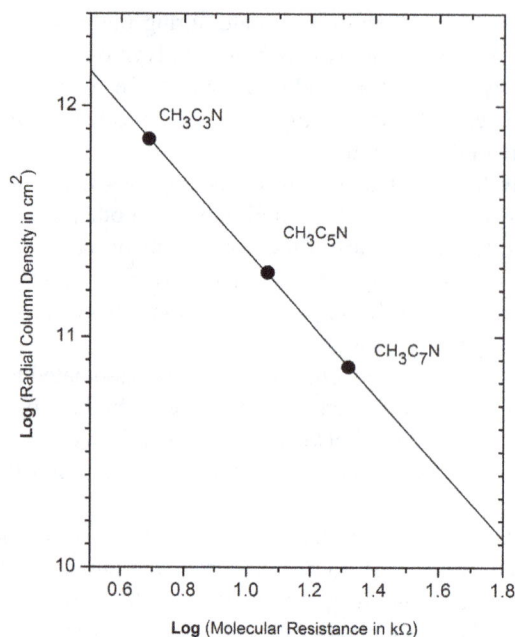

Figure 2. Observed trend between Radial Column Density (RCD) and Molecular Resistance ($R(L)$) in a logarithmic scale.

considered when chemical models are being used in a complex network of reaction schemes under LTE as well as in radial column density estimations of ISM. Effectively, our predictions about how the radial column density of long molecular wires declines with increasing molecular resistance, introduce a new criterion for radio searches of long molecules. The last question is important when a large amount of integration time will be necessary for observing these particular molecular systems.

Other linear molecular wires detected in circumstellar envelopes surrounding carbon-rich stars are being analyzed in our laboratory.

Acknowledgements

The authors acknowledge to the Centre for Environmental Sciences of the University of Chile for financial support.

References

[1] Müller, H.S.P., Goicoechea, J.R., Cernicharo, J., Agúndez, M., Pety, J., Cuadrado, S., Gerin, M., Dumas, G. and Chapillon, E. (2014) Revised Spectroscopic Parameters of SH⁺ from ALMA and IRAM 30 m observations. *Astronomy and Astrophysics*, **569**, L5-L9. http://dx.doi.org/10.1051/0004-6361/201424756

[2] Kahane, C., Ceccarelli, C., Faure, A. and Caux, E. (2013) Detection of Formamide, the Simplest but Crucial Amide, in a Solar-type Protostar. *Astrophysical Journal Letters*, **763**, L38-L42. http://dx.doi.org/10.1088/2041-8205/763/2/L38

[3] Jorgensen, J.K., Favre, C., Bisschop, S.E., Bourke, T.L., van Dishoeck, E.F. and Schmalzl, M. (2012) Detection of the Simplest Sugar, Glycolaldehyde, in a Solar-type Protostar with ALMA. *Astrophysical Journal Letters*, **757**, L4-L9. http://dx.doi.org/10.1088/2041-8205/757/1/L4

[4] Herbst, E. and van Dishoeck, E.F. (2009) Complex Organic Interstellar Molecules. *Annual Review of Astronomy and Astrophysics*, **47**, 427-480. http://dx.doi.org/10.1146/annurev-astro-082708-101654

[5] Pardo, J.R., Cernicharo, J. and Goicoechea, J.R. (2005) Observational Evidence of the Formation of Cyanopolyynes in CRL 618 through the Polymerization of HCN. *Astrophysical Journal*, **628**, 275-282. http://dx.doi.org/10.1086/430774

[6] Woods, P.M., Millar, T.J., Herbst, E. and Zijlstra, A.A. (2003) The Chemistry of Protoplanetary Nebulae. *Astronomy and Astrophysics*, **402**, 189-199. http://dx.doi.org/10.1051/0004-6361:20030215

[7] Millar, T.J., Herbst, E. and Bettens, R.P.A. (2000) Large Molecules in the Envelope Surrounding IRC+10°216. *Monthly Notices of theRoyal Astronomical Society*, **316**, 195-203.

[8] Cordiner, M.A. and Millar, T.J. (2009) Density-Enhanced Gas and Dust Shells in a New Chemical Model for IRC+10216. *Astrophysical Journal*, **697**, 68-78. http://dx.doi.org/10.1088/0004-637X/697/1/68

[9] Morales, R.G.E. and González-Rojas, C. (2005) Dipole Moments of Polyenic Oligomeric Systems. Part II. Molecular Organic Wire Resistivities: Polyacetylenes, Allenes and Polylines. *Journal of Physical Organic Chemistry*, **18**, 941-944. http://dx.doi.org/10.1002/poc.931

[10] McCarthy, M.C., Chen, W., Travers, M.J. and Thaddeus, P. (2000) Microwave Spectra of 11 Polyyne Carbon Chains. *Astrophysical Journal*, **129**, 611-623. http://dx.doi.org/10.1086/313428

[11] Bell, M.B., Feldman, P.A., Kwok, S. and Matthews, H.E. (1982) Detection of $HC_{11}N$ in IRC+10°216. *Nature*, **295**, 389-391. http://dx.doi.org/10.1038/295389a0

[12] Bell, M.B., Feldman, P.A., Travers, M.J., McCarthy, M.C., Gottlieband, C.A. and Thaddeus, P. (1997) Detection of HC11N in the Cold Dust Cloud TMC-1. *Astrophysical Journal Letters*, **483**, L61-L64. http://dx.doi.org/10.1086/310732

[13] Kroto, H.W., Kirby, C., Walton, R.M., Avery, L.W., Broten, N.W., MacLeod, J.M. and Oka, T. (1978) The Detection of Cyanohexatriyne, $H(CC_3)CN$, in Heiles' Cloud 2. *Astrophysical Journal Letters*, **219**, L133-L137. http://dx.doi.org/10.1086/182623

[14] Smith, D. (1992) The Ion Chemistry of Interstellar Clouds. *Chemical Reviews*, **92**, 1473-1485. http://dx.doi.org/10.1021/cr00015a001

[15] Carter, F.L. (1986) Molecular Electronic Devices. Marcel Dekker Inc., New York.

[16] Joachim, C. and Roth, S. (1997) Atomic and Molecular Wires. Kluwer Academic Publishers, Alphen aan den Rijn. http://dx.doi.org/10.1007/978-94-011-5882-4

[17] James, D.K. and Tour, J.M. (2005) Molecular Wires. *Topics in Current Chemistry*, **257**, 33-62. http://dx.doi.org/10.1007/b136066

[18] Hernández, C. and Morales, R.G.E. (1993) Bridge Effect in Charge-Transfer Photoconduction Channels. 1. Aromatic Carbonyl Compounds. *Journal of Physical Chemistry*, **97**, 11649-11651. http://dx.doi.org/10.1021/j100147a016

[19] Morales, R.G.E. and Hernández, C. (2012) Cyanopolyynes as Organic Molecular Wires in the Interstellar Medium. *International Journal of Astronomy and Astrophysics*, **2**, 230-235. http://dx.doi.org/10.4236/ijaa.2012.24030

[20] Morales, R.G.E. and González-Rojas, C. (1998) Dipole Moments of Polyenic Oligomeric Systems. Part I. A One-Dimensional Molecular Wire Model. *Journal of Physical Organic Chemistry*, **11**, 853-856. http://dx.doi.org/10.1002/(SICI)1099-1395(199812)11:12<853::AID-POC74>3.0.CO;2-Y

[21] González, C. and Morales, R.G.E. (1999) Molecular Resistivities in Organic Polyenic Wires. I. A One-Dimensional Photoconduction Charge Transfer Model. *Chemical Physics*, **250**, 279-284. http://dx.doi.org/10.1016/S0301-0104(99)00335-3

[22] Frisch, M.J., Trucks, G.W., Schlegel, H.B., Scuseria, G.E., Robb, M.A., Cheeseman, J.R., Scalmani, G., Barone, V., Mennucci, B., Petersson, G.A., Nakatsuji, H., Caricato, M., Li, X., Hratchian, H.P., Izmaylov, A.F., Bloino, J., Zheng, G., Sonnenberg, J.L., Hada, M., Ehara, M., Toyota, K., Fukuda, R., Hasegawa, J., Ishida, M., Nakajima, T., Honda, Y., Kitao, O., Nakai, H., Vreven, T., Montgomery Jr., J.A. Peralta, J.E., Ogliaro, F., Bearpark, M., Heyd, J.J., Brothers, E., Kudin, K.N., Staroverov, V.N., Kobayashi, R., Normand, J., Raghavachari, K., Rendell, A., Burant, J.C., Iyengar, S.S., Tomasi, J., Cossi, M., Rega, N., Millam, J.M., Klene, M., Knox, J.E., Cross, J.B., Bakken, V., Adamo, C., Jaramillo, J., Gomperts, R., Stratmann, R.E., Yazyev, O., Austin, A.J., Cammi, R., Pomelli, C., Ochterski, J.W., Martin, R.L., Morokuma, K., Zakrzewski, V.G., Voth, G.A., Salvador, P., Dannenberg, J.J., Dapprich, S., Daniels, A.D., Farkas, O., Foresman, J.B., Ortiz, J.V., Cioslowski, J. and Fox, D.J. (2009) Gaussian 09, Revision A.01. Gaussian, Inc., Wallingford.

6

Determination of Velocity and Radius of Supernova Remnant after 1000 yrs of Explosion

Baha T. Chiad[1], Lana T. Ali[2], Abdhreda S. Hassani[1]

[1]Department of Physics, College of Science, University of Baghdad, Baghdad, Iraq
[2]Department of Astronomy and Space, College of Science, University of Baghdad, Baghdad, Iraq
Email: Lana_talib@yahoo.com

Abstract

Supernova explosions are described as very violent events which transfer a significant amount of energy to interstellar media and are responsible for a large variety of physical processes. This study does not discuss the actual explosion mechanisms but follows the behavior of the dynamical evolution of some selected type I and type II supernova remnant and particularly after a thousand years from their explosion and shows how the density of the medium affects the evolution and the lifetime of each remnant. By studying such behaviors, a simplified model has been proposed here for the velocity and radius of the remnant after thousand years of explosion that depends only on the density of the medium and age of the remnant. It has been found that all types of supernova remnants have similar behaviors after a thousand years from their explosion despite their origin formation. Moreover, it is demonstrated that, when those selected remnants have entered or will enter into their radiative phase, an idea on their physical properties will be obtained.

Keywords

Supernova, Supernova remnant, Radiative Phase

1. Introduction

The mass of the star plays a significant role in the determination of its lifetime. Once the star reaches the giant stage, its thermonuclear reaction will speed up and its fate will be determined. A star with mass < 8 $M_\odot$ will reach the carbon/oxygen reaction and cease since it has insufficient temperature to ignite further reaction and it ends its life expelling its outer envelope in a planetary nebula and leaving behind a carbon/oxygen white dwarf

[1]. However, the white dwarf is not always the end product in the collapse of a mid-sized star. If a white dwarf is in a contact binary system it will accrete material from the companion star, and this will continue until its mass exceeds the Chandrasekhar mass limit which is about 1.4 $M_\odot$ [2]. At this moment the density and the temperature in the center of the white dwarf become so severe that carbon starts fusing explosively and within one second it will undergo a thermonuclear explosion and is completely destroyed, leaving nothing behind except remnant, producing type Ia supernova [1]. While on the other hand, more massive stars (≥ 8 $M_\odot$) will go beyond carbon/oxygen reaction until they reach the iron reaction stage. At this point the thermonuclear reactions are no longer exothermic, and the core cannot support itself against gravity [2]. The core collapses in a few milliseconds, and an energetic shock propagates out through the envelope of the star to produce an energetic display of electromagnetic radiation marking the catastrophic death of this massive star as a core-collapse supernova [3]. At the same time, almost (99%) of the explosion energy will be released in the form of energetic neutrinos; the remaining energy is converted into kinetic energy, accelerating the stellar material to speed up greater than the speed of sound and causing a shock wave (moving outwards from the central star) which will compress heat and sweep the surrounding ambient gas as it expands [4]. With time, this expanding material from the explosion in addition to the material that blast wave collects as it travels through the interstellar medium, will form a Supernova Remnant (SNR) [5]. According to Green (Green 2009), more than 274 Galactic SNRs (Green 2009) have been recorded, discovered, and detected and well known examples of these are the remnants of Tycho's SN, Kepler's SN, the Cygnus loop, the Crab nebula, and SN1987A that provide delicate visual indication to their violent births [6].

The subject of supernova plays an important role across a wide range of fields in physics and astronomy as they mark the end point in the life of stars and at the same time their remnants mark the formation of the compact objects and the most massive black holes in the universe as well as they play a significant role in enriching the interstellar medium with heavy elements (up to iron). Moreover, they are the source of much of the energy that heats up the interstellar medium and interacting with it [4]. For these reasons, it is essential to study this remarkable subject. In this paper, we investigate the influence of the density of the medium that surrounds some type Ia and type II supernova remnants such as 1006 SNR, G1.9+0.3, Tycho, Kepler, SN1987A, CasA, Crab nebula, Cygnus loop, and IC443 on their evolution. In addition, we provide a model for the velocity and radius of the remnant when they enter the radiative phase (after thousand years of explosion) that depends only on the density of the medium and age of the remnant.

2. Evolutionary Stages of Supernova Remnant

Soon after the supernova explosion takes place, its new life as supernova remnant starts. All remnants from the supernovae explosion seem to go through the same evolutionary phases regardless their differences in the original spectra of supernovae when they all expand and evolve through a series of phases [7]. The first phase of those evolutionary phases will begun almost immediately after the explosion when the shock wave from the supernova expands supersonically into the surrounding medium which will accelerate and drive the interstellar material away from the supernova site with a velocity less than the shock wave's initial velocity which is typically in the order of 30,000 km/sec [4]. This phase in which the ejecta are expected to be mostly expanding freely known as the free expansion phase which will last for a few hundred years before the remnant enters the second evolutionary stage of its life as supernova remnant which is the adiabatic phase or the Sedov- Tyler phase. At this phase the remnant will sweep up mass equal to its initial mass that ejected from the explosion which made the shock wave to decelerate from 30,000 km/sec to about 10,000 km/sec, while the radius will increase by $R_s \propto t^{2/5}$ [8].

After about (100 - 1000) years from the beginning of the Sedov-Tyler phase, the temperature of the matter left behind the shock front will be decreased to about (10^3 - 10^6) K, which allows to the recombination process of the nuclei of abundant elements like C, N and O to recombine with electrons. Consequently, the radiation losses go up (via electron transition) [9] and the remnant experiences its forth evolutionary stage which is the radiative expansion phase that begins at [10]:

$$t = 1.9 \times 10^4 E_{51}^{3/14} n^{-4/7} \ (\text{yr})$$

(1)

and with a radius:

$$R = 16.2 E_{51}^{2/7} n^{-3/7} \ (\text{pc}) \tag{2}$$

where E_{51} represents the explosion energy in unit of 10^{51} ergs and n_o denotes the hydrogen number density in the surrounding interstellar medium in the unit of 10 atom per cm^3.

Once radiative losses become significant, the plasma flow of the remnant will rapidly change over: the gases quickly cool down in the region of the shock front, and it forms a cold dense thin shell behind the radiative shock [9]. This thin dense shell which is formed through the ISM is filled with a low-density, high-pressure interior so it will be driven by the pressure of the hot, roughly isobaric interior in addition to its own momentum. Subsequently, its velocity will be reduced which makes the post shock fluid velocity to approach shock velocity, so the expanding remnant will passes from its blast wave phase to a snowplow phase which is also said to be the pressure-driven snowplow (pds) stage of evolution [11]. The transition from the blast wave phase to the snowplow phase is illustrated in **Figure 1** [12].

The remnant will stay in the radiative phase for tens of thousands of years, until the shock has radiated all its energy and the remnant enters the last evolutionary stage of its life as a remnant which is the Dissipative phase. In this phase, the expansion rate slows down; seeding the local neighborhood with heavy elements and the expansion velocity drops to below the local ISM dispersion velocity which is about 10 km/sec and becomes indistinguishable from the surrounding medium [13]. In the next section we will obtain model for the expansion velocity and the radius of the remnant for type Ia and type II supernova when it reaches the radiative phase (after about 1000 years from the supernova explosion) which is independent on the supernova explosion energy.

3. Calculations and Discussion

Supernova remnants display a variety of shapes and features that are often remarkably influenced by the environment into which the ejected material is expanding. In this paper some remnants of type Ia and type II supernovae such as: 1006 SNR, G1.9+0.3, Tycho, Kepler, IC443, Cygnus loop, Crab nebula, CasA, and SN1987A, the physical properties of which are listed in **Table 1**, have been taken in order to investigate how the density of the region that surrounds it affects on the expansion rate of the remnant. In order to establish this, the Counting Pixels Method (see ref. [20]) was applied on the images of the remnants that mentioned in **Table 1** to determine their current radius which in turn will be substituted in the following equation in order to calculate the explosion energy (E) of each remnant [33]:

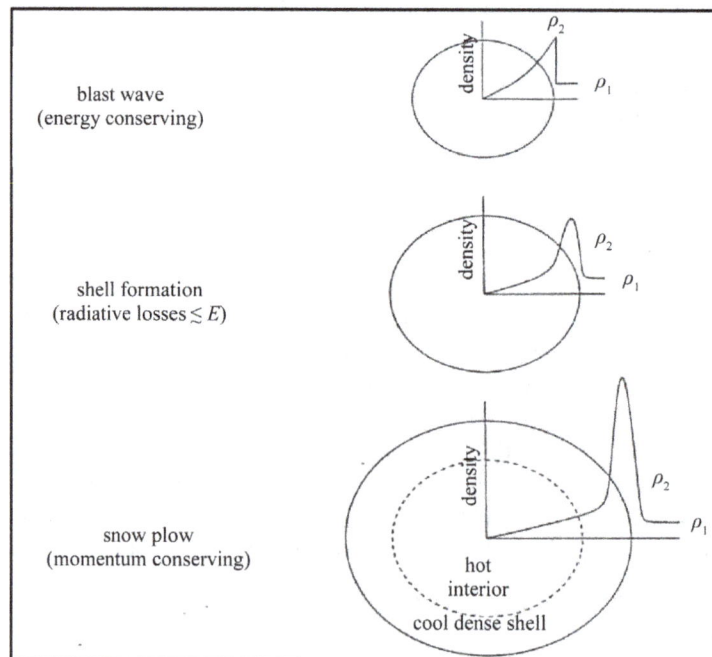

Figure 1. An expanding supernova remnant makes the transition from the blast wave phase (upper image) to the snowplow phase (lower image) [12].

Table 1. The physical properties of the selected remnants.

SNR	Age (year)	Distance (pc)	ISM no. density (cm^{-3})	Radius (pc)	Velocity (Km·s^{-1})
G1.9+0.3	100	8500 [14]	0.03 [14]	2.2 - 4.7 [14]	14000 [14]
Kepler	410	5000 - 6400 [15]	0.1 [16]	2.5 - 3.8 [15]	1550 - 2000 [17]
Tycho	442	1500 - 3100 [18]	0.2 [18]	3.7 [18]	1500 - 2800 [18]
1006	1000	2180 ± 80 [19]	0.1 [15]	7.1 - 7.5 [19]	2890 ± 100 [19]
SN 1987A	26	51000 [20]	10 [21]	0.39 [20]	4000 [21]
Cas A	358	3400 [22]	1.5 [22]	2.6 [22]	1000 - 1500 [22]
Crab	958	2000 [23]	0.5 [23]	3.4 [24]	1400 - 1500 [25]
Cygnus loop	17000	770 [26]	0.1 - 0.2 [27]	21.5 - 27 [28]	200 - 300 [29]
IC443	30000	1500 [30]	10 - 20 [30]	9.6 [31] - 15 [32]	65 - 100 [30]

$$E = 3.2 \times 10^{51} n_o R^5 t^{-2} \tag{3}$$

where n_o represents the number density of the surrounding medium, R is the radius of the remnant in unit of pc and t is the age of the remnant in unit of year.

The output results of this method are listed in **Table 2**. In terms of the radius, the velocity (v) and the mss (m) of the remnant can also be found by using the following equations:

$$v = \frac{R}{t} \tag{4}$$

$$m = \frac{4}{3}\pi\rho R^3 \tag{5}$$

where ρ is the mass density of the interstellar medium.

Moreover, in order to demonstrate the behavior of the radius and the expansion velocity of each remnant with time (which is taken from 1000 years to 30×10^3 years from the explosion) each of Equations (3) and (4) have been applied and the attainable results are plotted in **Figure 2(a1)** & **Figure 2(a2)**) for type Ia SNR and **Figure 3(a1)** and **Figure 3(a2)** for type II SNR.

On studying the behavior of **Figure 2(a1)** and **Figure 2(a2)**) and **Figure 3(a1)** and **Figure 3(a2)** of those selected remnants, we came up with new model of the radius and the expansion velocity of supernova remnant that depend only on the density of the surrounding medium and the age of the remnant which works after thousand year from the explosion, as follow:

$$R = 0.34332 t^{0.4} n^{-0.2} \tag{6}$$

$$V = 149666.25 \eta t^{-0.601} n^{-0.2} \tag{7}$$

where (η) is the expansion factor which is given in the following equation [34]:

$$\eta = \frac{v_b}{R_b/t} \tag{8}$$

where v_b, and R_b are the velocity and the radius of the blast wave respectively.

Subsequently, the results of applying Equations (6) and (7) have been plotted (**Figure 2(a1)** and **Figure 2(a2)** for type Ia SNR and **Figure 3(a1)** and **Figure 3(a2)** for type II SNR) and compared with those obtained from Equations (3) and (4) and it was found that, our results are in reasonable agreement with the observed results more than that obtained from applying CPM as shown in **Table 3**. As well as the results of **Table 3** also shows that, our model do not apply to the early evolutionary stages of the SNR as shown in the results of SN1987A and CasA since in these stages the explosion energy is dominant and it accelerates the remnant to higher velocity but with time as the mass of the remnant increases the influence of the explosion energy decreases and the remnant

Table 2. The result of applying counting pixels method on the selected remnants.

	G1.9+0.3	Kepler	Tycho	1006	SN 1987A	Cas A	Crab nebula	Cygnus loop	IC443
Radius (pc)	4.7	4	5.6	18	0.49	5.055	3.41	9.6	11.1
Explosion energy (erg)	0.2×10^{50}	0.02×10^{50}	0.18×10^{50}	6×10^{50}	1.34×10^{51}	0.86×10^{51}	0.015×10^{51}	0.26×10^{51}	0.06×10^{51}

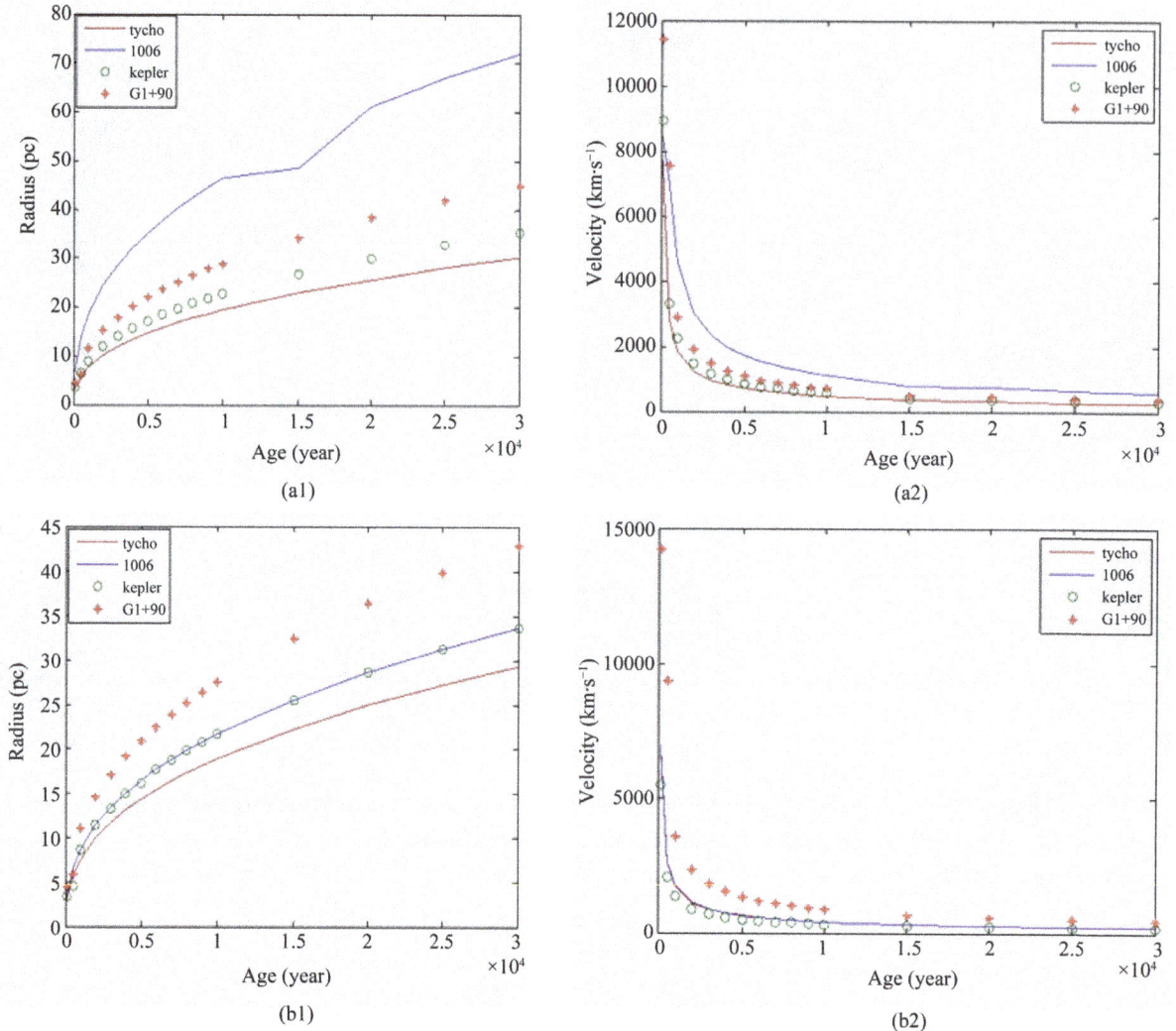

Figure 2. Variation of the radius and velocity of Type Ia SNRs with age. (a1) and (a2) by applying Equations (3) and (4), (b1) and (b2) by applying the present model (Equations (6) and (7)).

is decelerated according to one major parameter which is the density of the surrounding interstellar medium as a main parameter affecting the development and determination of the lifetime of each remnants for example the entrance of IC443 to the dens molecular cloud (that lie in the northern portion of the remnant) made the remnant experience an increasing in its mass which may reach a value of 1400 $M_\odot$. Consequently, its expansion velocity has been reduced to a value of 83 $Km \cdot s^{-1}$ and with current radius 13.4 pc. In addition to that, each of Equations (6) and (1) can be used in order to find when each remnants of type II supernova enters to radiative phase and what its physical properties will be. When those equations were applied on our selected remnants, we came up with the results that listed in **Table 4** which show that (for example) the remnants of IC 443 undergoes into a diative phase sooner than the other remnants even sooner than SN1987A even though they have the same inraterstel-

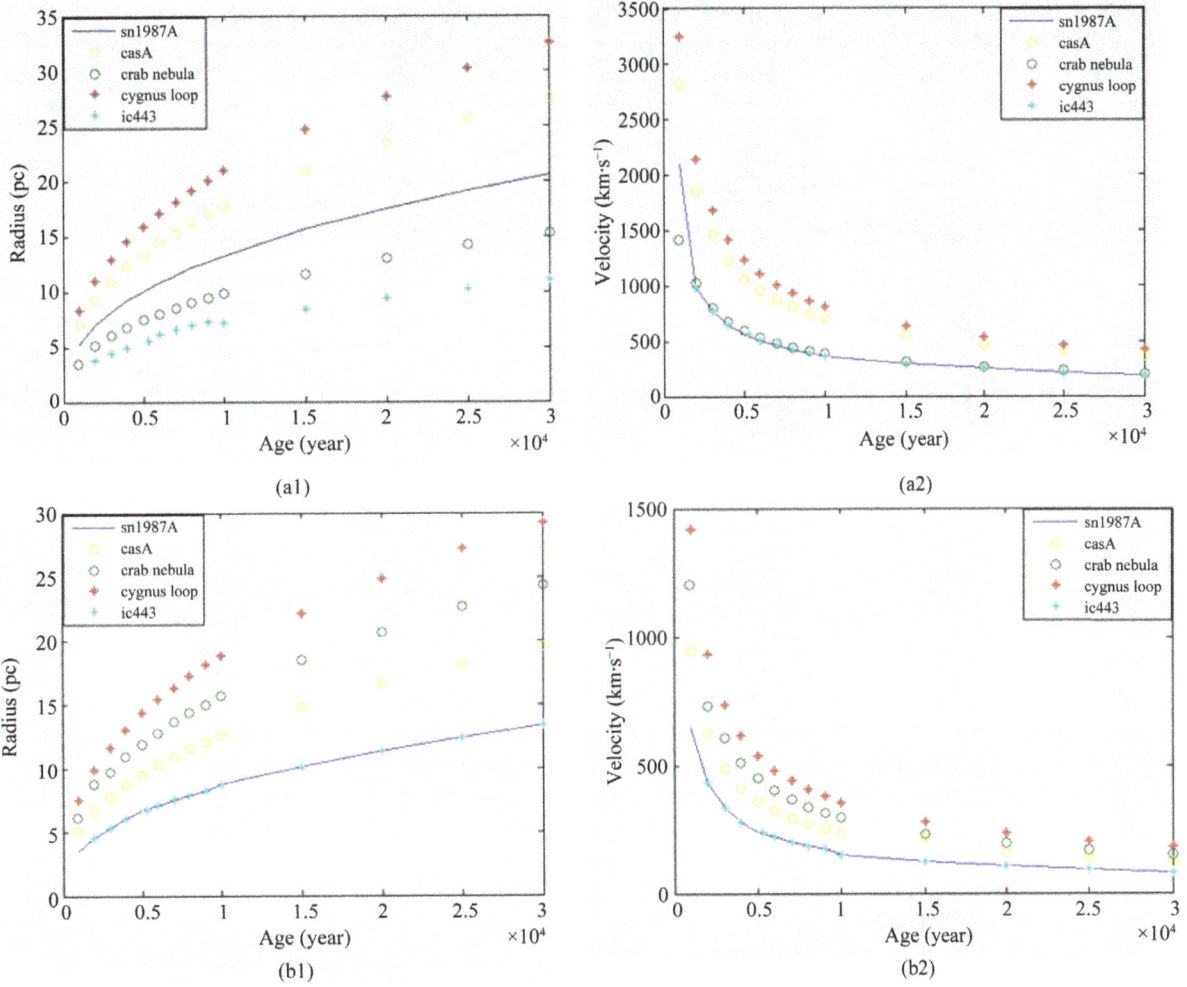

Figure 3. Variation of the radius and velocity of Type II SNRs with age. (a1) and (a2) by applying Equations (3) and (4) while (b1) and (b2) by applying the present model (Equations (6) and (7)).

Table 3. Comparison between the present CPM. and our model results and others published elsewhere.

SNR	Velocity (Km·s⁻¹)			Radius (pc)		
	Our	**CPM.**	**Researchers**	**Our**	**CPM.**	**Researchers**
G1.9+0	14200	11700	14000 [14]	4.37	4.7	4.7 [14]
kepler	2360	2420	1550 - 2000 [17]	6	4	4 [15]
Tycho	2442	3147	1500 - 2800 [18]	5.4	5.6	3.7 [18]
1006	1755	4436	2890 ± 100 [19]	8.6	18	7.1 - 7.5 [19]
SN 1987A	6246	4113	4000 [21]	0.78	0.49	0.39 [20]
Cas A	1755	3872	1000 - 1500 [22]	3.32	5.055	2.637 [22]
Crab	1220	1414	1500 [25]	6.1	3.41	3.4 [24]
Cygnus loop	260	224	200 - 300 [29]	23	9.6	20 - 25 [28]
IC443	83	101	65 - 100 [30]	13.4	11.1	9.6 [31] - 15 [32]

Table 4. The physical properties of the selected remnants at the beginning of the radiative phase.

Parameter	Unit	G1.9+0	kepler	Tycho	1006	SN 1987A	Cas A	Crab nebula	Cygnus loop	IC443
$t_{\text{beginning of radiative phase}}$	Yr	22.7×10^4	7×10^4	7.5×10^4	24×10^4	2×10^4	1.46×10^4	4.28×10^4	3.57×10^4	1.04×10^4
R	Pc	63.9	19.7	27.5	100.7	17.6	13	17.62	22	7.25
V	Km·s^{-1}	280	279.5	364	417	394	355	163.6	244.6	277
T	Kelvin	1.8×10^6	1.77×10^6	3×10^6	3.9×10^6	3.5×10^6	2.86×10^6	0.6×10^6	1.35×10^6	1.7×10^6
M	$M_\odot$	255	25	135	3321	557	336	279.5	217.6	389

lar density. The reason behind that, is the relatively lower explosion energy of IC 443 made it entered to the radiative phase earlier than the remnant of SN1987A. Which means that, not only the density of the medium determine the transition time to the radiative phase but there is another is another important parameter that play a significant role in this transition which is the explosion energy as demonstrated in the result of the remnant of Crab nebula and Cygnus loop which they will take longer time in order to enter into radiative phase according to their explosion energy and interstellar density.

4. Conclusion

The aim of this study is to show that all the remnants of supernovae have the same behavior as each other after 1000 years from the explosion but are shifted up or down depending on the one and the only parameter that effects the development and determination of the lifetime of each remnant which is the interstellar density in which they explode since after this time the explosion energy will not have the strength influence on the expansion and specially in the radiative phase. According to our model, we can calculate the expansion velocity and the radius of any supernova remnant (type Ia and type II) after 1000 years from its explosion by knowing only the density of the medium. When we applied our model on those selected remnant, we indicated that SNR which exploded under low density environment would expand freely and take almost a uniform shape (as almost all type Ia SNR and some of type II SNR such as Crab nebula, CasA). And it takes longer time in order to enter the radiative phase. On the other hand, the remnant that evolves primarily in the interclump medium, which has a density n = (5 - 25) H atoms per cm^3 such as IC443 (one of the best studied cases of supernova remnants interacting with surrounding molecular clouds) and SN1987A (that surrounded by a region with presupernova mass loss), will have ununiform shape. Since it interacts with the dense clumps, the molecular shock fronts is driven by a considerable overpressure (pressure driven snowplow) compared with the pressure in the rest of the remnant, and thus the remnant enters the radiative phase earlier than other remnants

References

[1] Katunari´c, J. (2009) Massive Stars: Life and Death. Dissertation, The Ohio State University, Columbus.

[2] Evans, J. (1998) Death of Stars. Vol. 103, Physics & Astronomy Department, George Mason University, Fairfax.

[3] Arny, T. (1998) Exploration: An Introduction to Astronomy. 2nd Edition, McGraw-Hill Companies, New York, 390-391.

[4] NASA'S HEASARC: Education and Public Information (2011) Introduction to Supernova Remnants.

[5] Roger, R. and Landecker, T. (1988) Supernova Remnants and the Interstellar Medium. Cambridge University Press, Cambridge, 12-20.

[6] Lee, J-J., Koo, B-C., Snell, R., *et al*. (2012) Identification of Ambient Molecular Cloud Associated with Galactic Supernova Remnant IC443.

[7] Theiling, M. (2009) Observation of Very High Energy Gamma Ray Emission from Supernova Remnants with VERITAS. Ph.D. Thesis, Clemson University, Clemson.

[8] Andrew, W. (2006) Filamentary Hα; Structure in the Milky Way. Ph.D. Thesis, University of Wollongong, Wollongong.

[9] Hnatyk, B., Petruk, O. and Telezhyns'kyi, I. (2007) Transition of Supernova Remnant from the Adiabatic Stage of Evolution to Radiative Stage. Analytical Description. *Kinematics and Physics of Celestial Bodies*, **23**, 137-146

[10] Candel, I. (2012) Search for Gamma-Ray Emission from Supernova Remnants with the Fermi/LAT and MAGIC Telescopes. Ph.D. Thesis, University of Aut`onoma, Barcelona.

[11] Cioffi, D., McKee, C. and Bertschinger, E. (1988) Dynamics of Radiative Supernova Remnants. *Astrophysical Journal*, **334**, 252-265. http://dx.doi.org/10.1086/166834

[12] Shu, F. (1992) The Physics of Astrophysics Volume II: Gas Dynamics. University Science Books, Mill Valley.

[13] Ali, L. (2011) Theoretical Study of the Physical Parameters of the Supernova 1987A. Master's Thesis, University of Baghdad, Baghdad.

[14] Ksenofontov, L.T., Voelk, H.J. and Berezhko, E.G. (2010) Non Thermal Properties of Supernova Remnant G1.9+0.3. http://arxiv.org/abs/1004.2555

[15] Patnaude, D.J., Badenes, C., Park, S. and Laming, J.M. (2012) The Origin of Kepler's Supernova Remnant. http://arxiv.org/abs/1206.6799

[16] Tang, Z.M. (1986) The Dynamic Evolution of the Kepler Supernova Remnant. *Astrophysics and Space*, **124**, 315-327. http://dx.doi.org/10.1007/BF00656043

[17] Blair, W.P., Long, K.S. and Vancura, O. (1991) A Detailed Optical Study of Kepler's Supernova Remnant. *Astrophysical Journal*, **366**, 484-494. http://dx.doi.org/10.1086/169583

[18] Hughes, J.P. (2000) The Expansion of the X-Ray Remnant of Tycho's Supernova (SN 1572). *The Astrophysical Journal*, **545**, L53-L56. http://dx.doi.org/10.1086/317337

[19] Winkler, P.F., Gupta, G. and Long, K.S. (2003) The SN 1006 Remnant: Optical Proper Motions, Deep Imaging, Distance, and Brightness at Maximum. *The Astrophysical Journal*, **585**, 324-335. http://dx.doi.org/10.1086/345985

[20] Chiad, B.T., Karim, L.M. and Ali, L.T. (2012) Study the Radial Expansion of SN 1987A Using Counting Pixels Method. *International Journal of Astronomy and Astrophysics*, **2**, 199-203. http://dx.doi.org/10.4236/ijaa.2012.24025

[21] Dwarkadas, V.V. (2006) Supernova Explosions in Winds and Bubbles with Applications to SN 1987A. http://arxiv.org/abs/astro-ph/0612665

[22] Hwang, U. and Laming, J.M. (2011) A Chandra X-Ray Survey of Ejecta in the Cassiopeia A Supernova Remnant. http://arxiv.org/abs/1111.7316

[23] Henry, R.C., Fritz, G., Meekins, J.F., Chubb, T.A. and Friedman, H. (1972) Absorption of Crab Nebula X-Ray. *Astrophysical Journal*, **174**, 389-397. http://dx.doi.org/10.1086/151498

[24] http://en.wikipedia.org/wiki/Crab_Nebula

[25] Bietenholz, M.F., Kronberg, P.P., Hogg, D.E. and Wilson, A.S. (1991) The Expansion of the Crab Nebula. *Astrophysical Journal*, **373**, L59-L62. http://dx.doi.org/10.1086/186051

[26] Blair, W.P., Sankrit, R., Raymond, J.C. and Long, K.S. (1999) Distance to the Cygnus Loop from *Hubble Space Telescope* Imaging of the Primary Shock Front. *The Astronomical Journal*, **118**, 942-947. http://dx.doi.org/10.1086/300994

[27] Martinez, A.P. (2010) The Cygnus Loop: A Weak Core-Collapse SN in Our Galaxy. *Astronomy & Astrophysics*, **527**, A55. http://dx.doi.org/10.1051/0004-6361/201015213

[28] Nemes, N. (2005) XMM-Newton Observation of the Northeastern Limb of the Cygnus Loop Supernova Remnant. Ph.D. Thesis, Osaka University, Osaka.

[29] Kirshner, R.P. and Taylor, K. (1976) High-Velocity Gas in the Cygnus Loop. *The Astronomical Journal*, **208**, L83-L86. http://dx.doi.org/10.1086/182237

[30] Rho, J., Jarrett, T.H., Cutri, R.M. and Reach, W.T. (2001) Near-Infrared Imaging and [O~I] Spectroscopy of IC 443 Using 2MASS and ISO. The Astrophysical Journal, **547**, 885-898. http://dx.doi.org/10.1086/318398 http://arxiv.org/abs/astro-ph/0010551

[31] Hnatyk, B. and Petruk, O. (1998) Supernova Remnants as Cosmic Ray Accelerators. SNR IC 443. *Condensed Matter Physics*, **1**, 655-667. http://dx.doi.org/10.5488/CMP.1.3.655

[32] Zhang, Z.Y., Gao, Y. and Wang, J.Z. (2010) CO Observation of SNR IC 443. *Science China*: *Physics, Mechanics & Astronomy*, **53**, 1357-1369.

[33] Lee, J.J., Koo, B.C., Yun, M.S., Stanimirović, S., Heiles, C. and Heyer, M. (2008) A 21 cm Spectral and Continuum Study of IC443 Using the Very Large Array and the ARECIBO Telescope. *The Astronomical Journal*, **135**, 796-808. http://dx.doi.org/10.1088/0004-6256/135/3/796

[34] Truelove, J.K. and Mckee, C.F. (1999) Evolution of Nonradiative Supernova Remnants. *The Astrophysical Journal Supplement Series*, **120**, 299-326. http://dx.doi.org/10.1086/313176

7

Progress in Physics of the Cosmos

M. Zafar Iqbal

Department of Physics, COMSATS Institute of Information Technology, Islamabad, Pakistan
Email: mziqbal@comsats.edu.pk

Abstract

Study of the Cosmos, at best, is considered a semi-scientific discipline, primarily because the laboratory for carrying out measurements and tests of theories (the Cosmos) has been largely inaccessible for centuries. The cosmic vista into the yonder, however, continued to fascinate humankind due to its inherent beauty and sheer curiosity. The invention of the optical telescope more than five centuries back, however, led to the opening of observational cosmology as a scientific discipline with firm experimental basis. However, the investigations based on visible light posed obvious limitations for the range of such observational cosmology. The advent of the radio telescope in the first half of the 20th century marked a fundamental new step in the progress of this branch of science. There has been no looking back in the march of knowledge in the discipline since then. A whole new vista was laid bare as a result of this development, leading to the discovery of altogether new celestial objects, such as quasars and pulsars and still newer galaxies. The parallel progress of the physics of fundamental constituents of the material world and their interactions led to an interesting merger of these two branches of physical sciences, yielding absolutely astounding knowledge of the nature and evolution of the Universe. New concepts of dark energy and dark matter thought to constitute the dominant share of the Universe were brought to light as a result of these new observations and theoretical ideas. This brief article aims to provide an overview of these exciting developments in the field of cosmology and the associated physics.

Keywords

Radio Astronomy, Cosmic Microwave Background, Dark Energy, Dark Matter, Swirling B Modes

1. Introduction

The birth and evolution of the Universe, of which we are one of the tiniest parts, have been an enigma and a challenge to unravel ever since humankind started taking a rational view of things around him. For a long time, Cosmology, as a branch of Physics, (Astrophysics) has been regarded more as soft science based on guesswork and heuristic ideas devoid of any solid scientific base, largely due to absence of concrete observational evidence. However, as the means to explore the Universe came within man's reach as a result of developments in various branches of physics and science, in general, more and more hard evidence on the different celestial objects and the universe beyond our immediate neighborhood in the cosmos started emerging and the revelation of extraordinary nature and properties of these objects fired the imagination of physicists and scientists, in general, to reach beyond our immediate galaxy, the Milky Way. A brief introduction to how it became possible to do so and with what scientific tools would be provided in the early part of this report. As amazing new facts about the nature of the astronomical bodies and systems inhabiting the cosmos emerged as a result of availability of these advanced means to access distances hitherto thought beyond our means to explore, a coherent picture of the origin and evolution of the universe started to emerge by the late 20th century, culminating in the so called *Standard Model of Cosmology*. This was greatly helped by exciting new developments in the realm of particle physics heralded by the advent of large and expensive accelerator machines, allowing exploration of the structure of the tiniest of the constituents of matter and hence of our universe. Although these developments led to a lot of clarity in our understanding of the science of cosmology, some new enigmas also emerged in their wake. Some of these unanswered questions will be touched upon towards the end of this article.

2. Historical Background

2.1. Early Explorations

Our understanding of the cosmos has, in general, been impeded by the lack of experimental means to observe it. These observations were limited to the range of the optical telescope invented by Galileo in 1609, to begin with. However, this scientific development, momentous though it was, could only extend our knowledge of the cosmos (at whatever rudimentary level possible), not far beyond our own solar system or the Milky Way galaxy, in which it lies. Our observations and, therefore, directly measured data, which are the basis of any scientific understanding of a system, could, at most, cover a miniscule fraction of the universe as long as we were limited to the light (radiation) spanning only the visible part of the optical spectrum. Clearly, we had to extend our capability of observation to the use of 'light' beyond this tiny band of wavelengths, if we intended to explore the elements of the Universe well beyond, since it was established soon after the discovery of the electromagnetic waves, which carry information about their origin and whatever lies in between the observer and such sources, that the cosmos was full of electromagnetic wave signals, harking for research and investigation to establish their nature and origins, just as visible light gave us a clue of what we were looking at around us and our earth. The development of electronic technology and the sensitive tools to detect even weak signals around us first brought us to the advent of radio astronomy. This made it possible for mankind to extend his sphere of exploration of the Cosmos to the 'invisible' (to the human eye or the optical telescope) part of the universe.

2.2. Radio Astronomy

Radio astronomy started with a serendipitous discovery in 1932 by Jansky [1], a scientist at the Bell Telephone Laboratories, of radio waves coming from space. Development of radar and allied technology during the 2nd World War led to a boost to this branch of science, bringing into limelight focused studies to look for radio wave sources from around the Universe, using specially designed high resolution radio telescopes, yielding remarkable results to enrich our knowledge of the universe. The power of radio astronomy is well demonstrated by the well-resolved structure, with much enhanced clarity, in the equatorial plane of our Milky Way galaxy, as observed by a radio telescope in comparison to that observed by visible light, as shown in **Figure 1**. Unlike visible light, radio waves are not absorbed by clouds, interstellar dust, or the atmosphere of the Earth. They, thus, provide a superior means of exploring distant celestial objects in the universe, as compared to the ordinary visible light waves. The original radio wave signal picked up by Jansky turned out to originate from the centre of the Milky Way. Large diameter dish antennae telescopes had to be constructed to achieve relatively high resolutions for meaningful exploration of the cosmos. Whereas radio wave emission from the Sun was first observed [2] in 1942, the first distant objects, *i.e.*, the galaxies Centaurus A and M87 and the Crab Nebula, shown in **Figure 2**, were identified as strong radio sources in 1949 [3].

Planetary science of our solar system benefitted from the development of the radio telescope greatly, since, although planets only reflect visible light, they may, however, emit radio waves, leading to their detailed studies

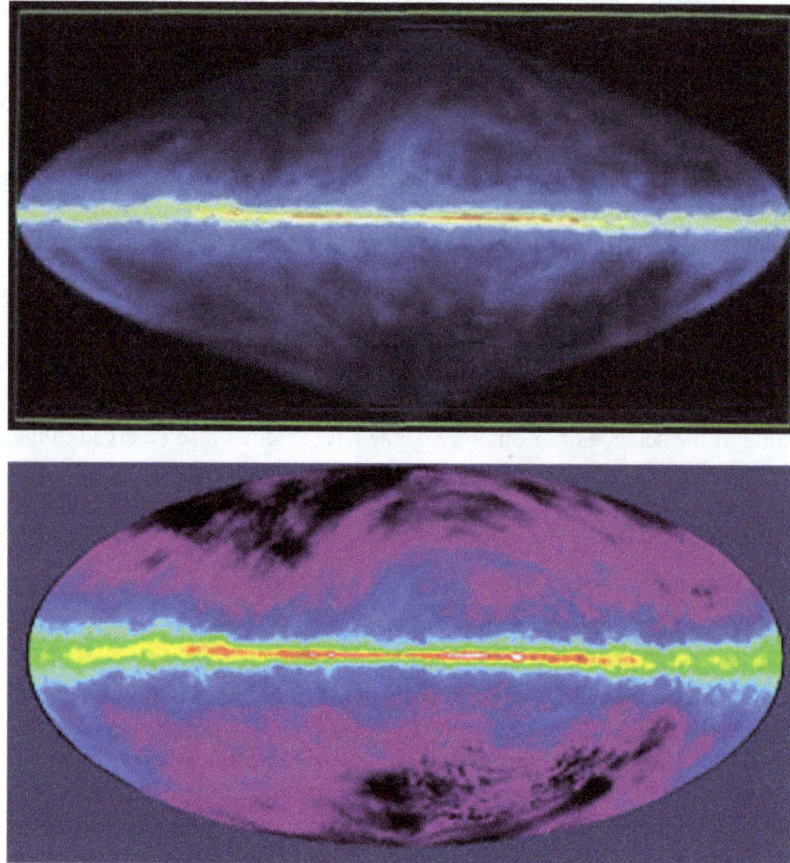

Figure 1. (Top) The Milky Way galactic plane under visible light. (Bottom) The structures of galactic plane become clear in 21 cm hydrogen spectral line observations. (Photo credit: NASA). Source: The National Radio Astronomy Observatory (NRAO). http://www.physics.hku.hk/~astro/hkusrt/radio_astro.html

using the new type of telescopes. The surface temperature of Venus was measured from such observations [4] as were some initial explorations made on radio wave emissions, both continuous and pulsed, from Jupiter [5]. Studies based on the 21 cm hydrogen line emission, first detected in 1951 [6] [7] have been extremely useful for the exploration of the interstellar matter, which largely consists of neutral hydrogen gas. Such studies, for example, led to the observation of the many spiral arms of our galaxy [8]-[11] finer than those of other galaxies. The formation and evolution of stars from regions of interstellar space containing other gases, which are the sources of radio waves, have been widely studied with radio telescopes, yielding exceptionally useful information on star birth. New galaxies, such as Cygnus A, have been found to be a million times brighter than Milky Way in radio-wave region [12] [13] of emission spectrum, with large lobes of radio-wave emission (**Figure 3**) around the central region emitting in the optical domain—detailed studies showed that the outer radio wave lobes are only 3 million years old as compared to the central region going back to 10 billion years. Radio astronomy has revealed altogether new facts about the Cosmos, such as the existence of quasi-stellar objects (quasars) in 1960 [14]-[16], which emit brightly both in the visible and the radio frequency region of the electromagnetic wave spectrum. These strange objects were at the centre of fascinating studies by astronomers, as well as physicists, during the 1960s; detailed redshift studies indicated them to be just about the farthest objects in the universe at the time, some 12 billion light years from Earth. As such, they held the promise of yielding information on the early stages of the Universe. The discovery of pulsars [17], following in 1967, showed how rotating neutron stars, some of the most densely packed material objects in the universe, could lead to the emission of pulses of radio waves at such an amazingly regular interval (33 milliseconds for the crab nebula pulsar), that they can be used as "astronomical clocks" with the highest precision. These wonderful lighthouse-like objects have inspired a great, in-depth understanding of the Universe by providing detailed insight into the life cycle of the stars. The

Figure 2. Hubble Space Telescope photograph of the Crab Nebula (2005)[1]. http://en.wikipedia.org/wiki/Crab_Nebula

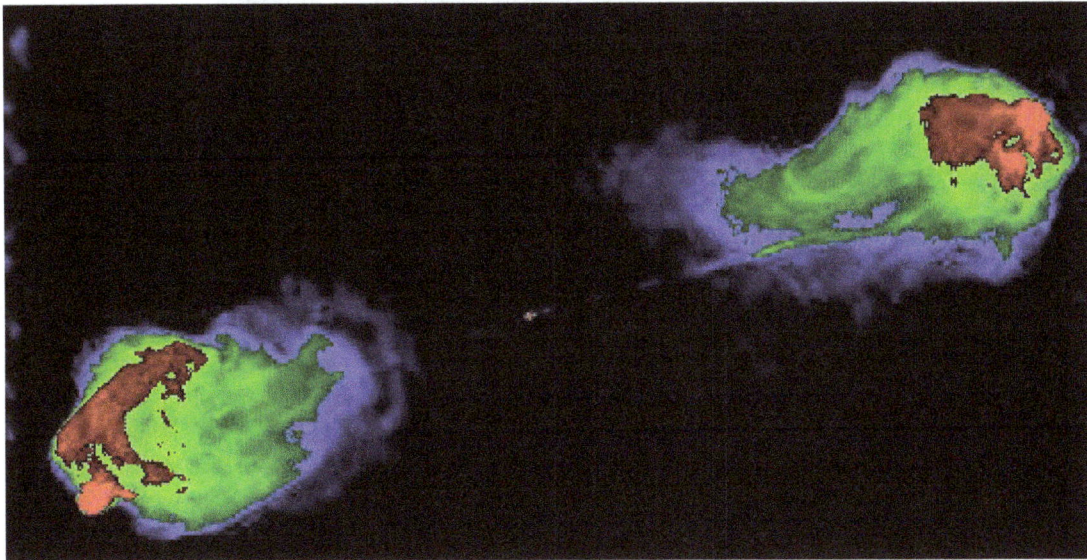

Figure 3. 5-GHz radio image of Cygnus A (3C405). http://en.wikipedia.org/wiki/Cygnus_A#cite_note-3

[1]This is a mosaic image, one of the largest ever taken by NASA's Hubble Space Telescope of the Crab Nebula, a six-light-year-wide expanding remnant of a star's supernova explosion. Japanese and Chinese astronomers recorded this violent event nearly 1000 years ago in 1054, as did, almost certainly, Native Americans. The orange filaments are the tattered remains of the star and consist mostly of hydrogen. The rapidly spinning neutron star embedded in the center of the nebula is the dynamo powering the nebula's eerie interior bluish glow. The blue light comes from electrons whirling at nearly the speed of light around magnetic field lines from the neutron star. The neutron star, like a lighthouse, ejects twin beams of radiation that appear to pulse 30 times a second due to the neutron star's rotation. A neutron star is the crushed ultra-dense core of the exploded star. The Crab Nebula derived its name from its appearance in a drawing made by Irish astronomer Lord Rosse in 1844, using a 36-inch telescope. When viewed by Hubble, as well as by large ground-based telescopes such as the European Southern Observatory's Very Large Telescope, the Crab Nebula takes on a more detailed appearance that yields clues into the spectacular demise of a star, 6500 light-years away. The newly composed image was assembled from 24 individual Wide Field and Planetary Camera 2 exposures taken in October 1999, January 2000, and December 2000. The colors in the image indicate the different elements that were expelled during the explosion. Blue in the filaments in the outer part of the nebula represents neutral oxygen, green is singly-ionized sulfur, and red indicates doubly-ionized oxygen.

development of radio astronomy, thus, expanded the horizon of our knowledge and understanding of the Universe from the elements of our own solar system and its planets to stars and other objects in our galaxy, to interstellar space and galaxies far beyond.

3. Cosmic Microwave Background

The discovery of cosmic microwave background (CMB) radiation in 1964 [18] is one of the most significant milestones in the development of the science of cosmology in the modern era, based on observations in radio astronomy. The serendipitous detection of this all pervading diffuse microwave radiation, found to have isotropic distribution over entire space, corresponding to a black body temperature of 2.73 Kelvin, although a mind boggling puzzle to start off with, was soon interpreted as the remnant of the massive explosion that accompanied the creation of the Universe, estimated to have taken place some 13.7 billion years ago. This event, heralding enormous release of energy, came to be dubbed as the *big bang* with time. The point in space of this event corresponded to a singularity from which radiation emanated in all directions, eventually transforming into matter (elementary particles of all descriptions constituting the material universe) as it cooled down. The material universe is believed to be expanding as per a law attributed to Hubble, but first derived from Einstein's equations of General Relativity by Georges Lemaître in 1927, which states that all cosmological bodies are receding from each other with a velocity proportional to their distance from each other. The proportionality constant which was first measured by Hubble in 1929 has come to be known as the Hubble constant and its most recent and accurate value is $H_0 = 74.3 \pm 2.1$ (km/s)/Mpc, pc being the symbol for a parsec, an astronomical unit of distance equal to 3.09×10^{13} km. The crab nebula, for instance, is receding from us at a rate of ~1500 km/s. The discovery of the cosmic microwave background radiation was the first definitive evidence for the big bang origin of our expanding universe.

4. Structure of the Universe

4.1. COBE

Whereas the uniformity of CMB in all directions in the cosmos seemed to be a striking feature of its discovery, detailed later studies based on COBE (Cosmic Background Explorer), a satellite in the Explorer series launched by NASA on November 18, 1989 in a sun-synchronous orbit, revealed strong evidence for an all important anisotropy in CMB, first announced on April 23, 1992 [19]. The famous full map of this anisotropic distribution of CMB obtained by COBE is shown in **Figure 4**. These fluctuations in CMB around the sky are extremely weak (about one part in 100,000 as compared to the 2.73 Kelvin average temperature of the radiation field), explaining why the initial measurements found a virtually isotropic microwave background, regardless of the position in the Cosmos. This anisotropy, clearly, hinted at the local density fluctuations of the Universe at or close to the time of the big bang, as evidenced today by the existence of galaxies interspersed by empty space. This was the first indication of the reason for the existence of today's structure in the Universe. This important observation led to very significant advances in our understanding of the formation of stars, galaxies of stars and their clusters and, hence, the evolution of the Universe. This significance of the discovery has been recognized by the award of the Nobel Prize for Physics in 2006 to John C. Mather of the NASA Goddard Space Flight Center and George F. Smoot at the University of California, Berkeley, two of COBE's principal investigators. According to the Nobel Prize committee, "the COBE-project can also be regarded as the starting point for cosmology as a precision science" [20].

4.2. WMAP

The COBE mission was followed by the Wilkinson Microwave Anisotropy Probe (WMAP) launched on June 30, 2001 with the objective of obtaining a more precise full sky map of the anisotropy of CMB distribution with a resolution of 13 arcminute - 45 times more sensitive and with 33 times the angular resolution of COBE. These precise data were expected to help understand the geometry, content, and evolution of the Universe, providing finer tests of the Big Bang model. WMAP's measurements played the key role in establishing the current *Standard Model* of cosmology and have led to the most precise value, till then, of the age of the Universe at 13.75 ± 0.11 billion years; the full timeline of the Universe according to the standard model is given in **Figure 5**.

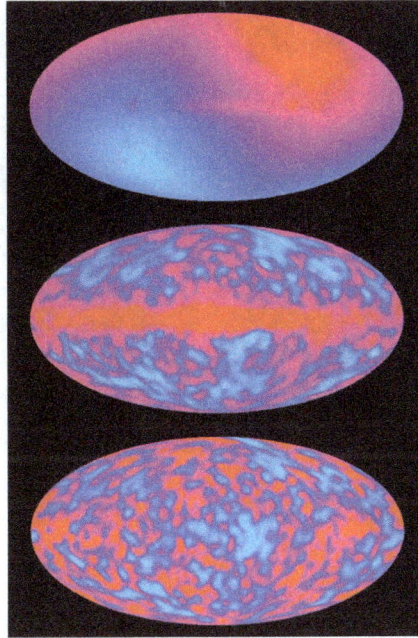

Uncorrected map

Temperature of the Cosmic Microwave Background (CMB) radiation spectrum as determined with the COBE satellite during the first two years of the Differential Microwave Radiometer (DMR) observation. The plane of the Milky Way Galaxy is horizontal across the middle of each picture: (top) uncorrected; (middle) corrected for the dipole term due to our peculiar velocity; (bottom) further corrected to remove the contribution of our galaxy. Note: This map is based on data collected over the two first years of the four-year COBE mission. Therefore, it has been superseded by the four-year map (shown below).

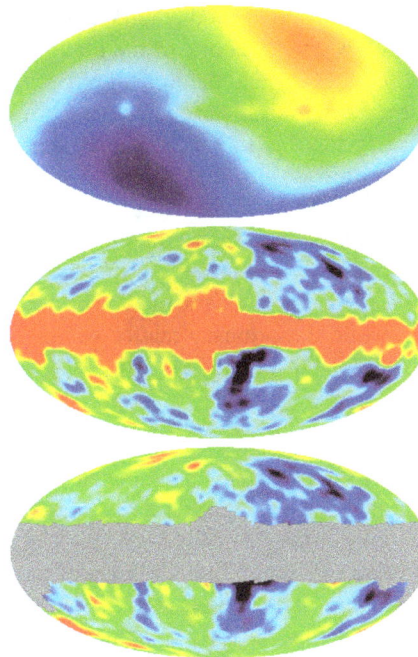

Four-year map

Temperature of the Cosmic Microwave Background (CMB) radiation spectrum as determined with the COBE satellite during the full four years of the Differential Microwave Radiometer (DMR) observation. The plane of the Milky Way Galaxy is horizontal across the middle of each picture. This map shows the 53 GHz channel: (top) prior to dipole subtraction; (middle) after dipole subtraction (due to the solar system movement); (bottom) after subtraction of a model of the Galactic emission.

Figure 4. Source: The COBE datasets were developed by the NASA Goddard space flight center under the guidance of the COBE science working group. http://en.wikipedia.org/wiki/Cosmic_Background_Explorer

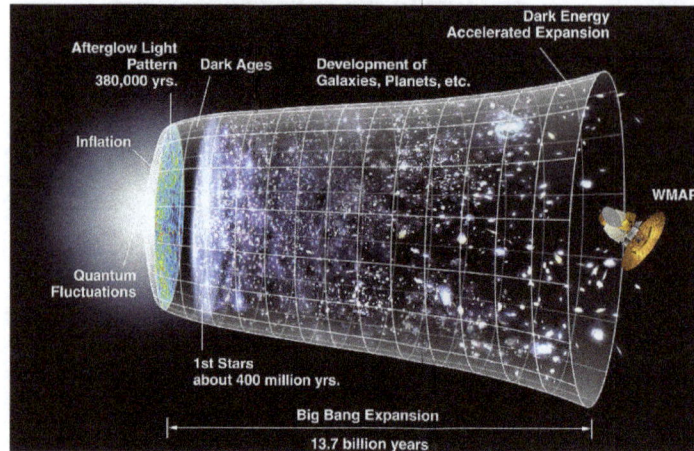

Figure 5. The timeline of the Universe, from inflation to the WMAP.
http://en.wikipedia.org/wiki/Wilkinson_Microwave_Anisotropy_Probe

5. Dark Energy and Dark Matter

This insight into the nature of the Universe has led to two remarkable new concepts. The first is to do with the expansion of the Universe referred to above. It has all along been expected that the speed of this expansion would slow down as we move farther and farther out in the Universe, since the galaxies would gravitationally pull each other. It was, therefore, to every body's astonishment that, to the contrary, this expansion was found to be accelerating rather than slowing down as we move outwards. This amazing recent discovery based on the observations of the distant Type 1a Supernovae—a late-in-life, dying state of a star—in 1998 [21] [22] caused a great stir among astronomers and physicists alike, leading to the award of the 2011 Nobel Prize in Physics. This novel phenomenon led to the idea of some mysterious "dark energy", which pulls the space apart [23]. The question of the precise nature of this dark energy is one of the hotly debated subjects in astrophysics and of ongoing research in both observational cosmology and theoretical physics.

Another subject of immense current research is an equally, if not more, mysterious object called the "dark matter". This is matter proposed to provide the gravitational glue binding the galaxies together in the cosmos. The original idea for this type of matter arose from the fact that galaxies in some clusters are found to move too fast to be allowed to hold together among themselves; even some stars within some individual galaxies move way too fast for gravity to hold them in these galaxies. Some mysterious invisible "dark matter" was, therefore, proposed to provide the missing gravitational pull in these systems, as early as 1933 [24]. Recent WMAP measurements provided a more direct evidence for this dark matter, however, suggesting that some 83% of the Universe constituted in "dark matter", while the ordinary matter we see and feel around us in the entire cosmos makes only about 4.56% of the mass of the Universe [25]-[28].

5.1. Observation of Dark Matter

The European Space Agency's (ESA) Planck mission, launched to study the cosmic microwave background (CMB), released the first results after the initial 15 months in March 2013 [29]. These observations revealed new structure in temperature fluctuations of the CMB on a far finer scale than known hitherto (**Figure 6**). Planck has scanned the entire sky in microwave and submillimeter range of the electromagnetic spectrum, materializing high resolution pictures of the temperature fluctuations since the Universe was very young. The slightly slower expansion rate of the Universe sensed by Planck has extended the age of the Universe slightly, in comparison with the previous estimates—to 13.81 ± 0.05 billion years from 13.77 billion years proposed by the WMAP mission and allied observations.

With the initial, opaque plasma state of the Universe after the Big Bang, the excessive energy present did not allow formation of stable atoms until a cooling period of 380,000 years, when an event known as *recombination* set in. The transparent universe followed with a huge outflow of photons, into the modern era. The initial ultraviolet photons resulting from recombination, however, converted to the microwave spectral region as the universe cooled further down to its current level. The power spectrum of the fluctuations over the sky, revealed by

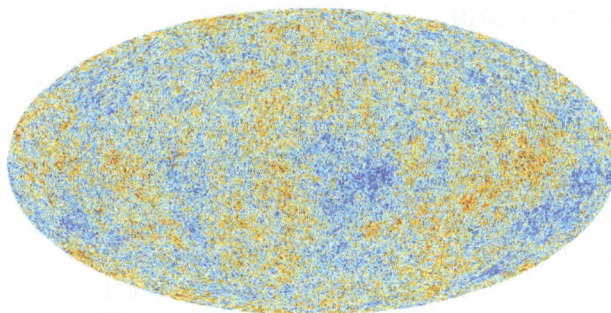

Figure 6. The cosmic microwave background—temperature fluctuations left over from 380,000 thousand years after the Big Bang. This new map is based on data from the Planck mission. ESA/Planck Collaboration/D. Ducros

Planck mission, **Figure 7**, yields crucial information on the structure and composition of the Universe. The most prominent fluctuations are due to the total energy content of the Universe, while the smaller ones are associated with the distribution of matter alone, both ordinary and dark. The relative magnitudes of the ordinary and dark matter and the conglomeration of the latter are revealed by the smaller fluctuations. Comparison of the Planck data with theoretical models led to finer, but significant adjustments in the hitherto known estimates of energy/matter composition of the Cosmos (**Figure 8**): dark energy lowered to 68.3% from 72.8%; dark matter up from 22.7% to 26.8%; ordinary matter also up from 4.5% to 4.9%.

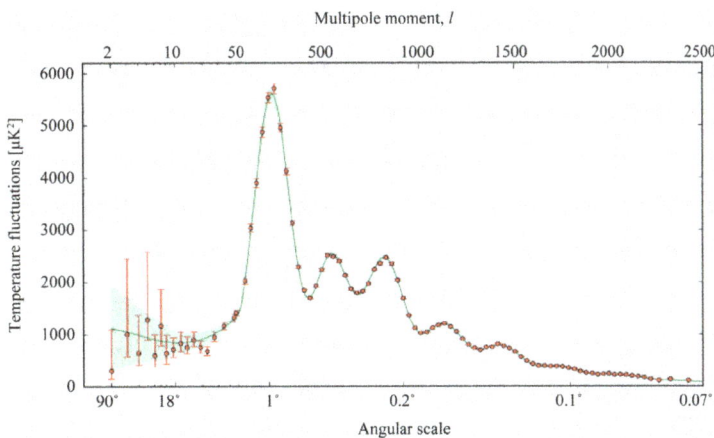

Figure 7. The power spectrum measured by Planck, showing the fluctuations in temperature at a range of size scales on the sky. The anomaly previously seen by WMAP lies at the left edge. The three major peaks show the relative contributions of dark energy, ordinary matter, and dark matter. ESA/Planck Collaboration/D. Ducros

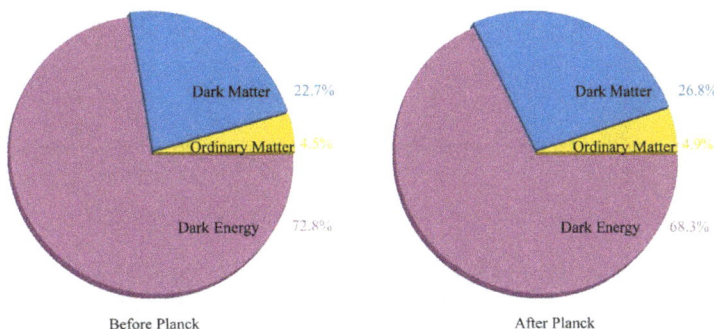

Figure 8. The composition of the cosmos, before and after Planck data release. ESA/Planck Collaboration/D. Ducros

5.2. Polarization Studies of CMB

The ultimate evidence for or against dark matter is expected to emerge from the polarization studies of the CMB radiation over the Universe and its anisotropy signature. A great excitement was recently generated by positive news on that count, coming from a telescope station located at the South Pole of the earth. On March 17, 2014 came the announcement of the observation of the so called primordial B-modes by BICEP2 detector of the telescope. These are swirling polarization patterns, **Figure 9**, in the tiny (1 part in 100,000) fluctuations in CMB temperature across the Universe. These miniscule variations in temperature are interpreted to be due to variations in the density of the primordial gas at the time of emission of the black body radiation, thus reflecting the evolutionary process of the Universe. These fluctuations are thought to be magnified by the gravitational pull of matter, leading to formation of galaxies and clusters of galaxies, now observed in the cosmos. The development of ultrasensitive detectors about a decade back made it possible to provide precise measurements of the polarization, calling in a new era in experimental cosmology and in studies of structure and evolution of the Universe. In particular, the swirling B modes would provide information on the inflationary development—proposed as early as 1980—of the very early Universe, during the first 10^{-36} to 10^{-32} seconds of the Big Bang, during which the Cosmos grew by a factor of 10^{26}. The gravitational waves during this period are supposed to generate this polarization pattern. Dust and magnetic fields in our galaxy were expected to easily mask the tiny polarization signal being sought after. Scales of $1°$ used for the initial measurement of B modes were concluded to be large enough to avoid interference due to signals from intervening galaxies. The sceptics found these signals to be far stronger than those predicted or even beyond the limits set by Planck mission on the power of the gravitational waves to generate the CMB temperature fluctuations.

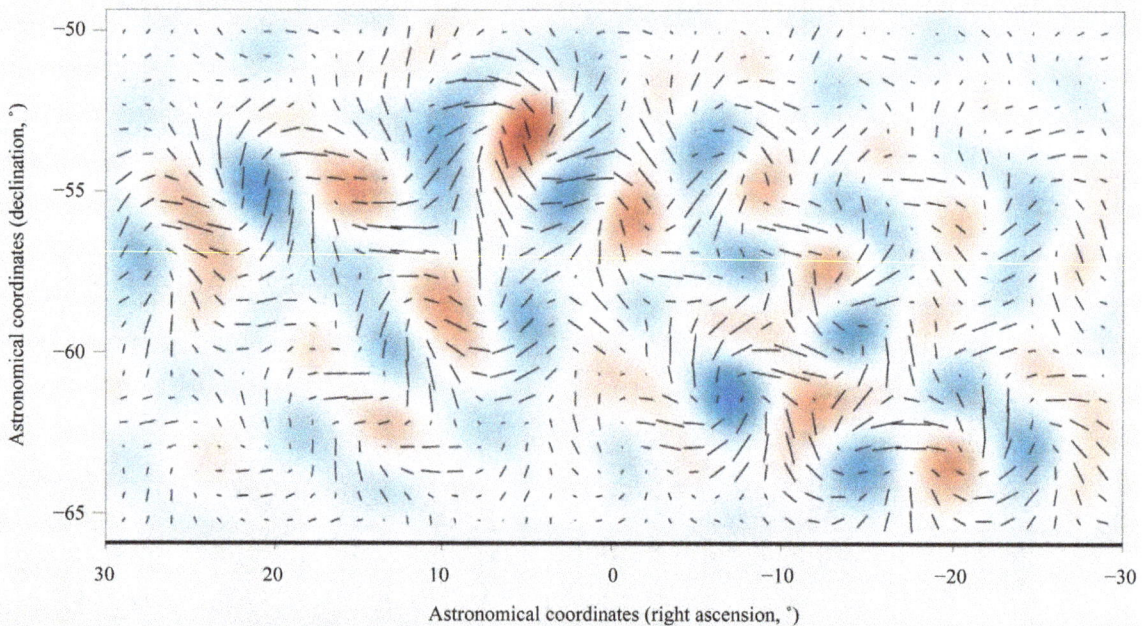

Figure 9. B-mode polarization in the Cosmic Microwave Background (source: BICEP2 Collaboration).
http://www.nature.com/news/telescope-captures-view-of-gravitational-waves-1.14876

However, it did not take too long to realize that the twisting CMB polarization pattern could easily be caused by the cosmic dust in the Milky Way galaxy [30]. It has been concluded that the BICEP2 team underestimated the contribution of this dust to the swirl polarization pattern, which could almost entirely be accounted for by the latest dust pattern data revealed by Planck mission.

This reversal of the evidence for the detection of primordial gravitational waves applies brakes to the progress in the quest for experimental proof of Big Bang inflationary model, to the idea of multiverse and, of course, to a definitive proof of the existence of dark matter. However, this may signal a temporary hiatus, since, at least, eight experiments, including BICEP3, the Keck Array and Planck, are already focused on this problem. Mean-while, a new perception has emerged regarding the inflationary model, which makes this theoretical model valid irrespective of the experimental evidence for the *primordial* gravitational waves—the paradigm of inflation would, in that perspective, appear to be unfalsifiable and, therefore, scientifically meaningless [31].

6. Concluding Remarks

The brief review above shows clearly how far we have travelled in unraveling the secrets of the Cosmos and the Universe in which we live, over less than a century. It is equally manifest that this rapid progress has been made possible by the enormous advancements in the experimental instrumentation and techniques available for observing the Cosmos. The huge progress in telescopy and the remarkable developments in space technology have made a tremendous impact on the achievable knowledge with high precision and accuracy. While the astronomical knowledge of the cosmos has witnessed impressive progress, the analysis of these observations has opened new frontiers in our understanding of the fundamental physics and nature and evolution of the Universe.

It is clear from the above brief description of the current state of our knowledge of the Universe that the age old maxim, *the more we know, less we conclude we actually know*, appears to be perfectly valid. However, such is perhaps the nature of all human knowledge. But how exciting the ever expanding frontiers of this field of fundamental physics are can be gauged from this brief narrative. It is also clear that our quest for the knowledge of the Universe will be a continuing chapter in the story of human search and discovery for the years to come and may yet raise more questions than it can answer. Thanks to the remarkable developments in technology, and hence the tools of observational astronomy, many new frontiers of the unknown may yet be opened for even more extensive exploration of the Cosmos than the territory conquered hitherto.

References

[1] Jansky, K.G. (1933) Radio Waves from Outside the Solar System. *Nature*, **132**, 66. http://dx.doi.org/10.1038/132066a0

[2] Hey, J.S. (1983) The Radio Universe. Vol. 1, Oxford Pergamon Press, Oxford, 253.

[3] Bolton, J.G., Stanley, G.J. and Slee, O.B. (1949) Positions of Three Discrete Sources of Galactic Radio-Frequency Radiation. *Nature*, **164**, 101-102. http://dx.doi.org/10.1038/164101b0

[4] Barath, F.T., Barrett, A.H., Copeland, J., Jones, D.E. and Lilley, A.E. (1963) Microwave Radiometers. *Science*, **139**, 908-909. http://dx.doi.org/10.1126/science.139.3558.908

[5] Burke, B.F. and Franklin, K.L. (1955) Observations of a Variable Radio Source Associated with the Planet Jupiter. *Journal of Geophysical Research*, **60**, 213-217. http://dx.doi.org/10.1029/JZ060i002p00213

[6] Muller, C.A. and Oort, J.H. (1951) Observation of a Line in the Galactic Radio Spectrum: The Interstellar Hydrogen Line at 1420 Mc./sec., and an Estimate of Galactic Rotation. *Nature*, **168**, 357-358. http://dx.doi.org/10.1038/168357a0

[7] Ewen, H.I. and Purcell, E.M. (1951) Observation of a Line in the Galactic Radio Spectrum. *Nature*, **168**, 356-358.

[8] Chen, W., Gehrels, N., Diehl, R. and Hartmann, D. (1996) On the Spiral Arm Interpretation of COMP^{TEL} ^ 26^ Al Map Features. *Astronomy and Astrophysics Supplement Series*, **120**, 315.

[9] Russeil, D. (2003) Star-Forming Complexes and the Spiral Structure of Our Galaxy. *Astron. Astrophys.-Berl.-*, **397**, 133-146.

[10] Levine, E.S., Blitz, L. and Heiles, C. (2006) The Spiral Structure of the Outer Milky Way in Hydrogen. *Science*, **312**, 1773-1777.

[11] Churchwell, E., Babler, B.L., Meade, M.R., Whitney, B.A., Benjamin, R., Indebetouw, R., Cyganowski, C., Robitaille, T.P., Povich, M., Watson, C. and Bracker, S. (2009) The Spitzer/GLIMPSE Surveys: A New View of the Milky Way. *Publications of the Astronomical Society of the Pacific*, **121**, 213-230. http://dx.doi.org/10.1086/597811

[12] Jennison, R.C. and Das Gupta, M.K. (1953) Fine Structure of the Extra-Terrestrial Radio Source Cygnus 1. *Nature*, **172**, 996-997. http://dx.doi.org/10.1038/172996a0

[13] Baade, W. and Zwicky, F. (1934) Cosmic Rays from Super-Novae. *Proceedings of the National Academy of Sciences*

of the United States of America, **20**, 259-263. http://dx.doi.org/10.1073/pnas.20.5.259

[14] Schmidt, M. (1963) 3C 273: A Star-Like Object with Large Red-Shift. *Nature*, **197**, 1040.
      http://dx.doi.org/10.1038/1971040a0

[15] Matthews, T.A. and Sandage, A.R. (1963) Optical Identification of 3c 48, 3c 196, and 3c 286 with Stellar Objects. *Astrophysical Journal*, **138**, 30.

[16] Greenstein, J.L. and Schmidt, M. (1964) The Quasi-Stellar Radio Sources 3c 48 and 3c 273. *Astrophysical Journal*, **140**, 1. http://dx.doi.org/10.1086/147889

[17] Hewish, A., Bell, S.J., Pilkington, J.D.H., Scott, P.F. and Collins, R.A. (1968) Observation of a Rapidly Pulsating Radio Source. *Nature*, **217**, 709-713. http://dx.doi.org/10.1038/217709a0

[18] Penzias, A.A. and Wilson, R.W. (1965) A Measurement of Excess Antenna Temperature at 4080 Mc/s. *Astrophysical Journal*, **142**, 419-421. http://dx.doi.org/10.1086/148307

[19] Boggess, N.W., Mather, J.C., Weiss, R., Bennett, C.L., Cheng, E.S., Dwek, E., Gulkis, S., Hauser, M.G., Janssen, M.A., Kelsall, T., *et al.* (1992) The COBE Mission—Its Design and Performance Two Years after Launch. *Astrophysical Journal*, **397**, 420-429. http://dx.doi.org/10.1086/171797

[20] Cosmic Background Explorer. Wikipedia, the free encyclopedia, 21 July 2014.

[21] Riess, A.G., Filippenko, A.V., Challis, P., Clocchiatti, A., Diercks, A., Garnavich, P.M., Gilliland, R.L., Hogan, C.J., Jha, S., Kirshner, R.P., *et al.* (1998) Observational Evidence from Supernovae for an Accelerating Universe and a Cosmological Constant. *Astronomical Journal*, **116**, 1009. http://dx.doi.org/10.1086/300499

[22] Perlmutter, S., Aldering, G., Goldhaber, G., Knop, R.A., Nugent, P., Castro, P.G., Deustua, S., Fabbro, S., Goobar, A., Groom, D.E., *et al.* (1999) Measurements of Ω and Λ from 42 High-Redshift Supernovae. *Astrophysical Journal*, **517**, 565. http://dx.doi.org/10.1086/307221

[23] Peebles, P.J.E. and Ratra, B. (2003) The Cosmological Constant and Dark Energy. *Reviews of Modern Physics*, **75**, 559. http://dx.doi.org/10.1103/RevModPhys.75.559

[24] Zwicky, F. (2009) Republication of: The Redshift of Extragalactic Nebulae. *General Relativity and Gravitation*, **41**, 207-224. http://dx.doi.org/10.1007/s10714-008-0707-4

[25] Ade, P.A.R., Aghanim, N., Armitage-Caplan, C., Arnaud, M., Ashdown, M., Atrio-Barandela, F., Aumont, J., Baccigalupi, C., Banday, A.J., Barreiro, R.B., *et al.* (2013) Planck 2013 Results. I. Overview of Products and Scientific Results. arXiv:1303.5062.

[26] Francis, M. (2013) First Planck Results: The Universe Is Still Weird and Interesting. *Ars Technica*, 21 March 2013. http://arstechnica.com/science/2013/03/first-planck-results-the-universe-is-still-weird-and-interesting/

[27] First Observational Evidence of Dark Matter. http://www.darkmatterphysics.com/Galactic-rotation-curves-of-spiral-galaxies.htm

[28] Dark Matter. Wikipedia, the free encyclopedia, 22 July 2014.

[29] Cowen, R. (2014) Telescope Captures View of Gravitational Waves. *Nature*, **507**, 281-283. http://dx.doi.org/10.1038/507281a

[30] Cowan, R. (2014) Big Bang Finding Challenged. *Nature*, **510**, 20. http://dx.doi.org/10.1038/510020a

[31] Steinhardt, P. (2014) Big Bang Blunder Bursts the Multiverse Bubble. *Nature*, **510**, 9. http://dx.doi.org/10.1038/510009a

8

Time's Arrow in a Finite Universe

Martin Tamm

Department of Mathematics, University of Stockholm, Stockholm, Sweden
Email: matamm@math.su.se

Abstract

In this paper, a simple model for a closed multiverse as a finite probability space is analyzed. For each moment of time on a discrete time-scale, only a finite number of states are possible and hence each possible universe can be viewed as a path in a huge but finite graph. By considering very general statistical assumptions, essentially originating from Boltzmann, we make the set of all such paths (the multiverse) into a probability space, and argue that under certain assumptions, the probability for a monotonic behavior of the entropy is enormously much larger then for a behavior with low entropy at both ends. The methods used are just very simple combinatorial ones, but the conclusion suggests that we may live in a multiverse which from a global point of view is completely time-symmetric in the sense that universes with Time's Arrow directed forwards and backwards are equally probable. However, for an observer confined to just one universe, time will still be asymmetric.

Keywords

Multiverse, Time's Arrow, Cosmology, Entropy

1. Introduction

The riddle of Time's Arrow is an outstanding problem in modern physics. How does it come that we can remember yesterday but not tomorrow? Why do we all grow older but never grow younger? We all know that what is possible in one direction of time may be quite impossible in the other. The riddle of time consists in the observation that this asymmetry disappears when we turn to the microlevel: here, the all processes seem to be equally possible in both directions of time. So, if the fundamental laws of physics are time-symmetric, where does time asymmetry come from?

Ever since the time of Boltzmann [1], it has been clear that this has something to do with entropy and the Second Law of Thermodynamics, although it is still not quite clear exactly what the relation is. In spite of a large number of different attempts to explain the asymmetry of time, there has been so far no agreement at all even about where to look for the solution: is it a question about the dynamical laws or about the boundary conditions of the universe? Is it a quantum mechanical problem or is it an essentially classical question about probabilities?

It is also not clear what it will mean to have an answer to the riddle. Of course, if it will be something simple, like an asymmetry in the dynamical laws which we have for some reason previously overlooked, then this will perhaps not be a problem. But if it is, as will be suggested in this paper, a probabilistic property of our universe, then the situation is different. In most cases, physical theories should try to describe as many details and factors as possible. In view of the enormous complexity of the universe, however, this may not be a possible strategy to start within the present case. Rather, we are forced by necessity to take the opposite path: to simplify as far as possible and discard everything which is not absolutely indispensable in order to reveal the underlying mechanism.

Another complication may be the concept of entropy itself. Under ordinary circumstances, entropy may be a fairly well understood concept. But it should always be remembered that it is developed in classical thermodynamics, and is best understood in stationary or quasi-stationary situations. This is very far from what we meet in cosmology, where some of our ordinary physical concepts get a new twist.

Still, another specific problem with Time's Arrow seems to be that our very human perspective tends to make us formulate the questions in the wrong way, and that many of our conclusions may in fact already be unconsciously built into our assumptions (see Price [2]). For this reason, it seems important to start the investigation by formulating some of these fundamental questions which underlie the discussions. In Sections 2 and 3, we, therefore, review three such assumptions. In Section 4, we study a very simple model for a semi-classical multiverse, and in Section 5 we apply Boltzmann's ideas to relate the dynamics of it to the concept of entropy. A fundamental problem in this context is that we really do not know what we should mean by a state of the multiverse and how we should count them. In Section 6, we then turn this multiverse into a probability space by adding some assumptions about probability weights, and in Section 7 we discuss a non-technical argument why a monotonic growth of the entropy should be more likely than a symmetric behavior with low entropy at both ends. This is related to the old discussion originating from Gold's famous paper [3] and continued by Hawking, Page, LaFlamme and others (see [4] [5]), but from a rather different perspective.

In any case, these arguments are just the beginning of what should be done. Finally, in Section 8, we discuss how these ideas can be further developed.

Some of the ideas in this paper have been discussed earlier in a more preliminary and technical form in [6], where some related ideas can also be found.

2. Our Underlying Assumptions

Examples of fundamental questions underlying the discussion are:
- Is temporal causality a fundamental property of nature?
- Is our universe unique? (instead of just one of many parallel ones?)
- Is our world an infinite structure?

Depending on what answers, yes or no, we give to these questions, we get eight different perspectives on physics in general and on Time's Arrow in particular. From our present state of knowledge, it is hard to argue that any one of these perspectives is obviously wrong or obviously the correct one. But it can still be very important to be aware of which perspective we choose and why. The starting point for this paper is to assume that the answers to these questions are all no.

Temporal causality refers to the idea that events are somehow caused by previous events in the backward light-cone of the given event, and hence that the latter should be considered to be a consequence of these. From a human perspective, this may all be very natural. But from the point of view of fundamental physics, this is quite problematic. As has many times been pointed out by Price [2], if we accept that the laws of physics are essentially time-symmetric, then there is no obvious reason why we could not equally well consider the given event as a consequence of events in the forward light-cone instead, since the equations of motion can in general equally well be solved backwards and forwards in time.

Temporal causality is very closely related to the question of Time's Arrow. In fact, assuming temporal causality is more or less equivalent to taking a preferred direction of time for granted, which will obviously not do for an explanation of Time's Arrow. There are several ways to try to get around this problem. First, we can of course deny the time-symmetry of the laws of physics. Or we can assume that the laws are symmetric, but that it is somehow a question about the boundary conditions of the universe. Both ways have been attempted by many authors, but so far without any obvious success (see e.g. [7] [8]). In this paper, we instead make the assumption that both laws and boundary conditions are time-symmetric, but it will be argued that this can still lead to what we perceive as a directed Arrow of Time.

We will also assume that our universe is just one of many, and that this is the most natural interpretation of quantum mechanics. This is a question where the scientific community is clearly divided. But in most cases, our descriptions and predictions will be equivalent. However, one can note that in this context, accepting or not accepting the multiverse really will imply a difference: the explanation for Time's Arrow in this paper does make natural sense in the multiverse, but it does not seem to do so if we insist on just a single universe.

As for the third question, the situation is still more complicated. One of the purposes of this paper is to argue that the different answers which we can choose to this question may generate very different pictures of reality in general and of Time's Arrow in particular. And that this aspect is often forgotten in the discussion.

The use of infinities has a long history in science. They can easily be traced thousands of years back to Greek philosophers, like Zeno and Aristotle, and even further. Obviously, no single experiment has ever produced a truly infinite value, so what is it that makes the belief in infinities so strong?

At least part of the answer seems to be that in many cases infinities really do simplify our description of the world. There is an abundance of examples in any course in calculus or classical physics where studying a limit situation simplifies the problem tremendously.

Also, there is a kind of scale invariance argument for accepting infinities: to say that space-time is infinitely large and divisible into infinitely small parts in a sense amounts to saying that what we see on our intermediate level is exactly what we would see on any level of magnification which creates a, possibly illusory, feeling of complete understanding.

During the development of modern science, mathematicians and philosophers have had a lot to say about infinities, however without really having reached a state of consensus. Within physics the situation is somewhat different. Here one tends to accept infinities as long as we can make useful and consistent computations with them. Often, they are even regarded as powerful instruments in the toolbox, rather than as problems.

A rather trivial situation where assuming infiniteness tends to hide the real problem is the simple Martingale betting system and the so called St Petersburg paradox (see [9]): suppose we play heads or tails and we win the stake if the coin comes up heads and lose it if the coin comes up tails. If we double our bet after every loss, the first win will recover all previous losses plus win an amount equal to the original stake. Thus, if we have access to an infinite amount of money, eventually heads will sooner or later come up and the Martingale betting strategy will always generate a gain in the long run if we go on in this way. But, of course, if the assumption about infinite wealth is removed then we will instead have to face the opposite fate of a certain ruin within a finite time. Clearly, assumptions about infiniteness in physics are usually much more subtle than in this example. Nevertheless, it can be that we are making similar mistakes in cosmology.

It may also be said that from a classical point of view, it was very natural to suppose the world to be infinite, simply because the perception of space-time was essentially based on the idea of real numbers. And once the infiniteness at one end of the scale is accepted, it is much easier to accept it at the other end too. With a quantum mechanical perspective, this is no longer true: within a finite volume, the quantum states will be discrete. Hence, the choice between finiteness and infiniteness essentially becomes the question whether the universe is closed or open.

3. Are the Underlying Assumptions in This Paper Compatible with Our Present Knowledge?

Concerning temporal causality, the present situation in science is that it is so deeply built into our way of thinking that it is very difficult to work without it. But on the fundamental level, temporal causality has been questioned by many. Some people even go as far as questioning the flow of time itself. Often, the starting point is the Wheeler-deWitt equation (see e.g. [10]-[12]). However, the approach in this paper is in fact rather independent of all such attempts.

When it comes to the multiverse interpretation, originating from the work of Everett [13], some people would say that simple physical experiments, where particles seem to interfere with themselves, should be considered as genuine proofs of the existence of parallel worlds. But of course, other interpretations are possible. Let us not enter this discussion here. Critics of the multiverse interpretation often focus on the tendency in recent times to expand this idea too far: there is an obvious risk that we will end up finding ourselves in a wonderland where everything is possible and nothing is certain. The point of view in this paper however, can rather be said to be the opposite one: an extremely simplified model where it should be possible to do computations and in the end prove rigorous results.

As for the third question about infinities, a possible interpretation of the situation is as follows: since the birth of modern cosmology about a hundred years ago, a number of different cosmological models have been proposed. For a long time, the closed Friedmann model (see [14] [15]) was the most wide-spread model, but towards the end of the 20th century, open models became increasingly popular. When it was discovered in 1998 that the expansion of the universe is actually accelerating (see [16] [17]), this was by many people interpreted as a very strong argument in favor of open models.

Although there are still critical voices, it is probably fair to say that the acceleration of the expansion is rather firmly established. But does this imply that the universe is open? This may in fact be another instance where our use of infinities mislead us. The conclusion is reached by extrapolating Einstein's field equations with a cosmological constant. But it can also be said that the very existence of the accelerating expansion itself seems to indicate that there might be something wrong with the field equations on a cosmological scale, so it might be rather dangerous to extrapolate them. In fact, an alternative explanation of the observed phenomena may be that it is the finiteness of the universe itself which causes that we now perceive as an accelerating expansion. For a more thorough discussion of these matters, see [18].

So far the conclusion must be that, for the time being, we must leave all these three questions open. The best thing we can do is to analyze different models, and in the end hope that our understanding and comparison with measurements will single out one preferred theory.

4. The Finite Multiverse

One of the main points of this paper is that if we answer no to all the three questions in Section 2, a rather natural perspective on cosmology emerges. On the other hand, this perspective is not so easy to fit together with a yes to any of the questions.

To start with, a finite universe will be bounded, both in space and time. In fact, simplifying somewhat, we may assume that there are only a finite number of moments of time between the Big Bang and the (inevitable) Big Crunch. To make things as simple as possible, let us choose a time-axis between $-T_0$ (Big Bang) and T_0 (Big Crunch), such that these discrete moments of time correspond to the integers. And for each such moment of time, there will be only a finite number of possible states of the multiverse.

What these states are like and in what sense they are related to quantum mechanical states is a non-trivial question. The problem is that different states of the multiverse cannot be treated as quantum mechanical states in the usual sense for the following simple reason: time-development is unitary and therefore maps quantum mechanical states at one moment of time onto quantum mechanical states at another moment of time in an isomorphic way. Thus, a state now will always correspond to a unique state in the future. But the most characteristic property of the multiverse is that any state now can give rise to many different states in the future (in the following, this will be referred to this as the branching property of the multiverse). In the model of this paper, this property is lost if we consider ordinary quantum states.

It is perfectly possible to interpret states in a classical way, e.g. by splitting phase space into lots of small cells and consider a state to be determined by specifying which particles belong to which cells. Nevertheless, the world is after all quantum mechanical, so something is obviously lost here too. But if the states of the multiverse are neither classical nor quantum states, what are they then? In fact, it may very well be that there is no reasonable such concept. Perhaps the best compromise for the time being is to think of them as distinguishable configurations. Still, it may be illuminating to study a semi-classical approximation of the multiverse where states do make sense. This is very much in the tradition of physics where discrete approximation to continuous systems or continuous approximations to discrete systems are frequently used. Certainly, it is by no means evident that this extremely simplified way of thinking is good enough to reflect the true character of the world. Nevertheless, in this paper it will be consider as an adequate model for discussing Time's Arrow.

As we pass from one moment of time to the next (or previous), the laws of physics will give different probabilities for a transition from one state to another. In general this is, on the scale of the whole multiverse, an enormously complicated process. But if we, following the philosophy in Section 1, simplify as much as we can, then we may just classify each transition as possible or impossible.

In this way, we obtain something which is actually a enormous but finite graph: a node is simply a possible state at some moment of time and an edge connects states at adjacent moments of time such that the transition between them is possible. Note that we will always suppose that if a transition is possible in one direction of time, then it is also possible in the other, which corresponds to assuming that the dynamical laws are time-symmetric.

Definition 1 *A universe U is a path, or a chain of states (one state U_t for each moment of time t), with the property that the transition between adjacent states is always possible.*

Definition 2 *The multiverse M is the set of all possible universes U in the sense of Definition 1 (see* **Figure 1**).

5. Boltzmann in the Multiverse

To be able to say something about Time's Arrow in the multiverse of the previous section, we also need to discuss entropy. My position in this paper is to try not to take more than necessary about entropy for granted. As a starting point, let us return to two well-known and fundamental ideas, both due to Boltzmann.

The first idea, as well as the foundation of the modern theory of entropy of a macro-state Ξ, is contained in Boltzmann's famous formula [1]:

$$S = k_B \log \Omega, \tag{5.1}$$

or alternatively

$$\Omega = W^S, \quad \text{where} \quad W = e^{1/k_B}. \tag{5.2}$$

Here, Ω denotes the number of micro-states compatible with Ξ, and k_B is Boltzmann's constant.

Although this formula is usually derived under circumstances which are very far from what we meet in cosmology, it still represents a very fundamental truth. In the following it will always be assumed that in whatever way we measure the entropy S, the number of states corresponding to that entropy is an exponentially growing function of S as in (5.2).

The other fundamental idea of Boltzmann is that the second law of thermodynamics is a manifestation of the fact that the universe, as time develops, passes from less probable states to more probable ones. This idea is by no means less important than the first one. But as it stands, it is unfortunately rather useless for explaining Time's Arrow, since it already has supposed direction of time built into it. Therefore, we need to reformulate Boltzmann's second idea in a time symmetric form. To do this, the following points must be taken into account:

- The dynamics must be time-symmetric.
- The branching property of the multiverse must be included.
- The relation to the entropy must be clear.

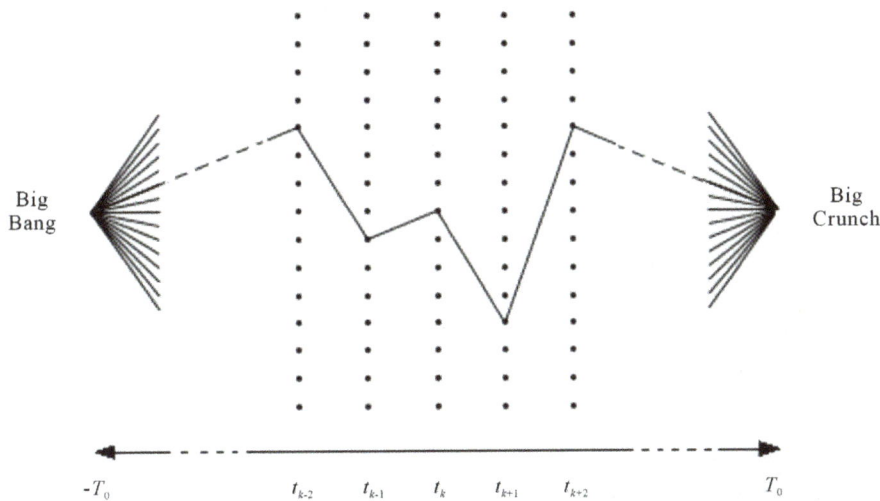

Figure 1. A schematic picture of a universe as one particular path from the Big Bang to the Big Crunch in the huge graph of all possible states.

The following formulation seems to include the essence of Boltzmann's idea in a time-symmetric way:

Statistical Assumption *Given any state of a universe, at any moment of time between the Big Bang and the Big Crunch, which is not very close to the end points and where the entropy is not close to being maximal, the number K of accessible states with higher entropy at the next/previous moment of time is very large, and the number of accessible states with lower entropy at the next/previous moment of time is very small (or rather, the probability p for finding any such accessible state is very small). In the following, K will be considered to be essentially independent of time and of the particular state.*

Also it will always be assumed in the following that $1 \ll K \ll W$.

Near the end points, the dynamics may be quite different from the rest of the time (see the next section). At the end points $-T_0$ and T_0, it will be assumed that there is only one state with zero volume and entropy.

If the entropy is maximal or close to, among all states at some moment of time t, this assumption obviously has to be modified somewhat. But let us explicitly agree to not consider the life-span of the multiverse to be long enough for the entropy to become anything near to maximal, except possibly close to the end-points. In fact, according to [19] [20], the entropy of our present universe is very far from maximal and may continue to be so for a very, very long time still.

Remark 1 *Once W and K are determined, the p in the Statistical Assumption above is also determined. To see this, consider for a given moment of time $t+1$ (or $t-1$), all states with entropy $S-1$. According to (5.2), there are W^{S-1} such states. For each of these states there are K accessible states with entropy S at time t. Assuming statistical independence, we see that there are $\sim W^{S-1} \cdot K$ accessible state with entropy S at time t. Since there are in total W^S such states, we conclude that only the ratio*

$$p \sim \frac{W^{S-1}K}{W^S} = \frac{K}{W} \ll 1 \tag{5.3}$$

of these states can be accessible from states with lower entropy.

6. The Multiverse as a Probability Space

We can now make the multiverse into a finite probability space, using the Statistical Assumption of the previous section.

Thus, let $[-T_0, T_0]$ be the time-interval from the Big Bang to the Big Crunch. It will be convenient in the following to split the life-span of the multiverse into three different phases: we choose symmetric moments of time $-T_1$ and T_1 close to the end-points of the interval $[-T_0, T_0]$ so that we can write

$$[-T_0, T_0] = [-T_0, -T_1] \cup [-T_1, T_1] \cup [T_1, T_0]. \tag{6.1}$$

Coarsely speaking, this may be thought of as a kind of idealized division into the extreme initial phase, the normal phase and the extreme final phase of the multiverse. It is difficult to have a definite opinion about how long the extreme phases should be, but it sounds reasonable to assume that the passage into the normal phase would occur somewhere close to the inflation.

We can now to each universe assign an (un-normalized) probability weight as follows:

$$\omega = \omega_{-1}\omega_0\omega_1. \tag{6.2}$$

Here ω_{-1} and ω_1 represent the weights for the development from $-T_0$ to $-T_1$ and from T_1 to T_0 respectively, whereas ω_0 refers to the normal phase in between.

During the normal phase, according to the idea in Section 4, each transition is just classified as either possible or impossible, *i.e.* has weight 1 or 0. It follows that ω_0 will be 1 for all universes. Let us also assume that we measure the entropy on an appropriate scale where it is reasonable, again as an extreme simplification, to assume that it can (during this phase) only change by ± 1 per each unit interval of time.

During the extreme phases, the situation is very different. The volume will be very small and quantum effects dominate. Here, in a sense, everything can happen, although of course not with the same probability. On the other hand, the duration of the extreme phases is very short, so we will assume that the most probable scenario is that very little happens at all. To make this somewhat more formal, let us assume that a universe, starting from the unique state at time $-T_0$ with zero entropy, can develop into any state at time $-T_1$. But the probability for such a development will be given by an exponential distribution

$$\sim E^{-S}, \tag{6.3}$$

where E is a large constant. In other words, the probability for the universe to enter the normal phase in a disordered state decreases very rapidly with the entropy. The symmetric situation applies to the interval $[T_1, T_0]$. We now have made the multiverse into a probability space. Note however, that there are still several unspecified parameters, in particular K, W and E.

7. Monotonic Universes Versus Universes with Low Entropy at the Ends

In this section, we finally turn to the question why the probability space in Section 6 should give rise to time asymmetry: the idea is that although the model itself is time symmetric, individual universes with a directed time could simply be much more probable than other relevant types. In other words, there could be an enormous number of different universes which share the same Big Bang and Big Crunch, and it could also be that in almost 50% of them the Arrow of Time points in the same direction as in our universe, and in almost 50% of them it points in the other direction. And the fact that we perceive an asymmetry of time could just reflect the fact that we can only observe the very small part of the multiverse to which we are confined.

To keep the discussion in this section as simple and non-technical as possible, let us just concentrate on comparing the total probability weight for universes with a monotonic behavior of the entropy (starting from approximately zero entropy at one end of the normal phase), with universes with low entropy at both ends of the normal phase. We let P_m be the total probability weight for a monotonic behavior and P_s be the corresponding probability weight for the-low-entropy-at-the-ends type of behavior. The task is to compute these two probability weights and then compare their sizes.

To start with P_m, the argument goes as follows: according to the statistical way of counting in Section 5 and the assumption that the entropy can only change by ± 1, a state with entropy 0 at time $-T_1$ can lead to K different states with entropy 1 one unit of time later. Each such state can again lead to K states with entropy 2 after two units of time, which gives a total of K^2 developments so far. Continuing in this way, we see that there are K^{2T_1} developments with monotonically increasing entropy leading from the state with zero entropy at $-T_1$ to states with entropy $S = 2T_1$ at time T_1. Each of these will according to (6.3) have a very small probability weight E^{-2T_1} during the last extreme phase. It can be noted that this method of counting gives the same weight if we start from the W^{2T_1} states with entropy $2T_1$ at time T_1 and use the p of Remark 1 to count backwards instead. Clearly, there are also equally many developments with decreasing entropy during the normal phase, so we arrive at

$$P_m \approx 2\left(\frac{K}{E}\right)^{2T_1} \tag{7.1}$$

Next we estimate P_s. According to our model, each universe with low entropy at both ends will be characterized by transitions which increases the entropy by 1 at T_1 moments of time and decreases it by -1 at equally many moments of time. As above, each transition of the first type will contribute to the number of universes by a factor K, and each transition of the second type will contribute (according to Remark 1) with a factor $p \approx K/W$. Hence we get the number

$$K^{T_1} \cdot \left(\frac{K}{W}\right)^{T_1} = \left(\frac{K^2}{W}\right)^{T_1} \tag{7.2}$$

of developments (in our case, this will give a very small number, thus the chance for finding any such universe is very small). The number of different choices of intervals where the entropy increases and decreases is obviously less than

$$\binom{2T_1}{T_1} < 2^{2T_1}, \tag{7.3}$$

which together with (7.2) gives

$$P_s < \left(\frac{4K^2}{W}\right)^{T_1}. \tag{7.4}$$

Again we note that we get the same result if we count in both directions of time. Hence,

$$\frac{P_m}{P_s} > 2\left(\frac{W}{4E^2}\right)^{T_1}. \tag{7.5}$$

Summing up the discussion, we arrive at

Claim 1 *If* $1 \ll K \ll W$ *and* $1 \ll E^2 \ll W$, *then* $P_m/P_s \gg 1$.

In other words, the chance for a behavior with low entropy at both ends in our probability space is very small compared to the chance for a monotonic behavior (**Figure 2**).

8. Discussion

The claim in Section 7 is deduced under extremely simplified premises and is obviously only the very beginning of what we should do if we want to pursue this path towards a deeper understanding of the Arrow of Time. In fact, recently there have been some progress in understanding and simulating Time's Arrow in classical N-body systems with N of intermediate size, e.g. ≈ 1000 (see [21]). This kind of system is certainly large enough to investigate time-asymmetry, although perhaps still not quite large enough to analyze entropy.

A natural question in connection with the model in this paper is: how about high entropy at both ends? It can of course be argued that this may not be directly relevant for the question of Time's Arrow. After all, it seems to be a well established observational fact that we live in a universe with low entropy at least at one end, so the question is really about how such universes behave. Universes with high entropy at both ends are probably quite un-inhabitable, so maybe we should not bother about them?

Nevertheless, if we take the multiverse perspective seriously, it should be interesting to understand whether monotonic universes are the most common among all possible universes or not. To construct models where this is indeed the case may be more difficult than to deal with the simple-minded approach in this paper, at least if we want to make realistic assumptions. The Statistical Assumption in Section 5 must perhaps be replaced by something more complicated. In particular, it should be noted that the assumption in this paper endows the multiverse with a kind of Markov property: whatever happens before a certain moment of time t is not important for predicting the entropy at time $t+1$ as long as we know the entropy of the present state. This is clearly

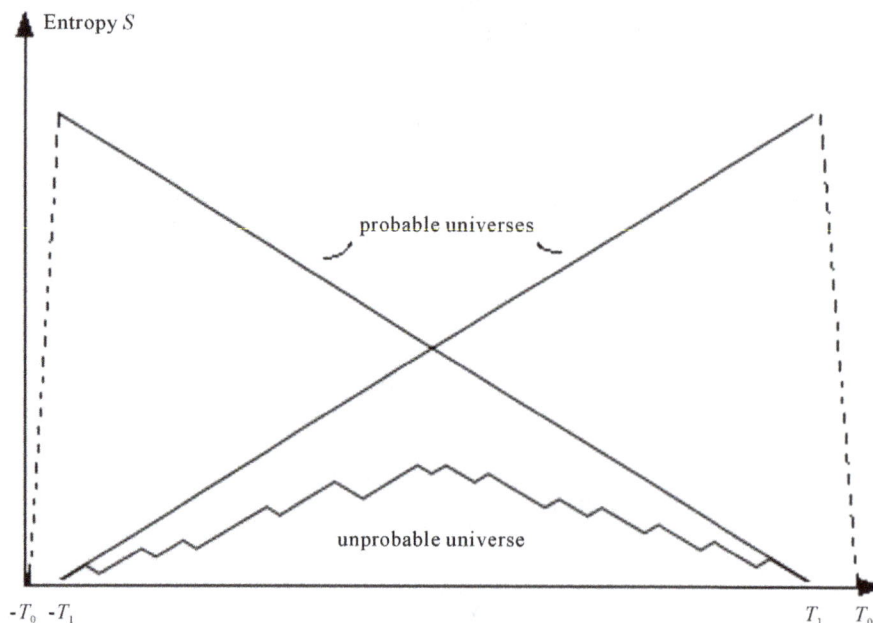

Figure 2. The two plots of entropy for probable universes represent the two kinds of scenarios included in P_m in (7.1). The plot of the unprobable universe represents one (of many) possible scenarios which are included in P_s in (7.4). The computation in (7.5) shows that in spite of the large number of different types of scenarios of type P_s, their total probability is very small under the assumption in Claim 1.

very unrealistic: many processes, e.g. light emanating from a source, can so to speak keep a memory of their origin for billions of years.

In addition to this, things like the inflation in the early universe may play an important role. Even more important is probably the question about what states are and how we count them. This is also closely related to the question whether the expansion of the universe contributes to the growth of entropy or not. On the one hand, it can be said that this is just a question about definitions: entropy is just what we define it to be. But the real question is of course how to define entropy in a way which is compatible with the second law. And the answer to this question may very well depend on which part of the history of our universe we consider: the inflationary phase, the early universe or the full-grown universe where galaxies have essentially already formed.

References

[1]   Boltzmann, L. (1974) Theoretical Physics and Philosophical Problems. Edited by Brian McGuinness. Trans. Paul Foulkes, Reidel Publishing Co., Dordrecht.

[2]   Price, H. (1996) Time's Arrow and Archimedes' Point. Oxford University Press, Oxford.

[3]   Gold, T. (1962) The Arrow of Time. *American Journal of Physics*, **30**, 403. http://dx.doi.org/10.1119/1.1942052

[4]   Hawking, S.W. (1985) Arrow of Time in Cosmology. *Physical Review D*, **32**, 2489. http://dx.doi.org/10.1103/PhysRevD.32.2489

[5]   Page, D. (1985) Will the Entropy Decrease If the Universe Recollapse? *Physical Review D*, **32**, 2496. http://dx.doi.org/10.1103/PhysRevD.32.2496

[6]   Tamm, M. (2013) Time's Arrow from the Multiverse Point of View. *Physics Essays*, **26**, 2.

[7]   Penrose, R. (1979) Singularities and Time-Asymmetry. General Relativity: An Einstein Centenary. Cambridge University Press, Cambridge.

[8]   Sakharov, A.D. (1967) ZhETF Pis'ma, 5, 32 (Sov. Phys. JEPT Lett., 6, 24).

[9]   Martin, R. (2004) The St. Petersburg Paradox. In: Zalta, E.N., Ed., *The Stanford Encyclopedia of Philosophy*, Summer 2014 Edition. http://plato.stanford.edu/archives/sum2014/entries/paradox-stpetersburg/

[10]   DeWitt, B.S. (1967) Quantum Theory of Gravity. I. The Canonical Theory. *Physical Review*, **160**, 1113-1148.

[11]   Zeh, H.D. (2001) The Physical Basis of the Direction of Time. 4th Edition, Springer-Verlag, Berlin.

[12]   Barbour, J. (1999) The End of Time. Oxford University Press, Oxford.

[13]   Everett, H. (1957) "Relative State" Formulation of Quantum Mechanics. *Reviews of Modern Physics*, **29**, 454. http://dx.doi.org/10.1103/RevModPhys.29.454

[14]   Friedman, A. (1922) Über die Krümmung des Raumes. *Zeitschrift für Physik*, **10**, 377-386.

[15]   Misner, C.M., Thorne, K.S. and Wheeler, J.A. (1973) Gravitation. W. H. Freeman and Company, San Francisco.

[16]   Riess, A.G., Filippenko, A.V., Challis, P., Clocchiatti, A., Diercks, A., Garnavich, P.M., *et al.*, Supernova Search Team (1998) Observational Evidence from Supernovae for an Accelerating Universe and a Cosmological Constant. *Astronomical Journal*, **116**, 1009. http://dx.doi.org/10.1086/300499

[17]   Perlmutter, S., Aldering, G., Goldhaber, G., Knop, R.A., Nugent, P., Castro, P.G., *et al.*, the Supernova Cosmology Project (1999) Measurements of Omega and Lambda from 42 High-Redshift Supernovae. *Astrophysical Journal*, **517**, 565. http://dx.doi.org/10.1086/307221

[18]   Tamm, M. (2015) Accelerating Expansion in a Closed Universe. *Journal of Modern Physics*, Special Issue on "Gravitation, Astrophysics and Cosmology", **6**, 239-251.

[19]   Adams, F. and Laughlin, G. (1997) A Dying Universe: The Long-Term Fate and Evolution of Astrophysical Objects. *Reviews of Modern Physics*, **69**, 337. http://dx.doi.org/10.1103/RevModPhys.69.337

[20]   Egan, C. and Lineweaver, C. (2010) A Larger Estimate of the Entropy of the Universe. *The Astrophysical Journal*, **710**, 1825. http://dx.doi.org/10.1088/0004-637X/710/2/1825

[21]   Barbour, J., Koslowski, T. and Mercati, F. (2014) Identification of a Gravitational Arrow of Time. *Physical Review Letters*, **113**, Article ID: 181101.

9

Use of an Energy-Like Integral to Study the Motion of an Axi-Symmetric Satellite under Drag and Radiation Pressure

Ahmed Mostafa

Department of Mathematics, Faculty of Science, Ain Shams University, Cairo, Egypt
Email: ahmed.mahmoud@guc.edu.eg

Abstract

The axi-symmetric satellite problem including radiation pressure and drag is treated. The equations of motion of the satellite are derived. An energy-like is given for a general drag force function of the polar angle θ, and then it is used to find a relation for the orbit equation of the satellite with initial conditions satisfying the vanishing of arbitrarily choosing higher derivatives of the velocity.

Keywords

Artificial Satellite, Drag Effect, Radiation Pressure

1. Introduction

The classical two body problem is one of the most important topics in the field of celestial mechanics, specially the applications of the theory of artificial satellites. Since Brouwer and Hori [1], so many works have been made to study the problem with different factors considered e.g. Mittleman and Jezewski [2] and Jezewski and Mittleman [3], Danby [4], Leach [5], Gorringe and Leach [6], McMahon and Scheeres [7] etc.

Marvaganis [8] studied the motion of an almost constant-speed two body problem under the effect of air resistance. The drag force was taken in the form of Danby's drag, while Marvaganis and Michalakis [9] studied the two body problem in the existence of Danby's drag and where the bigger body was radiating. They used a Laplace-like integral to derive the orbit equation. El-Shaboury and Mostafa [10] studied the problem of an axi-symmetric satellite under drag and radiation pressure by first neglecting the effect of axi-symmetry of the satellite, and then adding it as a perturbation to the problem.

In this work, an attempt is made to get a solution for the problem of an axi-symmetric satellite under drag and radiation pressures, which all the effects are included in the equation of motion from the beginning by using energy like integral. A relation for the orbit equation is derived first for a general air drag function and then for the case of Danby's drag. Finally, the solution of an almost constant speed satellite has been given.

2. The Equation of Motion and the Integral of Angular Momentum

The equation of motion of an axi-symmetric satellite under the gravitational force of a spherical body with an additional force due to the resistance force and radiation pressure can be modeled such as Mavraganis and Michalakis (1994), and El-Shaboury and Mostafa (2014).

$$\ddot{r} + R(r,\theta)\dot{r} + \left[\frac{(1-\beta)\mu}{r^3} + \frac{3\mu(C-A)(1-3\gamma^2)}{2r^5} \right] r = 0, \tag{1}$$

The air resistance is taken as a general function R of the polar coordinates r, θ, and the definitions of the involved parameters are as following:

μ is the gravitational constant.

β is the radiation constant, where $0 < \beta < 1$.

C, A are the principal moments of inertia of the satellite (C about the symmetric axis), and γ is the direction cosine of the radius vector with respect to the axis of the satellite. For simplification, we will rename

$$k = \frac{3\mu(C-A)(1-3\gamma^2)}{2}$$

Now, vector product Equation (1) with $\mathbf{r}$, and remembering that $\mathbf{H} = \mathbf{r} \times \dot{\mathbf{r}}$, we get,

$$\dot{\mathbf{H}} + R(r,\theta)\mathbf{H} = 0 \tag{2}$$

which gives immediately,

$$\mathbf{H} \times \dot{\mathbf{H}} = 0.$$

This expression admits a first vector which is the constant direction $\mathbf{e}_H = \mathbf{H}/H$ of the angular momentum $\mathbf{H}$. Therefore the motion is planar. This enables us to simplify Equation (2) by writing

$$\dot{H} + R(r,\theta)H = 0. \tag{3a}$$

Let the resistance be a general function of the polar angel $f(\theta)$ divided by the square of the radial distance r, i.e.

$$R = \frac{f(\theta)}{r^2} \tag{3b}$$

We get,

$$H = r^2\dot{\theta} = h - F(\theta). \tag{4}$$

where, h is the constant angular momentum in the absence of the drag force, and $F(\theta) = \int_{\theta_0}^{\theta} f(\theta)d\theta$.

3. The Energy-Like Invariant of Motion

Now, let ϕ be the angle between the radial distance r and the distance p of the origin O from the tangent (**Figure 1**).

The vector equation of motion is thus resolved to

$$\frac{dV}{dt} + RV - \left[\frac{(1-\beta)\mu}{r^2} + \frac{k}{r^4} \right] \sin\phi = 0, \tag{5}$$

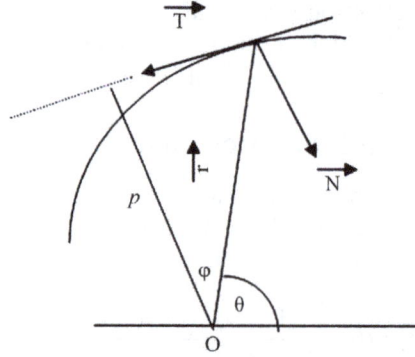

Figure 1. Geometrical meaning of the angle φ.

$$\frac{V^2}{\rho} - \left[\frac{(1-\beta)\mu}{r^2} + \frac{k}{r^4}\right]\cos\phi = 0,\tag{6}$$

where ρ is the radius of curvature. From the definition of the angle ϕ, we have

$$\sin\phi = -\frac{\mathrm{d}r}{\mathrm{d}s},\tag{7}$$

and using the substitution

$$\frac{\mathrm{d}}{\mathrm{d}s} = \frac{\mathrm{d}}{\mathrm{d}\theta}\frac{\dot\theta}{V},\tag{8}$$

Equation (5) becomes,

$$\frac{\mathrm{d}V}{\mathrm{d}\theta}\dot\theta + RV + \left[\frac{(1-\beta)\mu}{r^2} + \frac{k}{r^4}\right]\frac{\mathrm{d}r}{\mathrm{d}\theta}\frac{\dot\theta}{V} = 0,\tag{9}$$

This gives,

$$\frac{1}{2}\left(V^2 - V_0^2\right) + \int_{\theta_0}^{\theta}\frac{R}{\dot\theta}V^2\mathrm{d}\theta = \left[\frac{(1-\beta)\mu}{r} + \frac{3k}{r^3}\right] - \left[\frac{(1-\beta)\mu}{r_0} + \frac{3k}{r_0^3}\right]\tag{10}$$

where $r_0 = r(\theta_0)$. It is clear that the above equation gives the energy integral in the absence of resistance, radiation pressure and oblateness. However in the absence of resistance only, we will still have the invariant of motion

$$\frac{1}{2}\left(V^2 - V_0^2\right) = \left[\frac{(1-\beta)\mu}{r} + \frac{3k}{r^3}\right] - \left[\frac{(1-\beta)\mu}{r_0} + \frac{3k}{r_0^3}\right]\tag{10a}$$

4. A Relation for the Orbit Equation

Taking R in the form $R = \dfrac{f(\theta)}{r^2}$, where $f(\theta)$ is an arbitrary function of the angle θ. Equation (10) gives,

$$\frac{(1-\beta)\mu}{r} + \frac{3k}{r^3} = \frac{(1-\beta)\mu}{r_0} + \frac{3k}{r_0^3} + \frac{1}{2}\left(V^2 - V_0^2\right) + \int_{\theta_0}^{\theta}\frac{f(\theta)}{r^2\dot\theta}V^2\mathrm{d}\theta,\tag{11}$$

which implies by using Equation (4)

$$\frac{(1-\beta)\mu}{r} + \frac{3k}{r^3} = \frac{(1-\beta)\mu}{r_0} + \frac{3k}{r_0^3} + \frac{1}{2}\left(V^2 - V_0^2\right) + \int_{\theta_0}^{\theta}\frac{f(\theta)}{h-F(\theta)}V^2\mathrm{d}\theta\tag{12}$$

In order to integrate the required integration, we expand V in Taylor series of the polar angle θ,

$$V = V_0 + \sum_{n \geq 1} a_n (\theta - \theta_0)^n, \tag{13a}$$

$$a_n = \frac{1}{n!} \frac{d^n V}{d\theta^n}\bigg|_{\theta_0} \tag{13b}$$

where θ_0 is the initial value of θ and $V_0 = V(\theta_0)$. Then we have

$$V^2 = V_0^2 + 2V_0 \sum_{n \geq 1} a_n (\theta - \theta_0)^n + \left[\sum_{n \geq 1} a_n (\theta - \theta_0)^n \right]^2$$

We write,

$$\left[\sum_{n \geq 1} a_n (\theta - \theta_0)^n \right]^2 = \sum_{j \geq 1} a_j (\theta - \theta_0)^j \sum_{k \geq 1} a_k (\theta - \theta_0)^k = \sum_{n \geq 2} b_n (\theta - \theta_0)^n$$

where,

$$b_n = \sum_{k \geq 1} a_k a_{n-k}$$

Thus,

$$V^2 = V_0^2 + 2V_0 a_1 (\theta - \theta_0) + \sum_{n \geq 2} c_n (\theta - \theta_0)^n, \tag{14a}$$

where,

$$c_n = a_n + \sum_{k \geq 1} a_k a_{n-k} \tag{14b}$$

Therefore, we have the integration,

$$\int_{\theta_0}^{\theta} \frac{V^2 f(\theta)}{h - F(\theta)} d\theta = V_0^2 \int_{\theta_0}^{\theta} \frac{f(\theta)}{h - F(\theta)} d\theta + 2a_1 V_0 \int_{\theta_0}^{\theta} \frac{(\theta - \theta_0) f(\theta)}{h - F(\theta)} d\theta + \sum_{n \geq 2} c_n \int_{\theta_0}^{\theta} \frac{(\theta - \theta_0)^n f(\theta)}{h - F(\theta)} d\theta$$

$$= -V_0^2 \ln \frac{h - F(\theta)}{h - F(\theta_0)} + 2V_0 a_1 \left[-(\theta - \theta_0) \ln(h - F(\theta)) + \int_{\theta_0}^{\theta} \ln(h - F(\theta)) d\theta \right] \tag{15}$$

$$+ \sum_{n \geq 0} c_n \int_{\theta_0}^{\theta} \frac{(\theta - \theta_0)^n f(\theta)}{h - F(\theta)} d\theta$$

Thus, using Equations (12), (14a), and (15) we get a relation for the orbit equation in the form,

$$\left[\frac{(1 - \beta)\mu}{r} + \frac{3k}{r^3} \right] = \left[\frac{(1 - \beta)\mu}{r_0} + \frac{3k}{r_0^3} \right] + V_0 a_1 (\theta - \theta_0) + \frac{1}{2} \sum_{n \geq 2} c_n (\theta - \theta_0)^n - V_0^2 \ln \frac{h - F(\theta)}{h - F(\theta_0)}$$

$$+ 2V_0 a_1 \left[-(\theta - \theta_0) \ln(h - F(\theta)) + \int_{\theta_0}^{\theta} \ln(h - F(\theta)) d\theta \right] + \sum_{n \geq 2} c_n \int_{\theta_0}^{\theta} \frac{(\theta - \theta_0)^n f(\theta)}{h - F(\theta)} d\theta \tag{16}$$

where c_n is given by Equation (14b), a_n is given by Equation (13b) and $F(\theta) = \int_{\theta_0}^{\theta} f(\theta) d\theta$

Equation (16) describes a relation for the orbit equation of an axi-symmetric satellite with oblateness coefficient k under radiation pressure of coefficient β and air drag whose function is given by $R = \dfrac{f(\theta)}{r^2}$, where $f(\theta)$ can be chosen arbitrary.

The convergence of the involved series is guaranteed for initial velocity satisfying the vanishing of $\left.\dfrac{d^n V}{d\theta^n}\right|_{\theta_0}$ for all $n > N$, where N can be chosen arbitrary.

4.1. The Case of Danby Drag

In the special case of Danby's drag (Dabny, 1962), $f(\theta) = \alpha$, and $F(\theta) = \alpha\theta$ where α is a constant, the required integrations reduce to:

$$\int_{\theta_0}^{\theta} \ln\big(h - F(\theta)\big)\,d\theta = \int_{\theta_0}^{\theta} \ln\big(h - \alpha\theta\big)\,d\theta = -(\theta - \theta_0) - \frac{h}{\alpha}\ln\frac{h - \alpha\theta}{h - \alpha\theta_0} + \theta\ln\big(h - \alpha\theta\big) - \theta_0\ln\big(h - \alpha\theta_0\big) \tag{17}$$

$$\int_{\theta_0}^{\theta} \frac{(\theta - \theta_0)^n f(\theta)}{h - F(\theta)}\,d\theta = \alpha\int_{\theta_0}^{\theta} \frac{(\theta - \theta_0)^n}{h - \alpha\theta}\,d\theta = -\int_{\theta_0}^{\theta} \frac{(\theta - \theta_0)^n}{\theta - \dfrac{h}{\alpha}}\,d\theta = -\int_{\theta_0}^{\theta} \frac{(\theta - \theta_0)^n}{(\theta - \theta_0) + \left(\theta_0 - \dfrac{h}{\alpha}\right)}\,d\theta \tag{18}$$

Substituting,

$$a = \theta_0 - \frac{h}{\alpha}, \quad y = \theta - \theta_0, \tag{19}$$

we get the integration (18) in the form $-\displaystyle\int \frac{y^n}{y + a}\,dy$. To evaluate this integral, we distinguish between two cases for n when it is even or odd,

When n is even, we write

$$\frac{y^n}{y + a} = \frac{y^n - a^n}{y + a} + \frac{a^n}{y + a}$$

and then we use the expansion

$$y^n - a^n = (y + a)\big(y^{n-1} - ay^{n-2} + a^2 y^{n-3} - \cdots + a^{n-2} y - a^{n-1}\big)$$

$$= (y + a)\sum_{i=0}^{n-1} (-1)^{i+1} y^i a^{n-1-i}$$

thus we get

$$\int \frac{y^n}{y + a}\,dy = a^n \ln(y + a) + \sum_{i=0}^{n-1} \frac{(-1)^{i+1}}{i+1} y^{i+1} a^{n-1-i}$$

And when n is odd, we write

$$\frac{y^n}{y + a} = \frac{y^n + a^n}{y + a} - \frac{a^n}{y + a}$$

then we use the expansion

$$y^n + a^n = (y + a)\big(y^{n-1} - ay^{n-2} + \cdots - a^{n-2} y + a^{n-1}\big)$$

$$= (y + a)\sum_{i=0}^{n-1} (-1)^{i} y^i a^{n-1-i}$$

thus we get

$$\int \frac{y^n}{y+a} dy = -a^n \ln(y+a) + \sum_{i=0}^{n-1} \frac{(-1)^i}{i+1} y^{i+1} a^{n-1-i}$$

We can collect the two cases together in one case to get,

$$\int \frac{y^n}{y+a} dy = (-1)^n \left[a^n \ln(y+a) + \sum_{i=0}^{n-1} \frac{(-1)^{i+1}}{i+1} y^{i+1} a^{n-1-i} \right] \tag{20}$$

Substituting from Equations (17 - 20) into Equation (16), we get after simplification

$$\left[\frac{(1-\beta)\mu}{r} + \frac{3k}{r^3} \right] = \left[\frac{(1-\beta)\mu}{r_0} + \frac{3k}{r_0^3} \right] - V_0 a_1 (\theta - \theta_0) + \left[2a_1 V_0 \left(\theta_0 - \frac{h}{\alpha} \right) - V_0^2 \right] \ln \frac{h - \alpha\theta}{h - \alpha\theta_0}$$

$$+ \sum_{n \geq 2} c_n \left[\frac{1}{2} (\theta - \theta_0)^n - (-1)^n \sum_{i=0}^{n-1} \frac{(-1)^{i+1}}{i+1} \left(\theta_0 - \frac{h}{\alpha} \right)^{n-i-1} (\theta - \theta_0)^i \right] \tag{21}$$

$$- \sum_{n \geq 2} (-1)^n c_n \left[\left(\theta_0 - \frac{h}{\alpha} \right)^n \ln \left(\theta - \frac{h}{\alpha} \right) \right]$$

Equation (21) gives a relation for the orbit equation of an axi-symmetric satellite under the gravitational effect of a radiating body and air resistance described by Danby's drag.

4.2. The Case of an Almost Constant Speed Satellite

If the satellite is of almost constant speed, then we assume that the first derivative is of small value, and all the higher derivatives to be zero (e.g. Mavraganis, 1991), we get the solution

$$\left[\frac{(1-\beta)\mu}{r} + \frac{3k}{r^3} \right] = \left[\frac{(1-\beta)\mu}{r_0} + \frac{3k}{r_0^3} \right] - a_1 V_0 (\theta - \theta_0) + \left[2a_1 V_0 \left(\theta_0 - \frac{h}{\alpha} \right) - V_0^2 \right] \ln \frac{h - \alpha\theta}{h - \alpha\theta_0} \tag{22}$$

Equation (22) is a special case of Equation (21) when the satellite is of almost constant speed.

5. Conclusions

In this paper, the motion of an axi-symmetric satellite under the effect of a radiating body in the presence of air drag is studied. An energy-like integral for the problem has been evaluated using a Taylor expansion for the velocity around the initial value of the polar angel. The convergence of the integral is guaranteed by the assumption that the derivative $\left. \frac{d^n V}{d\theta^n} \right|_{\theta_0} = 0$ for all $n > N$, for an arbitrary N.

The energy-like integral has been used to get a relation for the orbit equation of the satellite. The relation is derived first for a general air drag function and then for the case of Danby's drag. Finally, the solution of an almost constant speed satellite has been given.

References

[1]   Brouwer, D. and Hori, G. (1961) Theoretical Evaluation of Atmospheric Drag Effects in the Motion of an Artificial Satellite. *The Astronomical Journal*, **66**, 193-225. http://dx.doi.org/10.1086/108399

[2]   Mittleman, D. and Jezwski, D. (1982) An Analytic Solution to the Classical Two-Body Problem with Drag. *Celestial Mechanics and Dynamical Astronomy*, **28**, 401-413. http://dx.doi.org/10.1007/BF01372122

[3]   Jezwski, D. and Mittleman, D. (1983) Integrals of Motion for the Classical Two-Body Problem with Drag. *International Journal of Non-Linear Mechanics*, **18**, 119-124. http://dx.doi.org/10.1016/0020-7462(83)90039-2

[4]   Danby, G.M.A. (1962) Fundamentals of Celestial Mechanics. MacMillan, New York.

[5]   Leach, P.G.L. (1987) The First Integrals and Orbit Equation for the Kepler Problem with Drag. *Journal of Physics A*:

Mathematical and General, **20**, 1997-2004. http://dx.doi.org/10.1088/0305-4470/20/8/019

[6]   Gorringe, V.M. and Leach, P.G.L. (1988) Hamiltonlike Vectors for a Class of Kepler Problems with Drag. *Celestial Mechanics and Dynamical Astronomy*, **41**, 125-130. http://dx.doi.org/10.1007/BF01238757

[7]   McMahon, J. and Scheeres, D. (2010) Secular Orbit Variation due to solar Radiation Effects: A Detailed Model for BYORP. *Celestial Mechanics and Dynamical Astronomy*, **106**, 261-300. http://dx.doi.org/10.1007/s10569-009-9247-9

[8]   Mavraganis, A.G. (1991) The Almost Constant-Speed Two-Body Problem with Resistance. *Celestial Mechanics and Dynamical Astronomy*, **51**, 395-405. http://dx.doi.org/10.1007/BF00052930

[9]   Mavraganis, A.G. and Michalakis, D.G. (1994) The Two-Body Problem with Drag and Radiation Pressure. *Celestial Mechanics and Dynamical Astronomy*, **58**, 393-403. http://dx.doi.org/10.1007/BF00692013

[10]   El-Shaboury, S.M. and Mostafa, A. (2014) The Motion of Axisymmetric Satellite with Drag and Radiation Pressure. *Astrophysics and Space Science*, **352**, 515-519. http://dx.doi.org/10.1007/s10509-014-1975-y

10

Flat Space Cosmology as a Mathematical Model of Quantum Gravity or Quantum Cosmology

Eugene Terry Tatum[1], U. V. S. Seshavatharam[2], S. Lakshminarayana[3]

[1]760 Campbell Ln. Ste. 106 #161, Bowling Green, KY, USA
[2]Honorary Faculty, I-SERVE, Hyderabad, India
[3]Department of Nuclear Physics, Andhra University, Visakhapatnam, India
Email: ett@twc.com, seshavatharam.uvs@gmail.com, lnsrirama@gmail.com

Abstract

We review here the recent success in modeling our expanding universe according to the rules of flat space cosmology. Given only a few basic and reasonable assumptions and a single observational input, our model derives a variety of results which correlate with astronomical observations, including best estimates of the size, total mass, temperature, age and expansion rate of our observable universe. Considering the apparent success of our model, we attempt to explain why we think it works so well, including the fact that it incorporates elements of both general relativity and quantum mechanics. We offer this approach as a possible avenue towards understanding cosmology at the quantum level ("quantum gravity").

Keywords

Flat space Cosmology, Hubble Parameter, Hubble Radius, CMBR Redshift, Schwarzschild Formula, Hawking Black Hole Temperature Formula, Quantum Gravity

1. Introduction

Flat space cosmology is our newly introduced heuristic model of cosmology [1] [2]. Relying on only a few basic and reasonable assumptions, it has allowed us to derive a variety of numbers in close agreement with astronomical observations, including 2015 Planck survey results [3]. In our introductory paper [1], we derived Hubble parameter H_0, Hubble radius R_0, Hubble time H_T and total mass M_0 values for our current observable universe relying only on our basic assumptions and current CMB radiation temperature $T_0 = 2.725$ K. In the follow-up paper [2], we derived current Hubble radius R_0, Hubble time H_T (universal age), total mass M_0, cosmic temperature T_0 and CMBR redshift, knowing only the current Hubble parameter H_0 value of 68.6 km/sec/Mpc (2015 Planck survey upper limit). In this paper, we summarize the key mathematical relationships and attempt to explain why such a simple model works so well.

2. Our Key Model Assumptions

Our key model assumptions can be expressed as follows, for any scale from the Planck scale to the scale of our observable universe:

1) Cosmic radius R and total mass M_R follow the Schwarzschild formula $R \cong \dfrac{2GM_R}{c^2}$ at all times.

2) The cosmic event horizon always translates at speed of light c with respect to its geometric center. Thus, in our model, the Hubble parameter H_R can be expressed as c/R. And considering assumptions 1 and 2 together, the cosmic Hubble radius can be expressed as $R \cong \dfrac{2GM_R}{c^2} \cong \dfrac{c}{H_R}$ and Hubble time (universal age) can be expressed as $R/c \cong 1/H_R$ for any stage of cosmic expansion.

3) The possible range of cosmic linear velocity of rotation at the Planck scale can be assumed to be anywhere from zero up to the special relativity limit of c. If we start by assuming the maximum possible value (c), then $\omega_{pl} \cong c/R_{pl} \cong c^3/2GM_{pl} \cong H_{pl}$, where ω_{pl} is the Planck scale angular velocity, M_{pl} is the Planck mass, H_{pl} is the Planck scale Hubble parameter and R_{pl}, according to assumption 1, is the Planck radius and equals twice the Planck length. As such, $M_{pl} \cong \sqrt{\hbar c/G} \cong 2.176507949 \times 10^{-8}\,\text{kg}$, and $R_{pl} \cong 2G\sqrt{\hbar c/G}/c^2 \cong 3.23240045 \times 10^{-35}\,\text{m}$, and the Planck scale Hubble parameter $H_{pl} \cong \left(c/R_{pl}\right) \cong \left(c^3/2G\sqrt{\hbar c/G}\right) \cong 9.274607607 \times 10^{42}\,\text{sec}^{-1} \cong \omega_{pl}$ in rad·sec^{-1}.

4) Following thermodynamics of Hawking's black hole temperature formula [4], at any radius R cosmic temperature T is inversely proportional to the geometric mean of cosmic total mass M_R and the Planck mass M_{pl}.

$$k_B T_R \cong \frac{\hbar c^3}{8\pi G \sqrt{M_R M_{pl}}} \tag{1}$$

3. Our Model Formulae Based on These Assumptions

3.1. Relations between Cosmic Radius, Total Mass and Hubble Parameter

$$M_R \cong \frac{Rc^2}{2G} \cong \frac{c^2}{2G}\left(\frac{c}{H_R}\right) \cong \frac{c^3}{2GH_R} \tag{2}$$

$$\left(M_R\Big/\frac{4\pi}{3}R^3\right) \cong \frac{3c^2}{8\pi GR^2} \cong \frac{3H_R^2}{8\pi G} \tag{3}$$

where R, M_R and H_R represent cosmic radius, total mass and Hubble parameter, respectively. Average mass density equaling Friedmann's critical density is seen in the second line. *Hence, our cosmic model is constantly "flat" by the Friedmann formula.*

3.2. Relations between Temperature, Mass, Radius and Hubble Parameter (Thermodynamics)

$$k_B T_R \cong \frac{\hbar c^3}{8\pi G \sqrt{M_R M_{pl}}} \cong \frac{\hbar \sqrt{H_R H_{pl}}}{4\pi} \cong \frac{\hbar c}{4\pi \sqrt{R R_{pl}}} \tag{4}$$

$$R T_R^2 \cong \frac{1}{R_{pl}}\left(\frac{\hbar c}{4\pi k_B}\right)^2 \cong 1.0272646 \times 10^{27}\,\text{m·K}^2 \tag{5}$$

$$\frac{T_R^2}{H_R} \cong \frac{c}{R_{pl}}\left(\frac{\hbar}{4\pi k_B}\right)^2 \cong \omega_{pl}\left(\frac{\hbar}{4\pi k_B}\right)^2 \tag{6}$$

$$\frac{H_R}{T_R^2} \cong \frac{1}{H_{pl}}\left(\frac{4\pi k_B}{\hbar}\right)^2 \cong 2.918356766 \times 10^{-19}\,\mathrm{K}^{-2}\cdot\sec^{-1} \tag{7}$$

where T_R represents the cosmic temperature for a given cosmic radius R.

4. Our Derivations of Current Cosmological Parameters

Using only our basic assumptions and the equations they generate above, derivations of current values for our observable universe are as follows:

Relations between universal current radius R_0, current temperature T_0, current Hubble parameter H_0, current total mass M_0, and current average mass density:

$$R_0 \cong \frac{1}{R_{pl}}\left(\frac{\hbar c}{4\pi k_B}\right)^2\left(\frac{1}{T_0}\right)^2 \cong \frac{1}{R_{pl}}\left(\frac{\hbar c}{4\pi k_B}\right)^2\left(\frac{1}{2.72548}\right)^2 \tag{8}$$

$$\cong 1.3829177 \times 10^{26}\,\mathrm{m}$$

$$H_0 \cong \frac{1}{H_{pl}}\left(\frac{4\pi k_B}{\hbar}\right)^2 T_0^2 \cong \frac{1}{H_{pl}}\left(\frac{4\pi k_B}{\hbar}\right)^2 (2.72548)^2 \tag{9}$$

$$\cong 2.167826 \times 10^{-18}\,\sec^{-1} \cong 66.89\,\mathrm{km/sec/Mpc}$$

$$M_0 \cong \frac{R_0 c^2}{2G} \cong \frac{c^3}{2GH_0} \cong 9.311752 \times 10^{52}\,\mathrm{kg}. \tag{10}$$

$$\left(\rho_0\right)_{average} \cong \left(\rho_0\right)_{critical} \cong \frac{3c^2}{8\pi GR_0^2} \cong \frac{3H_0^2}{8\pi G} \cong 8.4053137 \times 10^{-27}\,\mathrm{kg}\cdot\mathrm{m}^{-3} \tag{11}$$

where $\left(\rho_0\right)_{average}$ is the average mass density and $\left(\rho_0\right)_{critical}$ is Friedmann's critical average mass density.

The above-derived radius and mass values simulate a current observable universe with a radius of 14.6 billion light-years and roughly 2×10^{22} visible stars plus $5 \times$ dark matter, roughly 10^{53} kg.

Derived current cosmological values are consistent with 2015 Planck survey data. As per the 2015 Planck survey data, the current value of the Hubble parameter H_0 is reported to be:

$$\left.\begin{array}{l}\mathrm{Planck}\ TT + \mathrm{low}\ P : (67.31 \pm 0.96)\,\mathrm{km/sec/Mpc} \\[4pt] \mathrm{Planck}\ TE + \mathrm{low}\ P : (67.73 \pm 0.92)\,\mathrm{km/sec/Mpc} \\[4pt] \mathrm{Planck}\ TT, TE, EE + \mathrm{low}\ P : (67.7 \pm 0.66)\,\mathrm{km/sec/Mpc}\end{array}\right\}$$

As per the 2015 Planck data, the current value of CMBR temperature T_0 is reported to be:

$$\left.\begin{array}{l}\mathrm{Planck}\ TT + \mathrm{low}P + \mathrm{BAO} : (2.722 \pm 0.027)\,\mathrm{K} \\[4pt] \mathrm{Planck}\ TT; TE; EE + \mathrm{low}\ P + \mathrm{BAO} : (2.718 \pm 0.021)\,\mathrm{K}\end{array}\right\}$$

As per COBE/FIRAS, CMBR temperature T_0 is reported to be: $(2.7255 \pm 0.0006)\,\mathrm{K}$

From our recent analysis of conservation of angular momentum in flat space cosmology (pending publication), the maximum possible Planck scale angular momentum is:

$$\left.\begin{array}{l}A_{pl} \cong \left(M_{pl}R_{pl}^2\right)\omega_{pl} \cong \left(M_{pl}R_{pl}^2\right)\left(c/R_{pl}\right) \cong 2GM_{pl}^2/c \\[6pt] \cong 2\hbar \cong 2.10912261 \times 10^{-34}\,\mathrm{kg}\cdot\mathrm{m}^2\cdot\sec^{-1}\end{array}\right\} \tag{12}$$

where $\begin{cases} M_{pl} \cong 2.176507949 \times 10^{-8} \text{ kg}, R_{pl} \cong 3.23240045 \times 10^{-35} \text{ m}, \\ \text{and } c \cong 2.99792458 \times 10^8 \text{ m} \cdot \text{sec}^{-1} \end{cases}$

Assuming angular momentum conservation, maximum possible current angular momentum is:

$$A_0 \cong \left(M_0 R_0^2 \right) \omega_0 \leq 2\hbar \tag{13}$$

Hence, the maximum possible current angular velocity is:

$$\omega_0 \leq \frac{A_0}{M_0 R_0^2} \leq \frac{2\hbar}{M_0 R_0^2} \leq 1.2 \times 10^{-139} \text{ rad} \cdot \text{sec}^{-1} \tag{14}$$

Based upon this extremely small derived maximum theoretical value, our pending angular momentum paper concludes "this number is well beyond our ability to observe cosmic rotation effects in the present" or, presumably, even at the universal age at recombination.

Thus, our model predicts that rotational effects are unlikely to be detectable in higher resolutions of the cosmic microwave background radiation (CMBR).

As our cosmic model is always assumed to be expanding at light speed, from the beginning of the Planck scale, cosmic age can be expressed as follows:

$$t \cong \frac{\left(R - R_{pl} \right)}{c} \tag{15}$$

For the current case, since $\left(R_{pl} \right)$ is very small and $\left(R_0 - R_{pl} \right) \cong R_0$

$$t_0 \cong \frac{R_0}{c} \cong \frac{1}{H_0} \tag{16}$$

5. Our Model Correlations with Observed Cosmic Redshifts

We have proposed simple model equations (relations 17 - 23) for observed and predicted cosmic redshifts [2], including the CMBR redshift. One particularly simple model equation under current study is:

$$Z \cong \sqrt{\frac{R_0}{R_x} - 1} \cong \sqrt{\frac{2GM_0}{c^2 R_x} - 1} \tag{17}$$

where $R_x < R_0$ and $M_0 \cong c^3 / 2GH_0$.

and where R_0 and R_x represent current and past cosmic radii, respectively. With reference to the assumed equivalent cosmic temperature [2], redshift term Z can be obtained in the following way:

$$Z \cong \sqrt{\exp\left\{\left[1 + \ln\left(\frac{R_0}{R_{pl}}\right)\right] - \left[1 + \ln\left(\frac{R_x}{R_{pl}}\right)\right]\right\} - 1}$$

$$\cong \sqrt{\exp\left\{\left(\ln\left(\frac{R_0}{R_{pl}}\right) - \ln\left(\frac{R_x}{R_{pl}}\right)\right)\right\} - 1} \cong \sqrt{\exp\left\{\ln\left(\frac{R_0}{R_x}\right)\right\} - 1} \cong \sqrt{\frac{R_0}{R_x} - 1} \tag{18}$$

From relation (5) it is clear that the cosmic radius is inversely proportional to the squared cosmic temperature. The above relation (17) can be expressed, approximately, as follows:

$$Z \cong \sqrt{\frac{R_0}{R_x} - 1} \cong \sqrt{\frac{T_x^2}{T_0^2} - 1} \text{ where } T_x > T_0 \tag{19}$$

For past higher cosmic temperatures,

$$Z \cong \sqrt{\frac{T_x^2}{T_0^2} - 1} \cong \frac{T_x}{T_0} \quad \text{where } T_x \gg T_0 \tag{20}$$

This can be compared with the famous relation familiar to modern cosmologists:

$$Z + 1 \cong \frac{T_t}{T_0} \tag{21}$$

Given our stated basic assumptions, our expanding cosmic model shows average mass-energy density to be inversely proportional to R^2. See relation (11). In a very real sense, the deeper an observer from Earth looks into space (and time), the greater the temperature stage and average mass energy density stage of the cosmos one is observing. Thus, each progressively higher temperature stage and average mass density stage of the cosmos is associated with higher gravitational field strengths. So, there must be associated gravitational time dilation effects. This conclusion is firmly grounded in general relativity. Thus, *it appears likely that at least a portion of the progressively higher redshift we observe with increasing look-back distance is a manifestation of gravitational time dilation.* In addition, because of this inverse square relationship over very long distances, plots of proximal galactic redshifts per unit of distance observed would be expected to look relatively linear (as seen by the weaker telescopes of the 1920's and 1930's) and *deep space galactic redshifts per unit of distance observed would be expected to clearly fall away from linearity, along with decreasing luminosity (as redshifts extend into the infrared range), similar to the 1998 Type Ia supernovae observations [5]. Such an effect may possibly create an illusion of dark energy where there is none.* This is for further study.

The following graph (**Figure 1**), according to above relation (17), shows expected cosmic redshift as a function of the log ratio of current cosmic radius (R_0) to past cosmic radius (R_x) pertaining to a particular astronomical observation. In this manner, increasingly greater redshifts would be expected to correspond with more distant galactic observations. However, notice the apparent near-linearity below the radius ratio of about 20 (Log 1.3), and the increasingly nonlinear appearance with deeper space observations. The authors propose that something like this mathematical relationship could be useful in modeling the results of progressively deeper space observations. For data, see **Table 1**. The last row of **Table 1** correlates cosmic radius (R_x), Log (R_x/R_0), redshift and universal age corresponding to a temperature of 3116 K.

Thus, simple model formula relations (17) and (20) closely approximate the recombination temperature of 3000 K and CMBR redshift of 1093 believed to be related to formation of the first hydrogen atoms.

Of course, since we are modeling flat space cosmology, we are also testing a slightly more complex model formula, Minkowski's relativistic Doppler formula for flat space:

$$Z + 1 \cong \sqrt{\frac{[1 + (v/c)]}{[1 - (v/c)]}} \tag{22}$$

In order to keep scaling similar to **Figure 1**, the velocity term v in the Minkowski formula can be substituted with $[1 - (R_x/R_0)]c$ where $R_x < R_0$. The reduced relation can be expressed as:

$$Z \cong \sqrt{\frac{2 - (R_x/R_0)}{(R_x/R_0)} - 1} \tag{23}$$

Figure 2 shows the Minkowski relativistic Doppler redshift term Z as a function of decreasing log cosmic radius ratio (R_x/R_0) pertaining to progressively deeper space observations. The reader should note that the CMBR redshift of 1093 correlates with the place on the horizontal axis corresponding to log value −5.777. Perhaps more importantly, however, the reader's attention is directed to the place on the horizontal axis corresponding to the log value of −1.5. A greatly magnified portion of this region of **Figure 2** would show the nonlinearity corresponding to the earliest visible galaxies, which appear to be receding at up to about 0.95 c.

Thus, a combination of gravitational time dilation (**Figure 1**) and flat space relativistic Doppler effect (**Figure 2**) may possibly provide an explanation for the nonlinearity of deep space Type Ia supernovae observations currently being attributed to "dark energy." *The reader will, of course, object that dark energy is actually causing cosmic acceleration. However, there are recent credible arguments that the 1998 Type Ia supernovae data are*

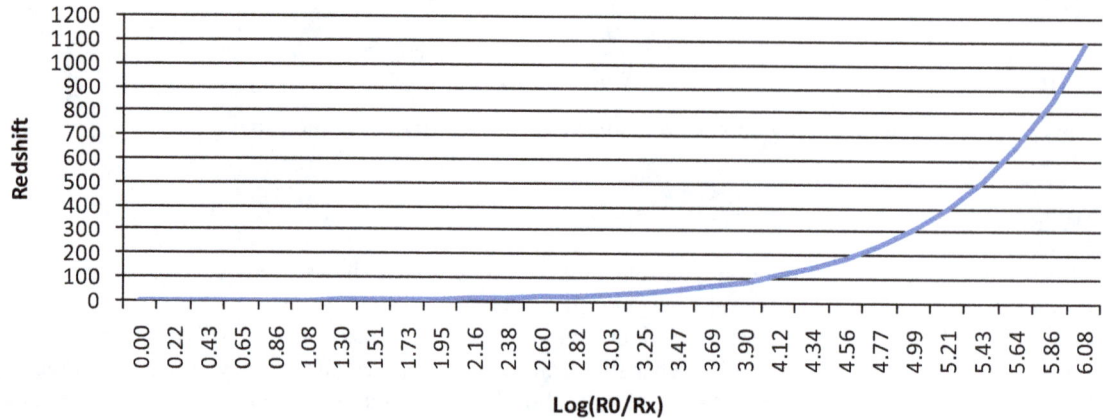

Figure 1. Cosmic Redshift vs Log Ratio of Current and Past Cosmic Radii.

Table 1. Cosmic Physical Parameters Obtained with above relations.

Assumed Cosmic radius (m)	Hubble parameter (sec^{-1})	Log (R_0/R_x)	Redshift	Temperature (K)	Age (Years)
1.34848E+26	2.22319E−18	0.00	0.0	2.7	1.4253E+10
8.19745E+25	3.65714E−18	0.22	0.8	3.5	8.6647E+09
4.98325E+25	6.016E−18	0.43	1.3	4.5	5.2673E+09
3.02933E+25	9.89632E−18	0.65	1.9	5.7	3.2020E+09
1.84154E+25	1.62795E−17	0.86	2.5	7.4	1.9465E+09
1.11948E+25	2.67797E−17	1.08	3.3	9.5	1.1833E+09
6.80533E+24	4.40526E−17	1.30	4.3	12.1	7.1932E+08
4.13698E+24	7.24665E−17	1.51	5.6	15.6	4.3728E+08
2.51488E+24	1.19207E−16	1.73	7.3	20.1	2.6582E+08
1.52417E+24	1.96692E−16	1.95	9.4	25.8	1.6110E+08
9.23739E+23	3.24542E−16	2.16	12.0	33.2	9.7639E+07
5.59842E+23	5.35495E−16	2.38	15.5	42.7	5.9175E+07
3.39298E+23	8.83566E−16	2.60	19.9	55.0	3.5864E+07
2.05635E+23	1.45788E−15	2.82	25.6	70.8	2.1736E+07
1.24627E+23	2.40551E−15	3.03	32.9	91.1	1.3173E+07
7.55318E+22	3.96909E−15	3.25	42.2	117.2	7.9837E+06
4.57768E+22	6.549E−15	3.47	54.3	150.9	4.8386E+06
2.77435E+22	1.08058E−14	3.69	69.7	194.2	2.9325E+06
1.68143E+22	1.78297E−14	3.90	89.5	249.9	1.7773E+06
1.01905E+22	2.94189E−14	4.12	115.0	321.6	1.0771E+06
6.17604E+21	4.85412E−14	4.34	147.8	413.9	6.5281E+05
3.74305E+21	8.0093E−14	4.56	189.8	532.7	3.9564E+05
2.26852E+21	1.32153E−13	4.77	243.8	685.6	2.3978E+05
1.37486E+21	2.18053E−13	4.99	313.2	882.4	1.4532E+05
8.33248E+20	3.59788E−13	5.21	402.3	1135.7	8.8074E+04
5.04999E+20	5.9365E−13	5.43	516.7	1461.6	5.3378E+04
3.06060E+20	9.79522E−13	5.64	663.8	1881.2	3.2351E+04
1.85491E+20	1.61621E−12	5.86	852.6	2421.2	1.9606E+04
1.12419E+20	**2.66675E−12**	6.08	**1095.2**	**3116.2**	**1.1883E+04**

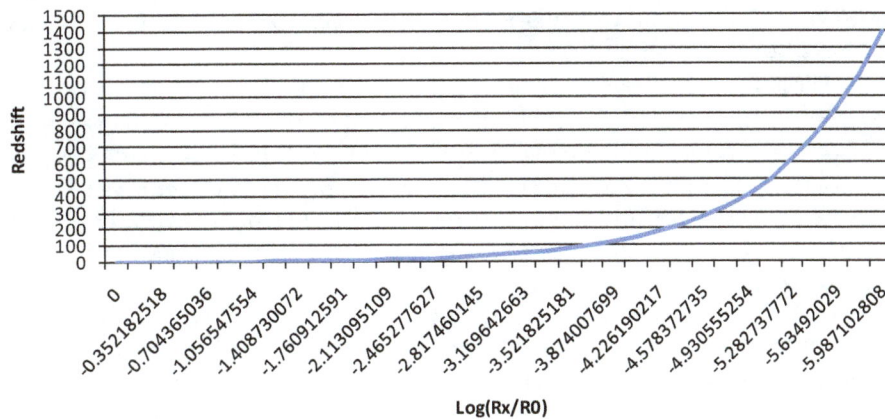

Figure 2. Redshift vs Decreasing Log (R_x/R_0).

not at all conclusive in this regard. Please see 2015 *references* [6] *and* [7]. *In particular, a very strong case has been made by Jun-Jie Wei, et al., that the Type Ia supernovae data are a better fit for a cosmic horizon coasting along at speed of light c* [7]!

This important question requires further study.

6. Summary Discussion

Given the apparent simplicity of our mathematical model, we have been pleasantly surprised by the excellent correlation between derivations from our model formulae and astronomical observations. However, upon closer inspection, the reason for this should be fairly obvious. Science progresses not by completely discarding reliable mathematical models from the past, but by refining them along with improvements in observation. Frequently, the older formulae exist as remnants within the newer refined formulae. Such will ultimately be the case when elements of general relativity and quantum mechanics come together as "quantum gravity".

In this tradition, our mathematical model incorporates a formula (Schwarzschild) derived from general relativity, combining it with Hubble's velocity-distance relation [8], conservation of angular momentum (to set a $2\hbar$ limit on cosmic rotation), and incorporating all of these relationships into formulae (relations 4) which closely resemble Hawking's black hole temperature formula [4]. Seeds in the development of these mathematical inter-relationships can be followed in references [9]-[14]. It is, therefore, not particularly unreasonable that our mathematical formulation has some relevance to our observable universe. While it may seem mysterious to some observers that quantum terms from the Planck scale, Boltzmann's constant, $\hbar$ and $2\hbar$ are creeping into our cosmological equations and derivations, we predict that further iterations of this process will ultimately lead us towards a more complete theory of "quantum gravity".

Acknowledgements

The authors express their thanks to Dr. Abhas Mitra for his kind and valuable suggestions in developing this subject. One of the authors, Seshavatharam U.V.S., is indebted to professors K.V. Krishna Murthy, Chairman, Institute of Scientific Research in Vedas (I-SERVE), Hyderabad, India and Shri K.V.R.S. Murthy, former scientist IICT (CSIR), Govt. of India, Director, Research and Development, I-SERVE, for their valuable guidance and great support in developing this subject. Author Dr. E. Terry Tatum would also like to thank Dr. Rudy Schild, Harvard Center for Astrophysics, for his support and encouragement in developing this subject.

References

[1]   Tatum, E.T., Seshavatharam, U.V.S. and Lakshminarayana, S. (2015) The Basics of Flat Space Cosmology. *International Journal of Astronomy and Astrophysics*, **5**, 116-124. http://dx.doi.org/10.4236/ijaa.2015.52015

[2]   Tatum, E.T., Seshavatharam, U.V.S. and Lakshminarayana, S. (2015) Thermal Radiation Redshift in Flat Space Cosmology. *Journal of Applied Physical Science International*, **4**, 18-26.

[3]   Planck Collaboration: Planck 2015 Results. XIII. Cosmological Parameters. http://arxiv.org/abs/1502.01589

[4]   Hawking, S.W. (1975) Particle Creation by Black Holes. *Communications in Mathematical Physics*, **43**, 199-220.
       http://dx.doi.org/10.1007/BF02345020

[5]   Riess, A.G., Filippenko, A.V., Challis, P., *et al.* (1998) Observational Evidence from Supernovae for an Accelerating
       Universe and a Cosmological Constant. *Astrophysical Journal*, **116**, 1009-1038. http://dx.doi.org/10.1086/300499

[6]   Nielsen. J.T., *et al.* (2015) Marginal Evidence for Cosmic Acceleration from Type Ia Supernovae. arXiv:1506.01354v2.

[7]   Wei, J.-J., Wu, X.-F., Melia, F., *et al.* (2015) A Comparative Analysis of the Supernova Legacy Survey Sample with
       ΛCDM and the $R_h = ct$ Universe. *The Astronomical Journal*, **149**, 102. http://dx.doi.org/10.1088/0004-6256/149/3/102

[8]   Hubble, E.P. (1929) A Relation between Distance and Radial Velocity among Extra-galactic Nebulae. *PNAS*, **15**, 168-
       173. http://dx.doi.org/10.1073/pnas.15.3.168

[9]   Seshavatharam, U.V.S. and Lakshminarayana, S. (2015) Primordial Hot Evolving Black Holes and the Evolved Pri-
       mordial Cold Black Hole Universe. *Frontiers of Astronomy, Astrophysics and Cosmology*, **1**, 16-23.

[10]   Seshavatharam, U.V.S. and Lakshminarayana, S. (2014) Friedmann Cosmology: Reconsideration and New Results.
       International Journal of Astronomy, Astrophysics and Space Science, **1**, 16-26.

[11]   Pathria, R.K. (1972) The Universe as a Black Hole. *Nature*, **240**, 298-299.
       http://dx.doi.org/10.1038/240298a0

[12]   Tatum, E.T. Could Our Universe Have Features of a Giant Black Hole? *Journal of Cosmology*, 25, 13061-13080.

[13]   Tatum, E.T. (2015) How a Black Hole Universe Theory Might Resolve Some Cosmological Conundrums. *Journal of
       Cosmology*, **25**, 13081-13111.

[14]   Melia, F. and Maier, R.S. (2013) Cosmic Chronometers in the $R_h = ct$ Universe. *Monthly Notices of the Royal Astro-
       nomical Society*, **432**, 2669. http://dx.doi.org/10.1093/mnras/stt596

The Efficiency of CP-Violating α^2-Dynamos from Primordial Cosmic Axion Oscillation with Torsion

L. C. Garcia de Andrade

Department of Theoretical Physics, State University of Rio de Janeiro (UERJ), Rio de Janeiro, Brazil
Email: garcia@dft.if.uerj.br

Abstract

Recently torsion fields were introduced in CP-violating cosmic axion α^2-dynamos [Garcia de Andrade, Mod Phys Lett A, (2011)] in order to obtain Lorentz violating bounds for torsion. Here instead, oscillating axion solutions of the dynamo equation with torsion modes [Garcia de Andrade, Phys Lett B (2012)] are obtained taking into account dissipative torsion fields. Magnetic helicity torsion oscillatory contribution is also obtained. Note that the torsion presence guarantees dynamo efficiency when axion dynamo length is much stronger than the torsion length. Primordial axion oscillations due to torsion yield a magnetic field of 10^9 G at Nucleosynthesis epoch. This is obtained due to a decay of BBN magnetic field of 10^{15} G induced by torsion. Since torsion is taken as 10^{-20} s^{-1}, the dynamo efficiency is granted over torsion damping. Of course dynamo efficiency is better in the absence of torsion. In the particular case when the torsion is obtained from anomalies it is given by the gradient of axion scalar [Duncan *et al.*, Nuclear Phys B 87, 215] that a simpler dynamo equation is obtained and dynamo mechanism seems to be efficient when the torsion helicity, is negative while magnetic field decays when the torsion is positive. In this case an extremely huge value for the magnetic field of 10^{15} Gauss is obtained. This is one order of magnitude greater than the primordial magnetic fields of the domain wall. Actually if one uses $t_{DW} \sim 10^{-4}$ s one obtains $B_{DW} \sim 10^{22}$ G which is a more stringent limit to the DW magnetic primordial field.

Keywords

Torsion Theories, Axion Dynamo, Primordial Magnetic Fields

1. Introduction

Earlier Mielke and Romero [1] have shown that Cartan spacetime torsion [2] in the chiral anomaly induces a dynamical axion coupled with gravitation. This torsion-induced pseudo-scalar is given in such way that its gradient yields the torsion vector. Earlier Campanelli et al. [3] have investigated the primordial oscillation of axions and respective magnetic fields when CP-violating dynamos [4] in QCD are present. In early universe torsion effects are stronger than usual [3] which justifies the introduction of torsion in the dynamo equation [5]. In this work we consider QCD era and we are very far any galaxy formation. Therefore instead of the usual classical Maxwell electrodynamics non-minimally coupled with photon-torsion coupling in the realm of quantum electrodynamics (QED) we use early universe electrodynamics. In this paper a FRW universe is given as background for axion dynamo equation with torsion. It is shown that torsion oscillations enhance cosmic axion oscillations computed by Campanelli and Gianotti [4]. Primordial axion oscillations due to torsion yields a magnetic field of 10^9 G at Nucleosynthesis epoch. This is obtained due to a decay of BBN magnetic field of 10^{15} G induced by torsion. In the last section of the paper a dynamo equation is obtained from axion scalar torsion string where huge magnetic fields are obtained from 10^{13} Gauss seed fields.

2. Axions, Photons and Torsion

Parity violation in gravity has been investigated recently by B Mukhopadhayaya, S Sen and Sur [6] which conclude that on a torsion-axion duality arising in a string scenario via Kalb-Ramond field leads to parity-violating interactions for spin-$\frac{1}{2}$ fermions. More recently the author [7] has investigated the role of parity violation in torsion has been used to built dynamo equation. By analogy photon-axion coupling may happen giving rise to magnetic fields that eventually may be amplified giving rise to the α^2-dynamos addressed in the next section.

3. Efficiency of CP-Violating Dynamos

In this section we shall consider the solution the CP-violation dynamos and its efficiency on a torsion background. Let us start by considering the dynamo equation as [1]

$$\dot{\boldsymbol{B}} = -2H\boldsymbol{B} + \frac{(\mathrm{div}\boldsymbol{S})\boldsymbol{B}}{\sigma a} + \frac{\nabla^2 \boldsymbol{B}}{\sigma a} + \alpha_{dyn}\frac{\nabla \times \boldsymbol{B}}{a} \tag{1}$$

where H is the Hubble parameter, a is the expansion of the universe, $\boldsymbol{S}$ represents the torsion vector and σ is the conductive. Here α_{dyn}

$$\alpha_{dyn} = \frac{\alpha_{em}}{2\pi}\frac{\dot{\Theta}}{\sigma} \tag{2}$$

the metric is given by

$$ds^2 = \left(dt^2 - a^2 d\boldsymbol{x}^2\right) \tag{3}$$

Here Θ represents the axion field primordial field. The Fourier analysis of first equation becomes

$$\dot{B}_{\pm} = -2HB_{\pm} - \frac{k^2 B_{\pm}}{\sigma a^2} \pm \frac{\alpha_{em}}{2\pi}\frac{\dot{\Theta}kB_{\pm}}{\sigma a} + \frac{ikS}{\sigma a}B_{\pm} \tag{4}$$

From this solution one notes that the torsion oscillating length is complex representing a true oscillation. Here k is the wave coherent scale number. The solution of this equation is

$$B_{\pm}(k,t) = B_{\pm}(k,t_i)\left(\frac{a_i}{a}\right)^2 \exp\left[-k^2 l_d^2 + k l_{\Theta} + ik l_S\right] \tag{5}$$

where the oscillation lengths are

$$l_S(t) = \int \frac{Sdt}{a\sigma} \tag{6}$$

$$l_d^2(t) = \int \frac{dt}{a\sigma} \tag{7}$$

$$l_{\Theta}(t) = -\frac{\alpha_{em}}{2\pi} \int \frac{\dot{\Theta} dt}{a\sigma} \tag{8}$$

Here l_{Θ} is the dynamo length [2]. Here $\eta \sim 1\sigma \sim 10^{29}$ cm$^2 \cdot$s^{-1} is the dissipation constant. Magnetic energy is then given by

$$\epsilon_B(k,t) = \epsilon_B(k,t_i)\left(\frac{a_i}{a}\right)^4 \exp\left(-2k^2 l_d^2\right)\cosh\left(2kl_{\Theta}\right)\cos\left(kl_S\right) \tag{9}$$

The magnetic helicity contribution of torsion oscillation is

$$\mathcal{H}_B = -\mathcal{H}_B^{max} \exp\left(-2k^2 l_d^2\right)\sinh\left(2kl_{\Theta}\right)\sin\left(kl_S\right) \tag{10}$$

Note that the magnetic helicity generation now has a contribution of the torsion oscillation length. Finally let us compute the torsion oscillation new term to compare it with the dynamo length to see if the torsion dissipative term may damp the dynamo length. Then

$$B_{\pm} = B_{seed}^{\pm} \sin\left(kl_S\right) \tag{11}$$

when the argument is small this expression reduces to

$$B_{\pm} = B_{seed}^{\pm} kl_S \tag{12}$$

and for Big Bang Nucleosynthesis (BBN)

$$B_{\pm} = B_{BBN}^{\pm} kl_S \sim 10^9 \text{ Gauss} \tag{13}$$

where we have used the seed field for $B_{BBN} \sim 10^{15}$ G. In the case of the seed field 10^{-21} G one obtains

$$B_{\pm} \sim 10^{-28} \text{ G} \tag{14}$$

which is able to seed galactic dynamos. Dynamo efficiency is given by $|l_{\Theta}| \gg l_S$. This ratio reads

$$\frac{\alpha_{em}}{2\pi} \sim 10^{-2} \gg S_{BBN} t_{BBN} \sim 10^{-20} \tag{15}$$

where torsion field $S \sim 10^{-20}$ s^{-1} [5] and $t_{BBN} \sim 1$ s. Here we have used $\Theta \sim 1$. Thus one may conclude that the torsion field is not high enough to avoid the dynamo efficiency of the dynamo by torsion damping of magnetic fields.

4. Dynamo Equation from Torsion Axion Anomalies

Earlier Duncan *et al.* [7] investigated axion hair anomalies in Riemann-Cartan spacetime where torsion was given by the gradient of axion scalar ϕ. In this section instead of using the α dynamo equation of the previous one we shall addopt this approach and obtain a simpler dynamo equation which solution gives rise to stronger magnetic fields starting from the magnetic field of 10^{13} G of the last section. The Maxwell equation obtained by Duncan *et al.* [7]

$$\nabla_{\mu} F^{\mu\nu} = 2\lambda \left(\nabla_{\mu}\phi\right)^* F^{\mu\nu} \tag{16}$$

where star in front of the Maxwell tensor $\boldsymbol{F}$ means that we are taking the dual of $\boldsymbol{F}$ given by $\epsilon^{\mu\nu\rho\sigma} F_{\rho\sigma}$. Taking the approximation of Minkowski space plus torsion to emphasize the torsion role one obtains the dynamo equation as

$$\partial_t \boldsymbol{B} - \nabla \times \boldsymbol{u} \times \boldsymbol{B} = -2\lambda S_0 \left(\boldsymbol{u} \times \boldsymbol{B}\right) \tag{17}$$

By taking the scaled version of this equation in Fourier space one obtains

$$\partial_t B = -2\lambda\left[S_0 - ik\right]uB \tag{18}$$

which solution yields

$$B \sim B_{seed} \exp\left[-2\lambda\left[\left(S_0 - ik\right)ut\right]\right] \tag{19}$$

From this expression one sees that the torsion helicity sign $S_0 u$ together with the coupling constant λ are fundamental to check for the efficiency of the dynamo; when torsion helicity is positive the magnetic field decays while when is negative dynamo effect is enhanced. By considering the approximation

$$|B| \sim B_{seed}\left[\left(s_0 - ik\right)ut\right]\lambda \tag{20}$$

Here the coupling constant $\lambda = \dfrac{e^2}{2 f_\phi \pi^2} = 10^{28}$ and f_ϕ is the axion decay constant. Taking the seed field as high as in the next section 10^{13} G one obtains for 1 kpc scale and a torsion field $S_0 \sim 10^{-17}$ cm^{-1}. Taking again $t_{BBN} \sim 1$ s a magnetic field as high as $10^{-3}\lambda$ Gauss. Which upon substitution of λ yields $B \sim 10^{25}$ Gauss which is one order of magnitude greater than the domain wall primordial magnetic field obtained by Cea and Tedesco [8]. Actually if one uses $t_{DW} \sim 10^{-4}$ s one obtains for the B-field $B_{DW} \sim 10^{30}$ G which is a much stronger limit for the DW primordial magnetic field. This is extremely stronger than the limit found by Kisslinger from walls obtained from chiral phase transition [9]. Thus one may conclude that the strenght of the magnetic field shall depend upon the strenght of the coupling constant.

5. Discussions and Conclusions

Torsion fields introduced in CP-violating cosmic axion α^2-dynamos by the author [2] in order to obtain Lorentz violating bounds for torsion are revisited. These time oscillating axion solutions of the dynamo equation with torsion modes [3] are obtained taking into account dissipative torsion fields. Magnetic helicity torsion oscillatory contribution is also obtained. Note that the torsion presence guarantees dynamo efficiency when axion dynamo length is much stronger than the torsion length. Primordial axion oscillations due to torsion yields a magnetic field of 10^9 G at Nucleosynthesis epoch. This is obtained due to a decay of BBN magnetic field of 10^{15} G induced by torsion [6]. Taking the Duncan et al. [7] torsion axion anomaly, the magnetic field strenght is huge depending on how strong is the coupling between torsion axions and photons.

Acknowledgements

We would like to express my gratitude to D. Sokoloff and A. Brandenburg for helpful discussions on the subject of this paper. I thank Prof. C. Sivaram for initiating me on the problem of dynamos and torsion. Financial support from CNPq. and University of State of Rio de Janeiro (UERJ) are grateful acknowledged.

References

[1] Campanelli, L. and Gianotti, M. (2005) Magnetic Helicity Generation from the Cosmic Axion Field. *Physical Review D*, **72**, Article ID: 123001. http://dx.doi.org/10.1103/PhysRevD.72.123001

[2] Garcia de Andrade, L.C. (2011) *Mod Phys Lett A.*, **26**, 2863.

[3] Garcia de Andrade, L.C. (2012) Primordial Magnetic Fields and Dynamos from Parity Violated Torsion. *Physics Letters B*, **711**, 143-146. http://dx.doi.org/10.1016/j.physletb.2012.03.075

[4] Widrow, L. (2001) *Rev Mod Phys*, **74**, 775; Turner, M. and Widrow, L. (1988) *Phys Rev D*; Prokopec, T., Tornkvist, O. and Woodward, R. (2002) *Phys Rev Lett.*, **89**, 101301; Ruzmakin, A., Sokoloff, D.D. and Shukurov, A. (1988) Magnetic Fields in Galaxies, Kluwer; Garcia de Andrade, L.C. (2011) *Nuclear Phys B*; Garcia de Andrade, L. (2011) *Phys Lett B*, **468**, 28; Ratra, B., Caltech preprint and Garcia de Andrade, L. (2011) Lorentz Violation Bounds from Torsion Trace and Radio Galactic Dynamos. *Phys Rev D (Brief Reports)*.

[5] Laemmerzahl, C. (1997) *Phys Lett A.*, **228**, 223.

[6] de Sabbata, V., Garcia de Andrade, L.C. and Sivaram, C. (1993) Torsion and Gauge-Invariant Massive Electrodynamics. *International Journal of Theoretical Physics*, **32**, 1523-1530. http://dx.doi.org/10.1007/BF00672853

[7] Duncan, M., Kaloper, N. and Olive, K.A. (1992) *Nucl Phys B*, **87**, 215.

[8] Cea, P. and Tedesco, L. (year) Dynamical Generation of the Primordial Magnetic Field by Ferromagnetic Domain Walls. arXiv:hep-th/9811221v1.

[9] Kisslinger, L.S. (2003) Magnetic Wall from Chiral Phase Transition and CMPR Correlations. arXiv: hep-ph/0212206v2.

12

Study of Principal Catalogues of Visual Double Stars

Yikdem Mengesha Gebrehiwot[1], Solomon Belay Tessema[2], Oleg Malkov[2]

[1]Astronomy and Astrophysics Research Division, Entoto Observatory and Research Center (EORC), Addis Ababa, Ethiopia
[2]Institute of Astronomy, Russian Academy Sciences, Moscow, Russia
Email: yikdema16@gmail.com, tessemabelay@gmail.com, malkov@inasan.ru

Abstract

The goal of this study is to develop a tool for detection of errors and misprints in principal catalogues of visual double stars. Modern statistical investigations of visual doubles (which are about 130 thousand, and they represent the most numerous observational type of binaries) are based almost exclusively on data, listed in those catalogues. So their correction is a challenging task, otherwise the statistics of visual binaries can be biased. Study of individual stars should also be based on correct data. We have presented tools to detect some of errors in the catalogues, especially, to indicate cases i) when positional information on components of a pair is contradictory, and ii) when a double star or its component is included in a catalogue twice, under different names.

Keywords

Visual Binary Star, Positional Inconsistency, Data Duplication, Reference Coordinates, Additional Component Coordinates

1. Introduction

Heggie in his investigation of formation and evolution of binaries identified two classes of binary stars [1]. He called these binaries as "soft" and "hard" binary stars. Actually this classification is analogues to "wide" and "close" binary stars respectively. There is no well defined boundary between close and wide binary systems. It differs from author to author according to the purpose and types of the data required. The operative parameter of a binary system is the semi-major axis a. Since its values vary over a wide range of several orders of magnitude, the exact form of semi-major axis distribution is not well known yet. According to [2]-[4] binary stars can be more or less classified as "close" when semi-major axis $a \leq 10$ AU and "wide" when semi-major axis $a \geq 100$ AU.

Close binary systems are very essential to study stellar properties to a high certainty. One of the types of close binary systems, eclipsing binary systems, are used to determine stellar parameters like masses and radii of components of binary systems [5]. On the other hand, wide binary systems are essential systems to study star formation processes as well as an exceptionally useful tracer of local potential and tidal fields through which they traverse. Specifically, they have been used to place constraints on the nature of halo dark matter [6] and to explore the dynamical history of the Galaxy [7]. Moreover, wide binaries used to constrain the mass of hypothetical MAssive Compact Halo Objects, so-called "MACHOs" [8].

Visual binaries represent the most numerous observational types of binary stars. Catalogues of visual binaries contain more than 10^5 pairs and provide us with astrometric, photometric and sometimes spectral information on the components. Consequently these data are extremely useful for statistical study and for construction of distributions on and relations between various observational and astrophysical parameters. However, catalogues of visual binaries are not error free. It is probably impossible to entirely fix the problem, but it is desirable to minimize the number of errors through various checks, to avoid any statistical investigations to be biased.

In this study we consider three principal catalogues of visual double stars, namely, The Washington Visual Double Star Catalog (WDS, [9]), the Catalog of Components of Double & Multiple stars (CCDM, [10]), and the Tycho Double Star Catalogue (TDSC, [11]). They contain information on about 130 thousand (*i.e.*, almost all known) visual double and multiple stars.

All catalogues contain annoying bugs and misprints, and this especially concerns large compilative catalogues WDS and CCDM. Some types of errors can be discovered without attraction of external data sources. These are, in particular, the cases (i) when positional information on an additional component in a pair is contradictory, and (ii) when a binary or a component is included in a catalogue twice, under different names.

Constructed tools and results of their application to the cases (i) and (ii) mentioned above are reported in Sections 2 and 3, respectively. Distribution of dublications as well as other errors found in the catalogues are discussed in Section 4. Finally, in Section 5 we make our conclusions.

2. Positional Inconsistency in the Catalogues

Catalogues of visual double stars provide us, among other data (brightness of the components, their spectral types, proper motion and cross-identification), with coordinates of the reference (α_1, δ_1) and the additional (α_2, δ_2) components of the pair. Also, the studied catalogues provide user with positional angle (θ) and separation (ρ) of the additional component in a pair, *i.e.*, its position relative to the reference component (with coordinates α_1, δ_1). From these data coordinates of the additional component (α_3, δ_3) can also be calculated according to

$$\alpha_3 - \alpha_1 = \frac{\rho \sin \theta}{\cos \delta_1}$$
$$\delta_3 - \delta_1 = \rho \cos \theta,$$

(1)

and compared with those, explicitly given in the catalogues (α_2, δ_2), using a formula for calculation of angular distance between two points on the sky:

$$d = \sqrt{\cos^2 \delta_1 (\alpha_2 - \alpha_1)^2 + (\delta_2 - \delta_1)^2}.$$

(2)

It can be argued that if the distance d exceeds certain selected value d_1, then the set ($\alpha_1, \delta_1, \theta, \rho, \alpha_2, \delta_2$) contains inconsistent data. The value of d_1 depends on observational conditions, and should be estimated experimentally for every studied catalogue.

2.1. WDS

It should be noted that WDS (unlike the other two catalogues) does not provide coordinates for the additional component (α_2, δ_2). It means that the detection of fictitious data may be performed only for those members of multiple (triple and higher multiplicity) systems, which appear as the reference component in one pair and as the additional component in another one. For example, WDS provides data for the following pairs of the system

WDS 00013+6021: AB, AC, AD, BD, and the task can be performed, consequently, only for component B.

The summary WDS catalog posted on-line provides two sets of θ, ρ values, for the first and the last observations (the full WDS database includes all measures). We used the latter one in this study.

For WDS we found $d_1 = 8$ arcsec, and there are about 340 pairs in the catalogue, demonstrating an inconsistency in positional data for additional components. The reasons for that inconsistency are large orbital motion in relatively close pairs (especially in a case of large difference in epochs of observation), large proper motion difference in optical pairs, confusion between two closely spaced objects, positional angle θ ambiguity by 180 degrees, large and consequently imprecisely defined value of separation ρ, misprints in the catalogue.

2.2. CCDM

CCDM, though being smaller than WDS, is based on a similar set of observational data (in fact it was based on an earlier version of the WDS). On the other hand, WDS is constantly updated, while CCDM keeps observational data collected before 2002. Also, unlike WDS, the detection of fictitious data may be performed for all catalogued objects (as coordinates are provided for every component in CCDM). That is why when we set d_1 for CCDM to be 8 arcsec, we found 1070 pairs demonstrating inconsistency between catalogued secondary coordinates and ones, calculated from primary coordinates, positional angle and separation.

Here we do not consider pairs, where θ is indicated roughly in CCDM, in the format like "SF" (to the South and following, $i.e.$, second quadrant) or "NP" (to the North and preceding, $i.e.$, fourth quadrant). Note that there are also pairs in the WDS with θ only published as N, S, SF, NP, etc. These have been converted to coarse θ values such as 0, 45, 90, etc. Usually these quadrant-only position angles were published only for very early measures, so the "last" angles are nearly always more accurately measured values.

2.3. TDSC

TDSC is a homogeneous catalogue, and it contains observations performed with the same instrument. As a result for the majority of objects we found d_1 to be about 1.2 arcsec. This limit is exceeded by eight very wide (ρ exceeds 10 arcmin) pairs only. For the extreme case $d = 27.4$ arcsec (TDCS 56356 = WDS 20452-3120AB, $\rho = 78$ arcmin).

3. Data Duplication in the Catalogues

Some star can happen to be a member of two different double (multiple) systems, included in a compilation catalogue of doubles. Moreover, a double star can be discovered by two different observers independently and, consequently, this pair appears twice in the catalogue, under different designations. This can be checked by coordinate comparison. Similarly to the case described in Section 2, angular distance between two points with celestial coordinates (α_x, δ_x) and (α_y, δ_y) can be calculated according to

$$d = \sqrt{\cos^2 \delta_x \left(\alpha_y - \alpha_x\right)^2 + \left(\delta_y - \delta_x\right)^2} \tag{3}$$

and compared to certain limiting value d_2. If d does not exceed certain d_2, one can conclude that (α_x, δ_x) and (α_y, δ_y) correspond to the same celestial object.

Again, d_2 value depends on observational conditions, and we should estimate it experimentally for every studied catalogue. Preliminary results were checked manually with the Binary star database (BDB, [12]), which allows us to visualize catalogued data, and SIMBAD.

3.1. WDS

As our analysis shows, $d_2(WDS)$ value is 21 arcsec, and all (altogether 306) cases with $d < 21$ arcsec indicate duplication of objects in WDS. Beside that, 26 other duplication cases are found, and they exhibit coordinate difference to be $21 < d < 41$ arcsec. At the same time we have found 59 cases when the angular distance between two really different objects is in the $21 < d < 41$ arcsec range.

The absence of a strict border between these two cases can be explained by variations of the positional accuracy of the catalogued data. In some cases coordinates of the objects are known with relatively low precision. It is true for overcrowded areas (e.g., Orion nebula), for very bright (e.g., α Cap) or very faint (e.g., WDS 19252-2900) stars, and for relatively wide ($\rho > 200$ arcsec) pairs (e.g., WDS 09479+1319). It also happens when K-band or other infrared magnitudes are given in WDS (e.g., vicinity of RK86 reflection nebula) instead of the visual photometry.

Duplication cases can be classified according to the following scheme.

• A star is included in two different double/multiple systems. E.g., WDS 00086-0418 A = WDS 00085-0419 A (**Figure 1**).

• Both components of a catalogued double star are included in another, also catalogued, multiple system. Those two stars can constitute (e.g., WDS 23243 + 6216 = WDS 23248 + 6217 CD, **Figure 2**) or cannot constitute (e.g., WDS 23288 + 0703 A and B are WDS 23287 + 0703 C and B, respectively, **Figure 3**) a pair in the multiple system.

• Two catalogued double stars represent in fact the same pair of objects (**Figure 4** and **Figure 5**). E.g., WDS 00281-2652 AB - WDS 00280-2654 AB.

All of these cases should be taken into account in statistical research of visual binaries. We have reported on these errors to the WDS authors, and necessary corrections were made in the WDS catalogue.

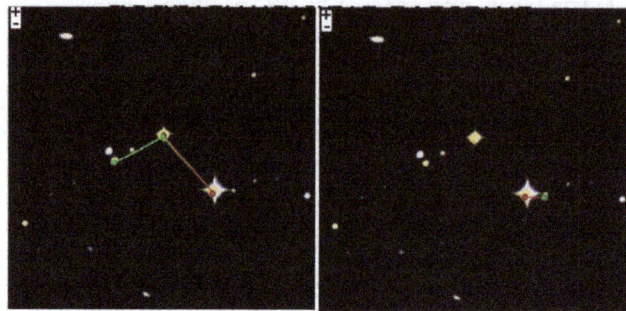

Figure 1. WDS 00086-0418 and WDS 00085-0419. Duplication case 1, a star is included in two different double/multiple systems.

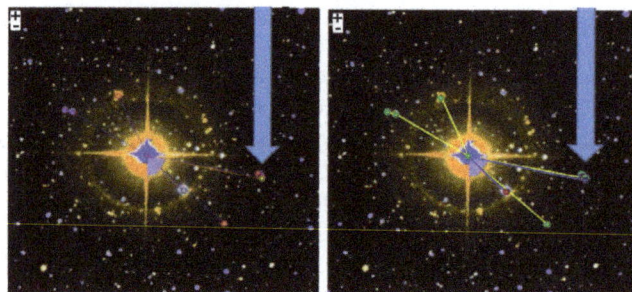

Figure 2. WDS 23243 + 6216 and WDS 23248 + 6217. Duplication case 2a, a catalogued double star are included in another system.

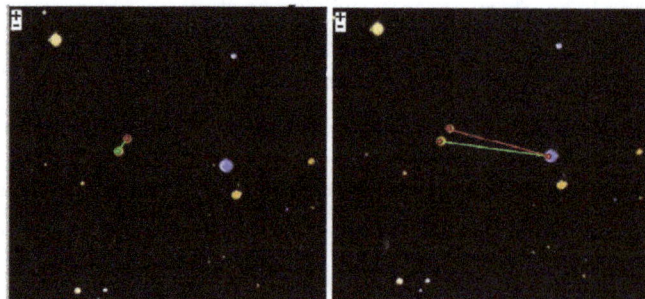

Figure 3. WDS 23288+ 0703 and WDS 23287 + 0703. Duplication case 2b, A catalogued double star is included in another system.

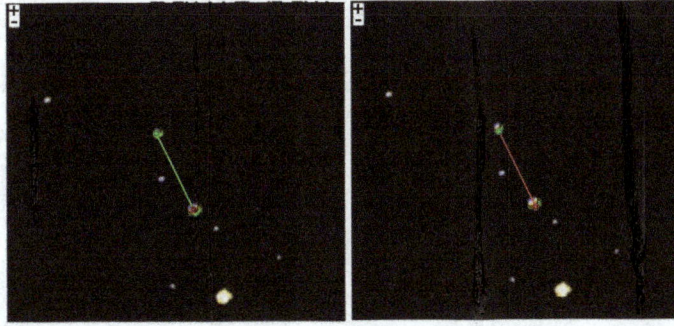

Figure 4. Duplication case 3. Two catalogued double stars are in fact the same pair of objects.

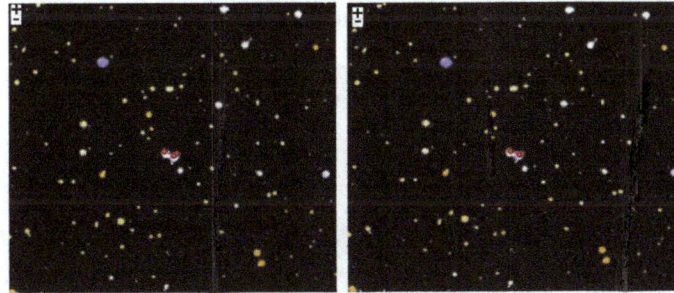

Figure 5. WDS 01450 + 6032 and WDS 01450 + 6038. Duplication case 3, two catalogued triple systems are the same.

3.2. CCDM

For CCDM we estimate d_2 to be about 11 arcsec, and in 19 cases components are included in two different binary/multiple systems in the catalogue (**Figure 6** and **Figure 7**).

3.3. TDSC

Owing to its homogeneity, we did not find duplications in TDSC, at least at the level of $d_2 \approx 50$ arcsec. The only exception is TDSC 29583 A = TDSC 29584 A.

4. Discussion

4.1. Data Duplication

It is instructive to compare distributions of duplication cases found in Sections 3.1 and 3.2 on component separation ρ. They are shown in **Figure 8** and **Figure 9**, respectively. One can see that the maximum of the WDS distribution is shifted to the smaller ρ in comparison with thr CCDM distribution, as WDS includes more modern and more precise observations and, consequently, can resolve closer pairs. Note also the difference of Y-axis scale in **Figure 8** and **Figure 9**.

4.2. Other Errors in the Catalogues

Sometimes catalogued positional angle θ can be ambiguous by 180 degrees. This is not always observer"s error. In some cases magnitude difference is too small for the observer to tell. It may also depend on the observing technique used- if the stars have different colors, the brighter star in B may not be the brighter star in R. Some of such cases can be indicated in WDS. As it was mentioned above, WDS provides two sets of (θ, ρ) values, for the first and the last observations. So, if the difference between θ values, obtained at the first and at the last observations, is close to 180 degree, it looks questionable. One can select pairs, where difference between 180 degree and $\left| \theta_{last} - \theta_{first} \right|$ does not exceed some d_3. To avoid pairs with high orbital motion, one

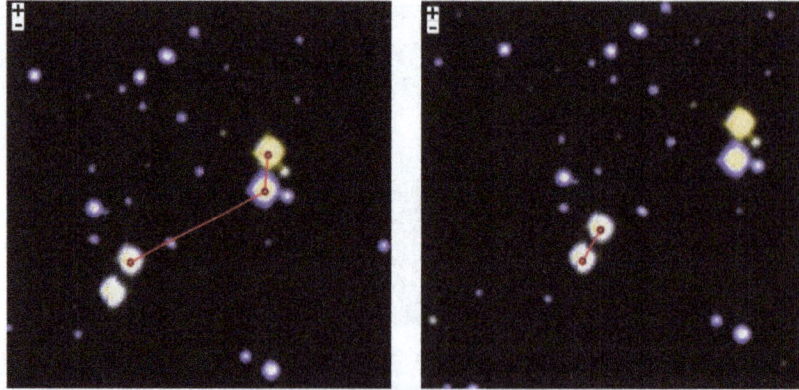

Figure 6. CCDM 00188 + 5924C and CCDM 00189 + 5923A, Duplication case 1. A star is included in two different double/multiple systems

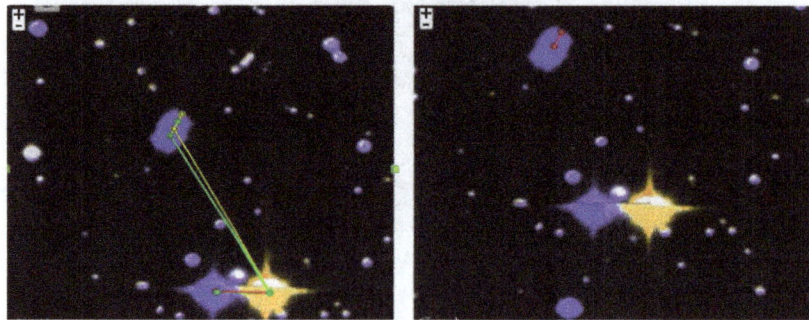

Figure 7. CCDM 01460 + 7142C and CCDM 01461 + 7143A, Duplication case 2. A catalogued double star is included in another system.

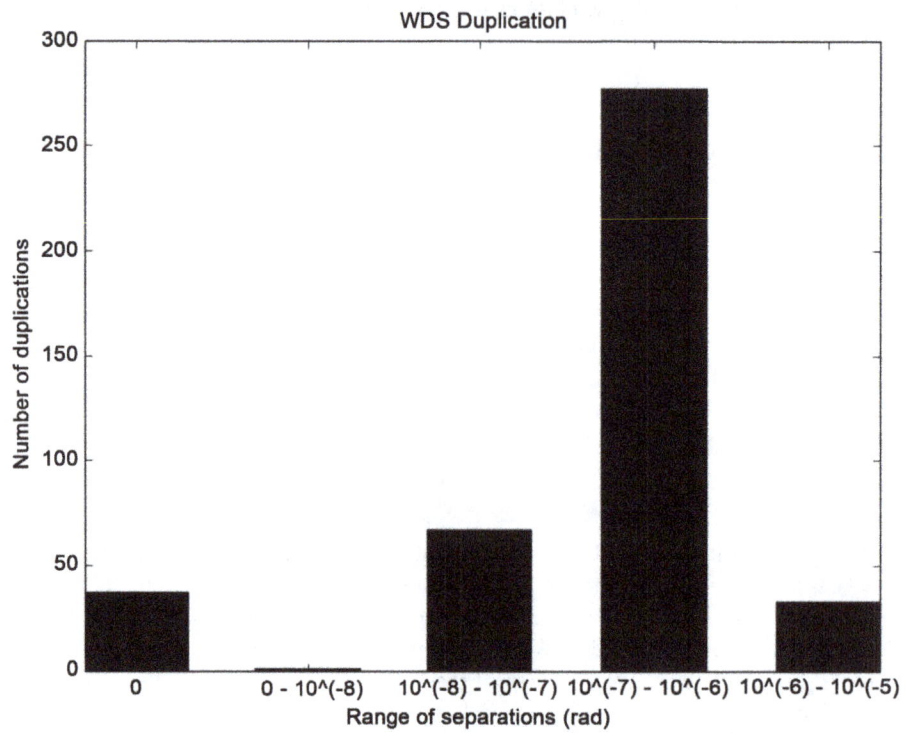

Figure 8. Distribution of duplications in WDS on the separation.

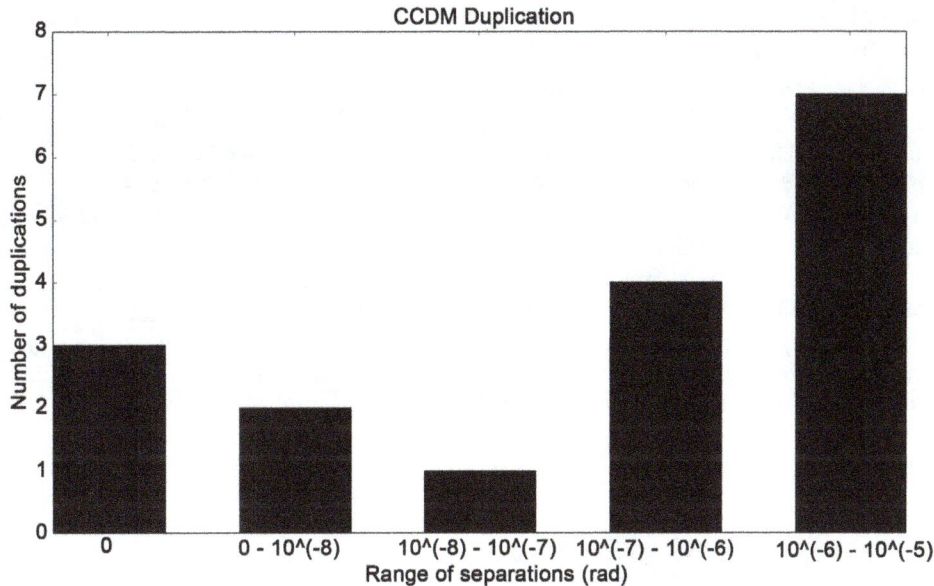

Figure 9. Distribution of duplications in CCDM on the separation.

can consider only pairs with $\rho > 10$ arcsec. If one takes $d_3 = 15$ degree, the resulting list contains 72 entries. Ten of them are happened to be pairs with large proper motions differences (and consequently with correct catalogued data), but other pairs contain erroneous data. To make the estimations more reliable, observation epoch difference can be taken into account.

All errors found in WDS were reported to the authors of this constantly updated catalogue.

TDSC contains cross-identification with WDS, but we have found that TDSC is apparently incorrect in identifying some objects with WDS objects, and, consequently, at least parts of the WDS names listed in TDSC are not correct. So we urge to be wary of WDS identifiers in TDSC. We note that TDSC contains data on 64894 systems, despite the fact that the latter number is 65210.

It should be noted also that our analysis has revealed about 70 errors (misprints) in CCDM, and the list of the errors is submitted to VizieR.

5. Conclusions

In this paper we developed methods of detection of errors in catalogues of double stars and presented results of their application to three principal catalogues of visual double and multiple stars, WDS, CCDM and TDSC.

Reports on the errors, detected in the studied catalogues, were sent to authors of the catalogues or to VizieR database. In some cases we could indicate a reason for the detected errors and suggest a way for correction. All lists of detected errors are available upon request.

Acknowledgements

We are grateful to William Hartkopf for helpful discussion and suggestions. We acknowledge the help of Negessa Tilahun, Seblu Humne, Tesfaye Dagne and Zeleke Amado. We thank Entoto Observatory and Research Center Astronomy and Astrophysics Research Division for giving such an opportunity and research facilities, and Mekele University for partly supported the research. This work was also partly supported by the Russian Foundation for Basic Researches (grants 15-02-04053 and 16-07-01162), by the Program of fundamental researches of the Presidium of RAS (P-41) and by the Program of support of leading scientific schools of RF (3620.2014.2). The use of the SIMBAD and WDS databases are acknowledged.

References

[1]   Heggie, D.C. (1975) Binary Evolution in Stellar Dynamics. *MNRAS*, **173**, 729-787.
       http://dx.doi.org/10.1093/mnras/173.3.729

[2]   Luyten, W.J. (1979) New Luyten Catalog of Stars with Proper Motions Larger than Two Tenths of an Arcsecond. Minneapolis: Univ. of Minnesota Press.

[3]   Chanamé, J. and Gould, A. (2004) Disk and Halo Wide Binaries from the Revised Luyten Catalog: Probes of Star Formation and MACHO Dark Matter. *The Astrophysical Journal*, **601**, 289. http://dx.doi.org/10.1086/380442

[4]   Salim, S. and Gould, A. (2003) Improved Astrometry and Photometry for the Luyten Catalog. II. Faint Stars and the Revised Catalog. *The Astrophysical Journal*, **582**, 1011. http://dx.doi.org/10.1086/344822

[5]   Andersen, J. (1991) Accurate Masses and Radii of Normal Stars. *The Astronomy and Astrophysics Review*, **3**, 91-126. http://dx.doi.org/10.1007/BF00873538

[6]   Yoo, J., Chanamé, J. and Gould, A. (2004) The End of the MACHO Era: Limits on Halo Dark Matter from Stellar Halo Wide Binaries. *The Astrophysical Journal*, **601**, 311. http://dx.doi.org/10.1086/380562

[7]   Allen, C., Poveda, A. and Hernańdez-Alcańtara, A. (2007) Halo Wide Binaries and Moving Clusters as Probes of the Dynamical and Merger History of our Galaxy. In: Hartkopf, B., Guinan, E. and Harmanec, P., Eds., *Binary Stars as Critical Tools and Tests in Contemporary Astrophysics*, *Proceedings of IAU Symposium No.* 240, Cambridge Univ. Press, Cambridge, 405.

[8]   Alcock, C., Allsman, R.A., Alves, D.R., Axelrod, T.S., Becker, A.C., Bennett, D.P., *et al.* (2001) MACHO Project Limits on Black Hole Dark Matter in the 1 -30 *M* Range. *The Astrophysical Journal Letters*, **550**, L169. http://dx.doi.org/10.1086/319636

[9]   Mason, B.D., Wycoff, G.L., Hartkopf, W.I., Douglass, G.G. and Worley, C.E. (2014) VizieR On-line Data Catalog: B/wds.

[10]   Dommanget, J. and Nys, O. (2002) VizieR On-line Data Catalog: I/274.

[11]   Fabricius, C., Høg, E., Makarov, V., Mason, B., Wycoff, G. and Urban, S. (2002) The Tycho Double Star Catalogue. *Astronomy & Astrophysics*, **384**, 180-189. http://dx.doi.org/10.1051/0004-6361:20011822

[12]   Kovaleva, D.A., Kaygorodov, P.V., Malkov, O.Yu., Debray, B. and Oblak, E. (2015) Binary Star DataBase BDB Development: Structure, Algorithms, and VO Standards Implementation. *Astronomy & Computing*, **11**, 119-125. http://dx.doi.org/10.1016/j.ascom.2015.02.007

13

Kinetic Electron-Ion Two Streams Instability in Space Dusty Plasma with Temperature Gradient

Sherif Mohamed Khalil[1,2]*, Weeam Saleh Albaltan[1]

[1]Physics Department, Faculty of Science, Princess Nora Bent Abdurrahman University, Riyadh, Saudi Arabia
[2]On Leave from Plasma Physics & Nuclear Fusion Department, N.R.C., Atomic Energy Authority, Cairo, Egypt
Email: *smkhalil@pnu.edu.sa

Abstract

The excitation, growing and damping of current instability is an important and vital subject for a lot of studies through its importance in communication for instance and in understanding the nature of space and the interpretation of many phenomena in space and astrophysics. Recent analytical and numerical works are presented to describe and investigate the excitation and growing of kinetic electron-ion two streams instability in anisotropic inhomogeneous dusty space plasmas. We elucidated the thermal effects of plasma species on the characteristics of such instability. It is found that the gradient of space plasma temperature, $\omega_{T\alpha} = \dfrac{k_{\perp} V_{T\alpha}^2}{\omega_{c\alpha}} \dfrac{\partial}{\partial x} \ln T_{\alpha}$, is a cause of interesting physical phenomena. Besides, different parameters, such as electron to ion temperature ratio $\dfrac{T_e}{T_i}$, magnetized plasma and dust grains, are also found to play a crucial role in the growth and depression of such instability.

Keywords

Electron-Ion Two Streams Instability, Anisotropic Space Plasma, Dusty Space Plasma

1. Introduction

The Buneman instability [1]-[7] is one of the current-driven plasma instabilities. The Buneman instability

---

*Corresponding author.

represents the stimulated Cherenkov radiation of the low-frequency. The Buneman (electron-ion two streams) instability takes place in a current-driven system where electron and ion beams drift at different velocities. Strong electron streaming leads to the growing of Buneman instability. This growing mode leads to strong localized electric fields. The resulting electron scattering produces strong enhanced resistivity and electron heating [8].

From the experimental point of view, a rapidly growing of the ultra-high-frequency (UHF) (>250 MHz) wave burst, identified as the Buneman two-stream instability, was detected at the beginning of a high-current plasma discharge [9]. The Buneman instability operates in both magnetized and unmagnetized plasmas near the electron plasma frequency (in the electron reference frame) and is excited when the electron drift velocity exceeds the electron thermal velocity [10]-[12].

For plasma turbulent heating, plasma heating in discharges and closed magnetic traps, acceleration of plasmoids, and for producing self-induced electric fields in plasmas, strong electric fields are used. This causes currents $u = u_e - u_i$ due to the relative motion of electrons and ions. If u is comparable to the electron thermal velocity V_{Te} ($V_{Te} = \sqrt{T_e/m_e}$), the current Buneman's instability arises.

This type of instability is excited kinetically [13] [14], however, it is excited hydrodynamically when u considerably exceeds a certain threshold value $\left(u > u_{cr} \approx V_{Te}\right)$. In this case, the development of the instability involves a displacement of the plasma regions and results in the variation of the spatial configuration of the plasma. On the other hand, near the threshold when $\Delta u = u - u_{cr}$, or when waves receive energy-due to Cherenkov resonance—from a small group of resonance particles, this instability becomes kinetic. In this case, the phase velocities of the waves are less by an order of magnitude than the thermal velocities of the particles.

As application, excluding the sheath regions, experiments in the magneto-plasmadynamic MPD thruster have shown that the electron drift velocity is only a small fraction of the electron thermal velocity, effectively stabilizing the Buneman instability. Streaming between reflected protons and upstream electrons gives rise to a strong Buneman instability [15] [16]. It is likely to attribute the Buneman instability and the modified two stream instability to play important roles in the production of high energy electrons at SNRs (supernova remnants). In astrophysical plasmas, the magnetic-reconnection Buneman instability may be excited [17].

Generally speaking, waves and instabilities occupy a significant part of modern plasma physics research because most properties of plasmas are related to the fundamental wave modes in laboratory and space plasma systems. In recent years, numerous studies showed interest in investigating dusty plasmas having electrons, ions, and charged dust grains of spherical shapes [18]-[23].

When dust grains are immersed in plasma, they become negatively charged because impinging electrons move faster than impinging ions. The negatively-charged dust grains can be considered as a third plasma species, so the plasma consists of electrons (negative), ions (positive), and dust grains (negative) [24].

In the presence of particle streaming, Jeans instability and the Buneman instability can overlap [25]. Further studies examined the influence of dust size distribution on the Jeans-Buneman instability [26]. Besides, the Buneman-type streaming instability may be developed in a low-temperature collisionless plasma in the presence of highly-charged impurities or dust and an ion flow. The important feature of the instability is that it takes place for supersonic as well as for subsonic velocities of the flow, thus being able to develop even in the pre-sheath and plasma bulk regions of low-temperature discharges where ion speeds are below the sound velocity [27].

Since charged dust in laboratory plasmas are generally levitated by electric fields, ions which acquire drifts due to these fields can stream through the dust, leading to various kinds of streaming instabilities, among them the Buneman instability. Recently, an ion-dust streaming instability with frequency less than the dust-neutral collision frequency was investigated [28]. The instability may have application to observations of waves in certain laboratory dc glow discharge dusty plasmas. Besides, dusty plasmas in the laboratory generally have finite spatial extent; therefore boundary effects may alter the properties of such ion-dust streaming instabilities. Recently, ion-dust streaming instability in plasma containing dust grains with large thermal speeds was considered using kinetic theory [29].

Buneman-type instabilities in a dusty plasma have been investigated before by many authors, e.g., ion-dust streaming instability in processing plasmas including collision effects [30], Buneman-type streaming instability in a plasma with dust particulates without collisions [27], and ion-dust two-stream instability in a collision including a magnetic field and collisions [31].

We extend previous work [13] [14] [32]-[37] for space plasma, with the following new consideration:

1) different temperature regimes (a mechanism for instability growth control),

2) plasma temperature inhomogeneity,
3) space dusty plasmas.

2. Space Plasma Kinetic Dispersion Relation

Let us consider an inhomogeneous space plasma immersed in a static magnetic field $H_0 = e_z H_0$ with cold weakly magnetized dusts. Plasma temperature inhomogeneity is perpendicular to H_0, i.e., directed along the x-axis.

In a dusty plasma, the quasi-neutrality condition should be adopted to include dust grains as:

$$\sum_{\alpha}^{e,i,d} e_\alpha n_{\alpha 0} = 0$$

For low β (the ratio of thermal to magnetic pressure), the kinetic dispersion relation governing the system is:

$$\varepsilon(k,\omega) = 1 + \sum_{\alpha}^{e,i,d} \chi_\alpha = 0 \tag{1}$$

$$\chi_\alpha = \frac{\omega_{p\alpha}^2}{k_\parallel^2 V_{T\alpha}^2} \left[1 + i\sqrt{\pi} \hat{l}_\alpha Z_\alpha W(Z_\alpha) A(\mu_\alpha) \right]$$

$k_\parallel$ is the wave number parallel to H_0, χ_α are the plasma susceptibilities, α is the type of particles, i.e., $\alpha = e,i,d$, for electrons, ions and dusts, respectively. The operator $\hat{l}_\alpha$ represents the effect of plasma inhomogeneity in density and temperature,

$$\hat{l}_\alpha = 1 - \frac{k_\perp V_{T\alpha}^2}{(\omega - ku_\alpha)\omega_{c\alpha}} \left[\frac{\partial}{\partial x} \ln n_\alpha + \frac{\partial}{\partial x} \ln T_\alpha \right]$$

Here, we are interested in the temperature inhomogeneity, i.e., $\partial \ln n_\alpha / \partial x = 0$. $A(\mu_\alpha)$ represents the magnetic field effect, $A(\mu_\alpha) = I_0(\mu_\alpha) e^{-\mu_\alpha}$, $\mu_\alpha = k_\perp^2 \rho_\alpha^2$, $k_\perp$ the wave number perpendicular to H_0, $\rho_\alpha = V_{T\alpha}/\omega_{c\alpha}$ is the Larmor radius, $\omega_{c\alpha} = e_\alpha H_0 / (m_\alpha c)$ is the cyclotron frequency, $V_{T\alpha} = \sqrt{T_\alpha / m_\alpha}$ the thermal velocity, $I_0(\mu_\alpha)$ is the modified Bessel function of the 0th order, and $\omega_{p\alpha} = \sqrt{4\pi e^2 n_\alpha(x)/m_\alpha}$ is the Langmuir frequency. $W(Z_\alpha)$ is the known probability integral with amplitude:

$$Z_\alpha = \frac{\omega - k \cdot u_\alpha}{\sqrt{2} k_\parallel V_{T\alpha}}$$

If $k \cdot u_i \approx k \cdot u_d$ ($u_i \approx u_d$), and we set $\dot{\omega} = \omega - k \cdot u_d = \omega - k \cdot u_i$, then Z_α for the different plasma different species are:

$$Z_e = \frac{\dot{\omega} - k \cdot u}{\sqrt{2} k_\parallel V_{Te}}, \quad Z_{(i,d)} = \frac{\dot{\omega}}{\sqrt{2} k_\parallel V_{T(i,d)}}$$

where $k \cdot u = ku \cos\theta$, and θ is the angle between k and $u = u_e - u_i$. We consider dusty plasma consisting of fairly massive grains, i.e., $m_d / m_i \gg 1$.

3. Threshold of Instability

For the excitation of instability we set

$$\dot{\omega} = \omega_0 + \Delta\omega, \ |\Delta\omega| \ll \omega_0, \ \Delta\omega = \omega_k + i\gamma_k$$

$$u = u_{cr} + \Delta u, \ \cos\theta = 1 - \frac{\theta^2}{2}, \ |\theta| \ll 1$$

The values $\Delta\omega, \Delta u, \theta^2$ are small perturbations, used for the expansion of

$$Z_\alpha = Z_{0\alpha} + \Delta Z_\alpha, \ |\Delta Z_\alpha| \ll Z_{0\alpha}; \ \alpha = e, i, d$$

where ω_k and γ_k are the frequency and growth rate of the instability, respectively, and u_{cr} and ω_0 are the

threshold values of current velocity and frequency. For such perturbations, we have $\dfrac{\Delta u}{u_{cr}}, \dfrac{\Delta \omega}{\omega}, \dfrac{\Delta k}{k}, \theta \ll 1$.

Now, for weakly magnetized dusts, $\mu_d \ll 1$, and under the frequency ranges $|\boldsymbol{k} \cdot \boldsymbol{u}_i, \boldsymbol{k} \cdot \boldsymbol{u}_d| \ll \grave{\omega}$, $|\boldsymbol{k} \cdot \boldsymbol{u}| \ll \grave{\omega}$, and validity of the parameters

$$n_d \ll n_e \approx n_i, \frac{n_d}{n_{e,i}} \le 1; Z_{0i} \approx Z_{0d} \gg Z_{0e}; \ Z_{0i} \gg 1, Z_{0e} \ll 1,$$

we get

$$A(\mu_i) \cong \frac{1}{\sqrt{2\pi\mu_i}}, \ A(\mu_e) \cong 1 - \mu_e, \ A(\mu_d) \cong 1$$

$$\hat{l}_e \approx 1 - \frac{\omega_{Te}}{\grave{\omega}}, \hat{l}_i \approx 1 - \frac{\omega_{Ti}}{\grave{\omega}}, \hat{l}_d \approx 1 - \frac{\omega_{Td}}{\grave{\omega}}$$

where

$$\omega_{T\alpha} = \frac{k_\perp V_{T\alpha}^2 \kappa_{T\alpha}}{\omega_{c\alpha}}, \ \kappa_{T\alpha} = \frac{\partial}{\partial x} \ln T_\alpha;$$

$$V_{T\alpha}^2 = \frac{T_\alpha}{m_\alpha}$$

Regardless the temperature ratio T_d/T_i, for a dusty plasma, we have:

$$\frac{\rho_d}{\rho_i} = \sqrt{\frac{T_d}{T_i}} \sqrt{\frac{m_d}{m_i}} \gg 1$$

In linear regime, and under the specific conditions mentioned above (which agrees with experiment), the dispersion relation (1) reads:

$$\begin{aligned}
& 1 + \frac{T_e}{T_i} + k^2 r_{de}^2 + i\sqrt{\pi}\left(1 - \frac{\omega_{Te}}{\grave{\omega}}\right) Z_e W(Z_e)(1 - \mu_e) \\
& + \frac{T_e}{T_i} i\sqrt{\pi}\left(1 - \frac{\omega_{Ti}}{\grave{\omega}}\right) Z_i W(Z_i) \frac{1}{\sqrt{2\pi\mu_i}} \\
& + \frac{T_e}{T_d} \frac{n_d}{n_e}\left[1 + i\sqrt{\pi}\left(1 - \frac{\omega_{Td}}{\grave{\omega}}\right) Z_d W(Z_d)\right] = 0
\end{aligned} \qquad (2)$$

where $r_{de} = V_{Te}/\omega_{pe}$ is the electron Debye radius.

Assuming in (2), $\omega' \rightarrow \omega_0$ and $u \rightarrow u_{cr}$, we obtain the threshold velocity of instability as:

$$\begin{aligned}
u_{cr} = & \sqrt{\frac{\pi}{8}} \frac{V_{T_i}\omega_0}{\omega_{ci}} (1 - \mu_e) e^{z_{0i}^2} \left(\frac{V_{T_i}}{V_{T_e}}\right)^3 \frac{m_i}{m_e}\left(1 - \frac{\omega_{Te}}{\omega_0}\right)\left(1 - \frac{\omega_{Ti}}{\omega_0}\right)^{-1} \\
& \cdot \left[1 + \left(\frac{V_{T_e}}{V_{T_d}}\right)^3 \frac{m_e}{m_d} \frac{n_d}{n_e} \frac{1}{(1 - \mu_e)} e^{-z_{0d}^2}\left(1 - \frac{\omega_{Td}}{\omega_0}\right)\left(1 - \frac{\omega_{Te}}{\omega_0}\right)^{-1}\right]^2
\end{aligned} \qquad (3)$$

It is clear that this instability depend strongly on the following parameters: plasma dust, temperature inhomogeneity (ω_{T_e}, ω_{T_i} and ω_{T_d}). **Figure 1** shows clearly these different dependencies.

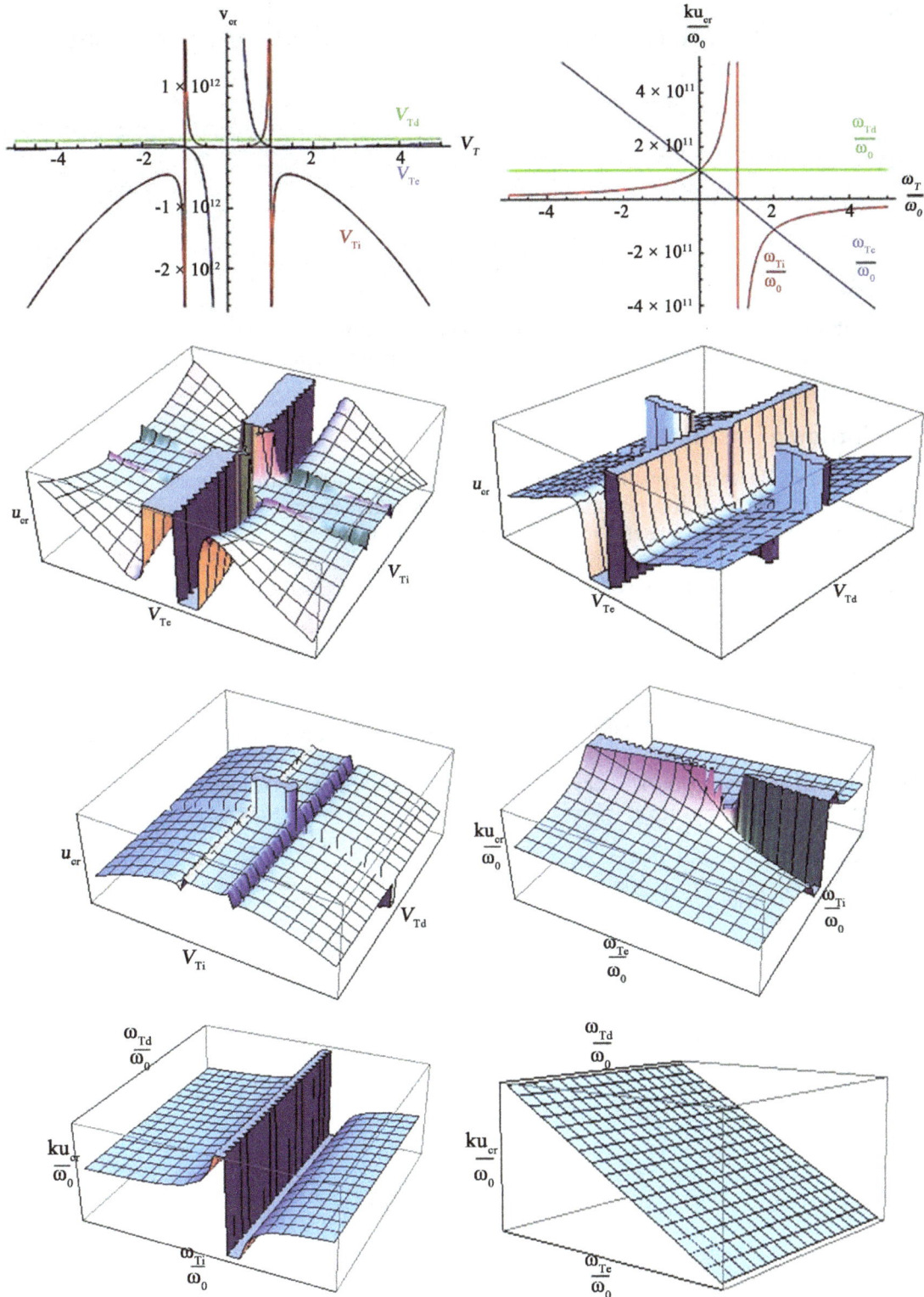

Figure 1. Critical velocity against ion and electron thermal velocities, and dusty particles; Critical current velocity against electrons and ions thermal velocities; Critical current velocity against electrons and dust thermal velocities; Critical current velocity against ions and dust thermal velocities, Critical current frequency against electrons and ions thermal frequencies due to inhomogeneity; Critical current frequency against ions and dust thermal frequencies due to inhomogeneity; Critical current frequency against electrons and dust thermal frequencies due to inhomogeneity.

Special cases

Table 1 declares the effect of temperature inhomogeneity on the threshold current velocity of the instability for different cases:

where

$$\aleph = \sqrt{\frac{\pi}{8}} \frac{V_{T_i} \omega_0}{\omega_{ci}} (1 - \mu_e) e^{z_{0i}^2} \left(\frac{V_{T_i}}{V_{T_e}}\right)^3 \frac{m_i}{m_e} \left(1 - \frac{\omega_{T_i}}{\omega_0}\right)^{-1}$$

$$\mathcal{M} = \left(\frac{V_{T_e}}{V_{T_d}}\right)^3 \frac{m_e}{m_d} \frac{n_d}{n_e} \frac{1}{(1 - \mu_e)} e^{-z_{0d}^2} \left(1 - \frac{\omega_{T_d}}{\omega_0}\right)$$

The table shows clearly the temperature inhomogeneity controlling, accordingly the initiation of current instability. For example, strong inhomogeneity in the dust temperature increases the u_{cr}, hence delaying appearance of instability.

4. Excitation of Instability

The excitation of Buneman instability put the plasma into a strongly turbulent state. Then a strong turbulent heating of the plasma leads to a rapid increase of the threshold current u_{cr} of the instability up to a value u, Therefore, it will be of great interest to investigate here this instability at current velocities close to the threshold values.

Accordingly, let us consider that the current velocity slightly exceeds the instability threshold, *i.e.,*

$$u = u_{cr} + \Delta u,$$

where $\frac{\Delta u}{u_{cr}} \ll 1$.

Now, for the excitation of instability $\gamma_k, \omega_k \neq 0$. Accordingly, the dispersion (1) takes the general form:

Table 1. Effect of temperature inhomogeneity on the threshold current velocity of the instability.

	Case	$|u_{cr}| =$
1	$\dfrac{\omega_{T_e}}{\omega_0} \ll 1$	$\aleph(1 + \mathcal{M})^2$
2	$\dfrac{\omega_{T_e}}{\omega_0} \gg 1$	$\aleph\left(\dfrac{\omega_{T_e}}{\omega_0}\right)\left(1 + \left(\dfrac{\omega_0}{\omega_{T_e}}\right)\mathcal{M}\right)^2$
3	$\dfrac{\omega_{T_i}}{\omega_0} \ll 1$	$\aleph\left(1 - \dfrac{\omega_{T_i}}{\omega_0}\right)^2\left(1 + \left(1 - \dfrac{\omega_{T_e}}{\omega_0}\right)^{-1}\mathcal{M}\right)^2$
4	$\dfrac{\omega_{T_i}}{\omega_0} \gg 1$	$\aleph\left(1 - \dfrac{\omega_{T_i}}{\omega_0}\right)^2\left(\dfrac{\omega_0}{\omega_{T_i}}\right)\left(1 + \left(1 - \dfrac{\omega_{T_e}}{\omega_0}\right)^{-1}\mathcal{M}\right)^2$
5	$\dfrac{\omega_{T_d}}{\omega_0} \ll 1$	$\aleph\left(1 - \dfrac{\omega_{T_e}}{\omega_0}\right)\left(1 + \left(1 - \dfrac{\omega_{T_e}}{\omega_0}\right)^{-1}\left(1 - \dfrac{\omega_{T_d}}{\omega_0}\right)^{-1}\mathcal{M}\right)^2$
6	$\dfrac{\omega_{T_d}}{\omega_0} \gg 1$	$\aleph\left(1 - \dfrac{\omega_{T_e}}{\omega_0}\right)\left(1 - \left(1 - \dfrac{\omega_{T_e}}{\omega_0}\right)^{-1}\left(1 - \dfrac{\omega_{T_d}}{\omega_0}\right)^{-1}\mathcal{M}\right)^2$

$$\sqrt{\pi}\left(1-\mu_e\right)\left(1-\frac{\omega_{Te}}{\omega_0}\right)\mathcal{R}_{0e}\frac{\omega'+\mathcal{W}}{\sqrt{2}k_\parallel v_{T_e}}+\frac{T_e}{T_i}\frac{1}{\sqrt{2\mu_i}}\left(1-\frac{\omega_{Ti}}{\omega_0}\right)\mathcal{R}_{0i}\frac{\omega'}{\sqrt{2}k_\parallel v_{T_i}}$$

$$+\frac{T_e}{T_d}\frac{n_d}{n_e}\sqrt{\pi}\left(1-\frac{\omega_{Td}}{\omega_0}\right)\mathcal{R}_{0d}\frac{\omega'}{\sqrt{2}k_\parallel v_{T_d}}+i\sqrt{\pi}\left(1-\mu_e\right)\frac{\omega_{Te}\omega'}{\omega_0^2}Z_{0e}e^{-z_{0e}^2} \tag{4}$$

$$-\left(1-\mu_e\right)\frac{\omega_{Te}\omega'}{\omega_0^2}+\frac{T_e}{T_i}i\frac{1}{\sqrt{2\mu_i}}\frac{\omega_{Ti}\omega'}{\omega_0^2}Z_{0i}e^{-z_{0i}^2}-\frac{T_e}{T_i}\frac{1}{\sqrt{2\pi\mu_i}}\frac{\omega_{Ti}\omega'}{\omega_0^2}+\frac{T_e}{T_d}\frac{n_d}{n_e}\frac{\omega_{Td}\omega'}{\omega_0^2}\left[i\sqrt{\pi}Z_{0d}e^{-z_{0d}^2}-1\right]=0$$

where $\mathcal{W}=\left[\dfrac{\Delta u}{u_{cr}}-\dfrac{\theta^2}{2}\right]$; $\omega'=\omega_k+i\gamma_k$ and $\mathcal{R}_{0\alpha}=i\left(1-2Z_{0\alpha}^2\right)e^{-z_{0\alpha}^2}-\dfrac{1}{\sqrt{\pi}Z_{0\alpha}}$

The dispersion relation (4) yields the growth rate and the frequency (high frequency) of Buneman instability in inhomogeneous dusty space plasma for different temperature regimes as follows:

$$\gamma_k=\left\{\mathbb{O}_1-\sqrt{\pi}Z_{0e}\mathbb{O}_2+\frac{n_d}{n_e}\left[\mathbb{N}_1+\mathbb{N}_2\right]\right\}\mathbb{R} \tag{5}$$

$$\omega_k=\left\{\mathbb{O}_2+\sqrt{\pi}Z_{0e}\mathbb{O}_1+\frac{n_d}{n_e}\left[\mathbb{N}_1+\mathbb{N}_2\right]\right\}\mathbb{R} \tag{6}$$

where,

$$\mathbb{R}=\left\{\frac{\dfrac{T_i}{T_e}\left(1-\mu_e\right)\sqrt{2\mu_i}\left(\dfrac{\omega_{Te}}{\omega_0}-1\right)\dfrac{1}{Z_{0e}}\omega_0\dfrac{ku_{cr}\mathcal{W}}{\sqrt{2}k_\parallel v_{T_e}}}{\mathbb{O}_1^2+\mathbb{O}_2^2+\dfrac{n_d}{n_e}\left[\dfrac{n_d}{n_e}\left(\mathbb{N}_1^2+\mathbb{N}_2^2\right)+2\left(\mathbb{O}_1\mathbb{N}_1+\mathbb{O}_2\mathbb{N}_2\right)\right]}\right\}$$

$$\mathbb{O}_1=Z_{0i}e^{-z_{0i}^2}\left[2Z_{0i}^2\left(1-\frac{\omega_{Ti}}{\omega_0}\right)+\frac{\omega_{Te}}{\omega_0}\left(1-\frac{\omega_{Ti}}{\omega_0}\right)\left(1-\frac{\omega_{Te}}{\omega_0}\right)^{-1}\right]$$

$$\mathbb{O}_2=-\frac{1}{\sqrt{\pi}}\left[1-\left(1-\frac{\omega_{Ti}}{\omega_0}\right)\left(1-\frac{\omega_{Te}}{\omega_0}\right)^{-1}\frac{\omega_{Te}}{\omega_0}\frac{Z_{0i}}{Z_{0e}}e^{-z_{0i}^2}\right]$$

$$\mathbb{N}_1=\frac{T_i}{T_d}\sqrt{2\pi\mu_i}Z_{0d}e^{-z_{0d}^2}\left\{\left(1-\frac{\omega_{Td}}{\omega_0}\right)\left[\frac{\omega_{Te}}{\omega_0}\left(1-\frac{\omega_{Te}}{\omega_0}\right)^{-1}+2Z_{0i}^2\right]-\frac{\omega_{Td}}{\omega_0}\right\}$$

$$\mathbb{N}_2=\frac{T_i}{T_d}\sqrt{2\mu_i}\left\{\left(1-\frac{\omega_{Td}}{\omega_0}\right)\left[\frac{\omega_{Te}}{\omega_0}\frac{Z_{0d}}{Z_{0e}}e^{-z_{0d}^2}\left(1-\frac{\omega_{Te}}{\omega_0}\right)^{-1}-1\right]-\frac{\omega_{Td}}{\omega_0}\right\}$$

The growth rate γ_k against the temperature gradients ω_{Td}, ω_{Te} and ω_{Ti} are shown clearly in **Figure 2**.

The growth rate and frequency of the instability for special cases of inhomogeneous magnetized dusty space plasma are investigated, *i.e.*, for:

1) Weak and strong electrons temperature inhomogeneity ($\dfrac{\omega_{Te}}{\omega_0}\ll1$; $\dfrac{\omega_{Te}}{\omega_0}\gg1$)

2) Weak and strong ions temperature inhomogeneity ($\dfrac{\omega_{Ti}}{\omega_0}\ll1$; $\dfrac{\omega_{Ti}}{\omega_0}\gg1$)

3) Weak and strong ions temperature inhomogeneity ($\dfrac{\omega_{Td}}{\omega_0}\ll1$; $\dfrac{\omega_{Td}}{\omega_0}\gg1$)

We obtained the following relations:

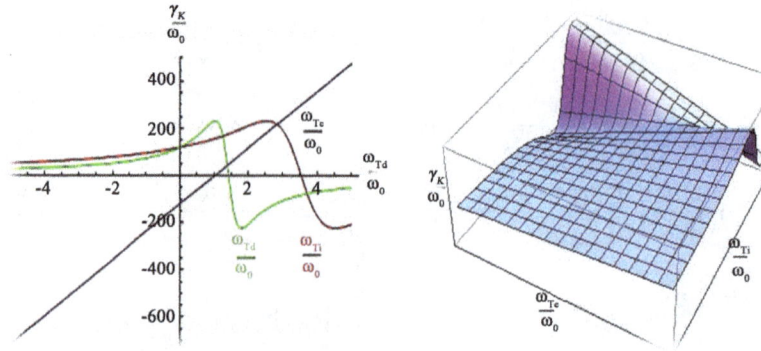

Figure 2. Growth rate γ_k against temperature inhomogeneity ω_{Td}, ω_{Te} and ω_{Ti}.

$$\frac{\gamma_k\left(\dfrac{\omega_{Te}}{\omega_0}\ll 1\right)}{\gamma_k\left(\dfrac{\omega_{Te}}{\omega_0}\gg 1\right)} < 1, \quad \frac{\gamma_k\left(\dfrac{\omega_{Ti}}{\omega_0}\ll 1\right)}{\gamma_k\left(\dfrac{\omega_{Ti}}{\omega_0}\gg 1\right)} > 1 \text{ and } \frac{\gamma_k\left(\dfrac{\omega_{Td}}{\omega_0}\ll 1\right)}{\gamma_k\left(\dfrac{\omega_{Td}}{\omega_0}\gg 1\right)} > 1 \tag{7}$$

5. Discussion and Conclusions

In this paper, analytical and numerical works are presented to describe and investigate the excitation and growing of kinetic electron-ion two streams instability in the anisotropic inhomogeneous dusty space plasmas. We elucidated the thermal effects of plasma species on the characteristics of such instability.

It is found that the gradient of space plasma temperature, $\omega_{Ta} = \dfrac{k_\perp V_{Ta}^2}{\omega_{ca}}\dfrac{\partial}{\partial x}\ln T_\alpha$, is a cause of interesting

physical phenomena. Besides, different parameters, such as electron to ion temperature ratio $\dfrac{T_e}{T_i}$, magnetized

plasma and dust grains, are found to play a crucial role in the growth and depression of such instability.

Thermal velocities of electrons and ions affect strongly the lagging and leading of instability according to the

relations, $\left(\dfrac{V_{Ti}}{u_{cr}}\ll 1\right)$; $\left(\dfrac{V_{Te}}{u_{cr}}\gg 1\right)$, respectively.

This result is also confirmed by the following relations

$$\frac{u_{cr}\left(\dfrac{\omega_{Te}}{\omega_0}\ll 1\right)}{u_{cr}\left(\dfrac{\omega_{Te}}{\omega_0}\gg 1\right)} > 1 \, ; \quad \frac{u_{cr}\left(\dfrac{\omega_{Ti}}{\omega_0}\ll 1\right)}{u_{cr}\left(\dfrac{\omega_{Ti}}{\omega_0}\gg 1\right)} > 1$$

which show that the frequencies due to temperature inhomogeneity ω_{Te}, ω_{Ti} influence the critical velocity u_{cr} of the instability.

ω_{Te}, ω_{Ti} and ω_{Td}, are found to play a crucial role in the growth rate of the instability (increasing or decreasing) according to the relations:

$$\frac{\gamma_k\left(\dfrac{\omega_{Te}}{\omega_0}\ll 1\right)}{\gamma_k\left(\dfrac{\omega_{Te}}{\omega_0}\gg 1\right)} < 1; \quad \frac{\gamma_k\left(\dfrac{\omega_{Ti}}{\omega_0}\ll 1\right)}{\gamma_k\left(\dfrac{\omega_{Ti}}{\omega_0}\gg 1\right)} > 1; \quad \frac{\gamma_k\left(\dfrac{\omega_{Td}}{\omega_0}\ll 1\right)}{\gamma_k\left(\dfrac{\omega_{Td}}{\omega_0}\gg 1\right)} > 1 \tag{8}$$

This shows that temperature inhomogeneity is a mechanism for growing or damping of the instability.

Another mechanism is the magnetic field. As in **Figure 3** the relations between the cyclotron frequencies ω_{ce},

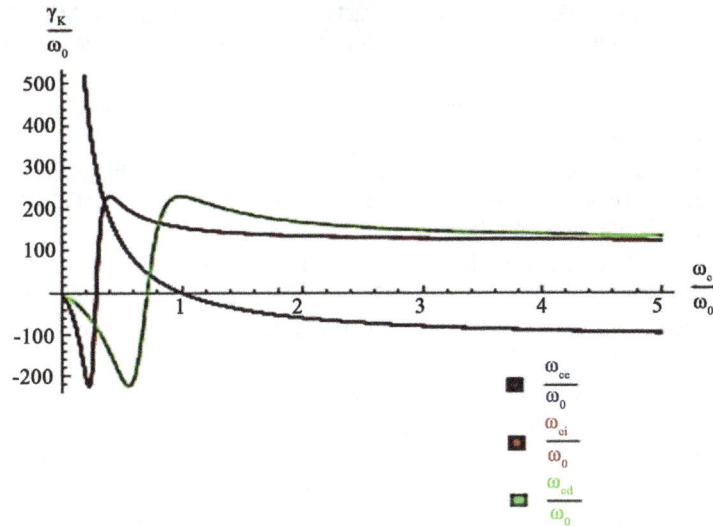

Figure 3. Growth rates γ_k against cyclotron frequencies ω_{ce}, ω_{ci}, ω_{cd}.

ω_{ci}, ω_{cd}, and growth rate γ_k are shown.

It is clear that magnetized electron is a good mechanism for depression and saturation of such instability. This mechanism is also applicable for extremely strong magnetized ions and dust.

Under investigation is the nonlinear and multi-ion species effect on the dispersion characteristics of such instability.

References

[1]   Buneman, O. (1958) Instability, Turbulence, and Conductivity in Current-Carrying Plasma. *Physics Review Letters*, **1**, 8. http://dx.doi.org/10.1103/PhysRevLett.1.8

[2]   Alexandrov, A.F., Bogdankevich, L.S. and Rukhadze, A.A. (1984) Principles of Plasma Electrodynamic. Springer-Verlag, Berlin. http://dx.doi.org/10.1007/978-3-642-69247-5

[3]   Raizer, Y.P. (1991) Gas Discharge Physics. Springer-Verlag, New York. http://dx.doi.org/10.1007/978-3-642-61247-3

[4]   Iizuka, S., Saeki, K., Sato, N. and Hatta, Y.J. (1985) Buneman Instability in a Bounded Electron Beam-Plasma System. *Journal of the Physical Society of Japan*, **54**, 146. http://dx.doi.org/10.1143/JPSJ.54.146

[5]   Shokri, B. and Vazifehshenas, T. (2001) Thermal Motion Effect on the Filamentation of a Strongly Collisional Current-Driven Plasma. *Physics of Plasmas*, **8**, 788. http://dx.doi.org/10.1063/1.1339231

[6]   Shokri, B., Khorashadi, S.M. and Dastmalchi, M. (2002) Ion-Acoustic Filamentation of a Current-Driven Plasma. *Physics of Plasmas*, **9**, 3355. http://dx.doi.org/10.1063/1.1490133

[7]   Shokri, B. and Niknam, A.R. (2003) Discharge Plasma Instabilities in the Presence of an External Constant Electric Field. *Physics of Plasmas*, **10**, 4153. http://dx.doi.org/10.1063/1.1609990

[8]   Drake, J. F., Shay, M.E., *et al.* (2001) Kinetic Treatment of Collisionless Reconnection. *Bull Am. Phys. Soc.*, **46**, 144.

[9]   Takeda, Y., Inuzuka, H. and Yamagiwa, K. (1995) Observations of Buneman Modes as Precursory Phenomena of a Solitary Potential Pulse. *Physics Review Letters*, **74**, 1998. http://dx.doi.org/10.1103/PhysRevLett.74.1998

[10]   Lemons, D.S. and Gary, S.P. (1978) Current-Driven Instabilities in a Laminar Perpendicular Shock. *Journal of Geophysical Research*, **831**, 1625. http://dx.doi.org/10.1029/JA083iA04p01625

[11]   Haihong Che, Goldman, M.V. and Newman, D.L. (2011) Buneman Instability in a Magnetized Current-Carrying Plasma with Velocity Shear. *Physics of Plasmas*, **18**, 5. http://dx.doi.org/10.1063/1.3590879

[12]   Bohata, M., Břeň, D. and Kulhánek, P. (2011) Generalized Buneman Dispersion Relation in the Longitudinally Dominated Magnetic Field. *ISRN Condensed Matter Physics*, **2011**, Article ID: 896321.

[13]   Hussein, A.M., Khalil, S.M. and Sizonenko, V.L. (1978) On the Theory of Turbulence in Weak Magneto-Active Plasma. *Plasma Physics*, **20**, 545-560. http://dx.doi.org/10.1088/0032-1028/20/6/005

[14]   El-Naggar, I.A., Khalil, S.M. and Sizonenko, V.L. (1978) On the Theory of Buneman's Instability in Hot Ion Plasma. *Journal of Plasma Physics*, **20**, 75-85. http://dx.doi.org/10.1017/S0022377800021371

[15] Drury, L. O'C., McClements, K.G., Chapman, S.C., Dendy, R.O., Dieckmann, M.E., Ljung, P. and Ynnerman, A. (2001) Computational Studies of Cosmic Ray Electron Injection. *Proceedings of the 27th International Cosmic Ray Conference*, Hamburg, 7-15 August 2001, 2096.

[16] Dieckmann, M.E., McClements, K.G., Chapman, S.C., *et al.* (2000) Electron Acceleration Due to High Frequency Instabilities at Supernova Remnant Shocks. *Astronomy and Astrophysics*, **356**, 377-388.

[17] Sakai, J.I., Sugiyama, D., Haruki, T., *et al.* (2001) Magnetic Field Energy Dissipation Due to Particle Trapping in a Force-Free Configuration of Collisionless Pair Plasmas. *Physical Review E*, **63**, Article ID: 046408. http://dx.doi.org/10.1103/PhysRevE.63.046408

[18] Salimullah, M., Shukla, P.K., Sandberg, I. and Morfill, G.E. (2003) Excitation of Dipole Oscillons in a Dusty Plasma Containing Elongated Dust Rods. *New Journal of Physics*, **5**, 40. http://dx.doi.org/10.1088/1367-2630/5/1/340

[19] Shukla, P.K. and Mamun, A.A. (2002) Introduction to Dusty Plasma Physics. Institute of Physics Publishing, Bristol. http://dx.doi.org/10.1887/075030653X

[20] Mendis, D.A. (2002) Progress in the Study of Dusty Plasmas. *Plasma Sources Science and Technology*, **11**, A219-A228. http://dx.doi.org/10.1088/0963-0252/11/3A/333

[21] Scales, W.A. (2004) Nonlinear Development of a Low-Frequency Hall Current Instability in a Dusty Plasma. *Physica Scripta*, **T107**, 107. http://dx.doi.org/10.1238/Physica.Topical.107a00107

[22] D'Angelo, N. and Merlino, R.L. (1996) Current-Driven Dust-Acoustic Instability in a Collisional Plasma. *Planetary and Space Science*, **44**, 1593-1598. http://dx.doi.org/10.1016/S0032-0633(96)00069-4

[23] Rosenberg, M. and Shukla, P.K. (1999) Instability of Ion Flows in Bounded Dusty Plasma Systems. *Physics of Plasmas*, **5**, 3786. http://dx.doi.org/10.1063/1.872743

[24] Bellan, P.M. (2004) A Model for the Condensation of a Dusty Plasma. *Physics of Plasmas*, **11**, 3368. http://dx.doi.org/10.1063/1.1740773

[25] Pandey, B.P. and Lakhina, G.S. (1998) Article title. *Pramana—Journal of Physics*, **50**, 191.

[26] Meuris, P., Verheest, F. and Lakhina, G.S. (1997) Influence of Dust Mass Distributions on Generalized Jeans-Buneman Instabilities in Dusty Plasmas. *Planetary and Space Science*, **45**, 449-454. http://dx.doi.org/10.1016/S0032-0633(96)00155-9

[27] Vladimirov, S.V. and Ishihara, O. (1999) Buneman-Type Streaming Instability in a Plasma with Dust Particulates. *Physica Scripta*, **60**, 370-372. http://dx.doi.org/10.1238/Physica.Regular.060a00370

[28] Rosenberg, M. (2002) A Note on Ion-Dust Streaming Instability in a Collisional Dusty Plasma. *Journal of Plasma Physics*, **67**, 235-242. http://dx.doi.org/10.1017/S0022377802001678

[29] Rosenberg, M., Thomas, E. and Merlino, R.L. (2008) A Note on Dust Wave Excitation in a Plasma with Warm Dust: Comparison with Experiment. *Physics of Plasmas*, **15**, Article ID: 073701. http://dx.doi.org/10.1063/1.2943218

[30] Rosenberg, M. (1996) Ion-Dust Streaming Instability in Processing Plasmas. *Journal of Vacuum Science & Technology A*, **14**, 631. http://dx.doi.org/10.1116/1.580157

[31] Mamun, A. A. and Shukla, P. K. (2000) Streaming Instabilities in a Collisional Dusty Plasma. *Physics of Plasmas*, **7**, 4412. http://dx.doi.org/10.1063/1.1315305

[32] Rosenberg, M. and Shukla, P.K. (2004) Ion-Dust Two-Stream Instability in a Collisional Magnetized Dusty Plasma. *Journal of Plasma Physics*, **70**, 317-322. http://dx.doi.org/10.1017/S0022377803002678

[33] Khalil, S.M. and Mohamed, B.F. (1989) Quasilinear Theory of Buneman's Instability in Hot Electron Plasma $T_e \gg T_i$. *Proceedings of the 16th European Conference on Controlled Fusion and Plasma Physics*, Venice, 13-17 March 1989, 1369.

[34] Khalil, S.M. and Mohamed, B.F. (1988) Electron-Ion Two-Stream Instability in Anisotropic Isothermal Plasma. *Journal de Physique*, **49**, 451-455. http://dx.doi.org/10.1051/jphys:01988004903045100

[35] Khalil, S.M. and Mohamed, B.F. (1991) Kinetic Buneman Instability in a Homogeneous Hot Electron Plasma. *Contributions to Plasma Physics*, **31**, 513-518. http://dx.doi.org/10.1002/ctpp.2150310506

[36] Khalil, S.M. and Sayed, Y.A. (1994) Kinetic Nonlinear Buneman's Instability in Field-Free Isothermal Collisonal Plasma. *Contributions to Plasma Physics*, **34**, 683-689. http://dx.doi.org/10.1002/ctpp.2150340508

[37] Khalil, S.M. and Al-Yousef, H. (2010) Kinetic Buneman Instability in Dusty Plasma of Different Temperature Regimes. *FIZIKA*, **A19**, 1.

14

If Quantum "Wave" of the Universe Then Quantum "Particle" of the Universe: A Resolution of the Dark Energy Question and the Black Hole Information Paradox

Mohamed S. El Naschie

Department of Physics, University of Alexandria, Alexandria, Egypt
Email: Chaossf@aol.com

Abstract

We start from a minimal number of generally accepted premises, in particular Hartle-Hawking quantum wave of the universe and von Neumann-Connes' pointless and self referential spacetime geometry. We then proceed from there to show, using Dvoretzky's theorem of measure concentration, that the total energy of the universe is divided into two parts, an ordinary energy very small part which we can measure while most of the energy is concentrated as the second part at the boundary of the holographic boundary which we cannot measure in a direct way. Finally the results are shown to imply a resolution of the black hole information paradox without violating the fundamental laws of physics. In this way the main thrust of the two opposing arguments and views, namely that of Hawking on the one side and Susskind as well as tHooft on the other side, is brought to a consistent and compatible coherent unit.

Keywords

Dvoretzky Theory, Wave-Particle Duality, Von Neumann Pointless and Self Referential Geometry, Cantorian Spacetime, Hartle-Hawking Quantum Wave of the Universe, Dark Energy, Black Hole Information Paradox, Connes Noncommutative Geometry

1. Introduction

Within the framework of Einstein's general relativity and the work of Hawking, Hartle, Penrose, Wheeler, Susskind and tHooft to mention only a few [1]-[20], the present work is mainly concerned with the problem of dark energy [8] [12] [15] and what turned out to be related to it, namely the famous information paradox of black holes [14]-[19].

We start from Hartle-Hawking quantum wave of the universe [1]-[5]. Using the wave-particle duality we reason that such a wave implies, by this well known duality, the existence of a corresponding quantum particle of the universe. Assuming the embedding spacetime of particle and wave to be akin to that of von Neumann-Connes' pointless geometrical-topological quasi manifold [9] [11], then it is a surprisingly simple task to calculate the entire energy density of the universe which is then shown by virtue of the celebrated theorem of Dvoretzky to split into two parts [12] [15] [16]. The first part is a measurable ordinary energy of about 4% to 4.5% of the total energy density and the second part is dark energy of about 95.5% to 96% of the total, which manifests its existence indirectly but cannot be measured directly [15] [16]. Noting the relation between energy and information via entropy we carry the aforementioned results over to the black hole information paradox and to our own great surprise resolves it without violating any fundamental laws of relativity or quantum theory [17]-[20].

2. If Quantum Wave Then a Colossal Quantum Particle as the Quantum Universe

We emphasize once more what we said earlier on, namely that there is a definite quantum wave-particle duality in quantum physics, but this by no means implies that there can be wave without particle or vice versa [1]-[7]. The duality says only that we can observe the one or the other and nothing more. This is in fact the most important result of our previous quantum set foundation of orthodox or historical quantum mechanics, namely that the quantum wave is simply the cobordism of the quantum particle. More accurately since the quantum pre-particle is the zero set of von Neumann-Connes' dimensional function [9] [15]

$$D = a + b\phi \tag{1}$$

where $a, b \in Z$ and $\phi = \left(\sqrt{5} - 1\right)/2$ then it follows that the empty set is the cobordism of the zero set [15]. In more intuitive mundane terminology, the particle is the zero set with $D \equiv (0, \phi)$ and its surface is the empty set with $D \equiv \left(-1, \phi^2\right)$. Here zero and -1 are Menger-Urysohn inductive topological dimensions and ϕ and ϕ^2 are the corresponding Hausdorff dimensions [21]. Now it is Hartle and Hawking who show that the Wheeler-de Witt equation [1]-[5] which is in essence a Schrödinger equation for the entire universe [1] can be solved to yield a wave function for the universe and consequently by the same fundamental duality, there must be a corresponding quantum particle also for the entire universe or we would not be here writing papers about dark energy and black hole information paradox [17]-[19]. That is why we gave this section the title we gave it and now we can proceed without further ado to the energy subdivisions [15].

3. Dvoretzky's Measure Concentration Theorem Applied to Cosmology of Dark Energy

In sufficiently high dimensional spaces, in particular a convex Banach space, Dvoretzky's theorem [12] states that the volume of a sphere will be concentrated at the surface [12]. More accurately the theorem implies that 96% of the volume would be at the surface while in the so called "false" bulk we will have only 4% [22] [23]. For our quantum particle of the universe surrounded by its quantum wave of the universe, the set up can be envisaged as embedded in an at least five dimensional spacetime akin to that of Klein-Kaluza unification theory [12]. Consequently from the 100 percent energy density we will have 96% concentrated on the surface, *i.e.* in the quantum wave of the universe where 4% are inside the "particle" which constitutes the quantum particle of the entire universe [1] [12]. This division and the corresponding energy densities agree in every respect with the essential measurements and observations of COBE, WMAP, Hubble and Planck [8] [12] [16]. In the next section we see how to make this estimate almost exact.

4. The Ordinary Energy of the Quantum Particle and the Dark Energy of the Quantum Wave

Let us reformulate our basic model of a zero set $(0, \phi)$ pre-particle [12] [15] [16] surrounded by its cobordism, *i.e.* its empty set $\left(-1, \phi^2\right)$ pre-wave surface, this time in a five dimensional space [15] [16]. The multiplicative volume of the zero set would be a "Hausdorff" information or a Hausdorff five dimensional volume equal [12] [15] [16]

$$V(O) = (\phi)(\phi)(\phi)(\phi)(\phi) = \phi^5 \tag{2}$$

while the corresponding additive Hausdorff volume of its dual empty set surface would be [12] [15] [16]

$$V(D) = \phi^2 + \phi^2 + \phi^2 + \phi^2 + \phi^2 = 5\phi^2 \tag{3}$$

From the above it follows that the corresponding energy density would therefore obviously be [12] [15] [16] [20]

$$E(O) = (1/2)(\phi^5)mc^2 \tag{4}$$

and

$$E(D) = (1/2)(5\phi^2)mc^2. \tag{5}$$

where $E(O)$ could be likened with the stationary potential energy while $E(D)$ could be likened with the kinetic energy of propagation in classical mechanics. Adding both energies together we found immediately the correct famous formula for the total energy of Einstein [15] [16] [20]

$$E = E(O) + E(D) = mc^2 \tag{6}$$

where $E(O)$ is exactly equal 4.508497197 percent of the total energy and agrees in an astounding degree with cosmic measurements and observations while $E(D)$ is equal to exactly $100 - E(O) = 95.4915028$ percent and equally in full agreement with measurements [8] [12] [16]. Needless to say this is also almost the same result obtained from applying the celebrated theorem of Dvoretzky [12] [23]. Now we are in a position to move on to considering the black hole information paradox [17]-[19] but we could mention on passing that equations 4 - 6 imply already a momentous conclusion, namely that dark energy is the global manifestation of the Casimir local effect [22] [23].

5. A Pointless and Self Referential Spacetime Resolution of the Black Hole Information Paradox

Let us examine the implicit assumptions in our previous exposition leading to the realization that dark energy is the energy of the quantum wave concentrated at the surface of the quantum particle universe. This was obtained using a model which de facto assumes that spacetime is a pointless and self referential fractal Cantorian ultimate L-like topological manifold where each point upon magnification reveals itself as an entire Cantor set, ergo an entire universe [12] [15] [16]. Thus our complete universe is a gigantic black hole in which we are peacefully living. Each point in this universe could be regarded as a quasi black hole seen from our end and there are of course "medium" sized black holes seen again from our view point as observer but self similarity or more accurately self affinity implies that these are again complete universes inside our universe seen again from our position as observer living in the gigantic black hole and so on ad infinitum [15] [21]. In short, the complexity of the situation uses the in-built self referentiality of Cantorian fractals [25]. That way infinity and zero are used and eliminated at the same time without getting entangled in computational or logical inconsistency. Finally invoking the well known relation between energy and information via entropy we see that the result obtained in our ordinary and dark energy investigation is applicable in duality to information. From our position the 96 percent information on the surface of the black hole could not be regarded in any way as lost, nor does it matter how small the black hole shrinks. Only the 4 percent information inside will not be directly accessible to us but it is not lost either. This is the perfect coincidentia oppositorium Hegelian resolution of the paradox while preserving a substantial part of the arguments of L. Susskind and Gerard tHooft [17]-[19] as well as some parts of Hawking's thesis [13] and most importantly, all that is obtained without violating the fundamental laws of physics thanks to the idea of a self affine pointless Cantorian fractal universe [17] [18] [24].

6. Conclusion

There is indeed a quantum particle universe corresponding to the Hartle-Hawking quantum wave of the universe.

Using the self referentiality [25] of our pointless Cantorian fractal spacetime and applying Dvoretzky's theorem to the compelling logic of this gigantic quantum particle universe, we can reason that $E = mc^2$ can be split into a quantum wave energy density $E(D) \cong mc^2 (21/22)$ concentrated at the boundary of the holographic boundary, that is to say the surface of the universe which we call dark energy and $E(O) \cong mc^2/22$ ordinary energy density of the core of the quantum particle universe which we can measure directly with relative ease, unlike $E(D)$ which we cannot measure easily in any direct way at least as far as the present time technology tells us. Thus by taking von Neumann-Connes' pointless quasi fractal-Cantorian spacetime seriously the preceding conclusion can be applied in dual form to the information paradox of black holes leading to a satisfactory resolution confirming that at a minimum approximately 96 percent of the black hole information will never be lost while 4 percent only will not be directly accessible to us, ergo the dual (in fact opposite) situation to dark energy. Consequently we may say that 96 percent of the information of a black hole is ordinary information like that of the electricity of a Faraday cage and the remaining 4 percent is dark information not directly detectable being essentially inside the black hole. In the above picture if it is right, and we are confident it is right, Hawking, Susskind, tHooft and the fundamental principles of theoretical physics can co-exist harmoniously together.

References

[1]   Hartle, J.B. and Hawking, S.W. (1983) Wave Function of the Universe. *Physics Review D*, **28**, 2960-2975. http://dx.doi.org/10.1103/PhysRevD.28.2960

[2]   Hawking, S. and Penrose, R. (1996) The Nature of Space and Time. Princeton University Press, Princeton, New Jersey.

[3]   Misner, C., Thorne, K. and Wheeler, J. (1973) Gravitation. Freeman, New York.

[4]   Hawking, S. (1993) Hawking on the Big Bang and Black Holes. World Scientific, Singapore.

[5]   Gibbons, G.W. and Hawking, S.W. (1993) Euclidean Quantum Gravity. World Scientific, Singapore. http://dx.doi.org/10.1142/1301

[6]   Hawking, S. and Israel, W. (1990) 300 Years of Gravitation. Cambridge University Press, Cambridge, UK.

[7]   Tayler, E.F. and Wheeler, J.A. (1966) Spacetime Physics. W.H. Freeman Company, New York, USA.

[8]   Amendola, L. and Tsujikawa, S. (2010) Dark Energy: Theory and Observation. Cambridge University Press, Cambridge, UK. http://dx.doi.org/10.1017/CBO9780511750823

[9]   Connes, A. (1994) Noncommutative Geometry. Academic Press, San Diego, USA.

[10] He, J.-H. (2014) A Tutorial Review on Fractal Spacetime and Fractional Calculus. *International Journal of Theoretical Physics*, **53**, 3698-3718. http://dx.doi.org/10.1007/s10773-014-2123-8

[11] El Naschie, M.S. (1998) Von Neumann Geometry and E-Infinity Quantum Spacetime. *Chaos, Solitons & Fractals*, **9**, 2023-2030.

[12] El Naschie, M.S. (2014) The Measure Concentration of Convex Geometry in a Quasi Banach Spacetime behind the Supposedly Missing Dark Energy of the Cosmos. *American Journal of Astronomy & Astrophysics*, **2**, 72-77. http://dx.doi.org/10.11648/j.ajaa.20140206.13

[13] Moskowitz, C. (2015) Stephen Hawking Hasn't Solved the Black Hole Paradox Just Yet. *Scientific American*, 27 August 2015.

[14] El Naschie, M.S. (2006) Fractal Black Holes and Information. *Chaos, Solitons & Fractals*, **29**, 23-35. http://dx.doi.org/10.1016/j.chaos.2005.11.079

[15] El Naschie, M.S. (2013) Topological-Geometrical and Physical Interpretation of the Dark Energy of the Cosmos as a "Halo" Energy of the Schrodinger Quantum Wave. *Journal of Modern Physics*, **4**, 591-596. http://dx.doi.org/10.4236/jmp.2013.45084

[16] El Naschie, M.S. (2013) What Is the Missing Dark Energy in a Nutshell and the Hawking-Hartle Quantum Wave Collapse. *International Journal of Astronomy and Astrophysics*, **3**, 205-211. http://dx.doi.org/10.4236/ijaa.2013.33024

[17] Susskind, L. and Lindesay, J. (2005) Black Holes, Information and the String Theory Revolution. World Scientific, Singapore.

[18] Susskind, L. (2008) The Black Hole War. Back Bay Books, New York.

[19] 't Hooft, G. (1985) On the Quantum Structure of a Black Hole. *Nuclear Physics B*, **256**, 727-745. http://dx.doi.org/10.1016/0550-3213(85)90418-3

[20] El Naschie, M.S. (2014) From $E = mc^2$ to $E = mc^2/22$—A Short Account of the Most Famous Equation in Physics and Its Hidden Quantum Entangled Origin. *Journal of Quantum Information Science*, **4**, 284-291. http://dx.doi.org/10.4236/jqis.2014.44023

[21] El Naschie, M.S. (2004) A Review of *E*-Infinity and the Mass Spectrum of High Energy Particle Physics. *Chaos, Solitons & Fractals*, **19**, 209-236. http://dx.doi.org/10.1016/S0960-0779(03)00278-9

[22] El Naschie, M.S. (2015) Banach Spacetime-Like Dvoretzky Volume Concentration as Cosmic Holographic Dark Energy. *International Journal of High Energy Physics*, **2**, 13-21. http://dx.doi.org/10.11648/j.ijhep.20150201.12

[23] Ball, K. (1997) An Elementary Introduction to Modern Convex Geometry. In: Levy, S., Ed., *Flavors of Geometry*, Cambridge University Press, Cambridge, 1-58.

[24] Baryshev, Y. and Teerikorpe, P. (2002) Discovery of Cosmic Fractals. World Scientific, Hackensack.

[25] Kauffman, L.H. (1987) Self-Reference and Recursive Form. *Journal of Social and Biological Structures*, **10**, 53-72. http://dx.doi.org/10.1016/0140-1750(87)90034-0

15

Calculations for Density of Quark Core Consisting of Mono Flavored Closely Packed Quarks inside Neutron Star

Jehangir A. Dar[1]*, Pawan Kumar Singh[2], Ram Swaroop[3]

[1]Plasma Waves and Particle Acceleration Laboratory, Department of Physics, Indian Institute of Technology Delhi, New Delhi, India
[2]Department of Physics, ARSD College, University of Delhi (South Campus), New Delhi, India
[3]North Bengal Science Centre, National Council of Science Museums, Ministry of Culture, Government of India, Siliguri, India
Email: *Jahangirahmad63@gmail.com

Abstract

The attempt has been taken to calculate the density of stars possessing quark matter core using sphere packing concept of crystallography. The quark matter has been taken as solid in nature as predicted in references 36 and 37, and due to immense gravitational pressure at the core of the star the densest packing of quarks as spheres has been assumed to calculate the packing fraction Φ, thus the density ρ of the matter. Three possible types of pickings—mono-sized sphere packing, binary sphere packing and ternary sphere packing, have been worked out using three possible types of quark matter. It has been concluded that no value about the ρ of quark matter can be calculated using binary and ternary packing conditions and for mono-sized packing condition different flavor quark matters of different values in the density have been calculated using results from the experiments done by HI, ZEUS, L3 and CDF Collaborations about the radius limit of quark. For example, for u quark matter ρ ranges from 4.0587×10^{48} - 7.40038×10^{48} MeV/c^2 cm^3 using results of L3 Collaboration, for s quark matter 15.91794×10^{48} - 17.6866×10^{48} MeV/c^2 cm^3, etc.

Keywords

Neutron Star, Packing Fraction, Quark Matter, Quark Gluon Matter, Kepler's Conjecture Theorem, Neutron

---

*Corresponding author.

1. Introduction

Neutron stars are one of the densest stars known in the cosmos. They are formed by the gravitational collapse of some massive stars having inner core mass above Chandrasekhar limit—1.4 M_o [1], but with mass lesser than the mass required to overcome neutron degeneracy pressure. The neutron degeneracy pressure is the pressure caused by degenerate neutron gas [2]. There is also a limit for neutron star above which the stars are said to form black hole [3] [4] which is a region of space time from which nothing can escape to infinity as predicted by General theory of relativity [5]. Up to 1967 the idea of neutron star was only theoretical but after the discovery of pulsar PSR B1919 + 21 by Joceyln Bell Burnell and Antony Hewish on November 28, 1967 [6], the idea was taken seriously. Pulsar is not more than a rotating neutron star that emits regular beams of electromagnetic radiation. Different scientists suggest different densities of the core of neutron star. The neutron star's density ranges from 1×10^{17} Kg/m^3 in the crust—exceeding with the depth—to above 6×10^{17} - 8×10^{17} g/m^3 that is denser than an atomic nucleus [7] [8]. We can say that the density of the core of neutron star is still mystery; it is still an unsolvable question in physics. In this paper to some extent this mystery has been tried to solve.

Moving towards its core composition of super dense matter, the composition of the matter is uncertain. According to different models it may be liquid or solid. One of the models says the core consists of super fluid-neutron—degenerate matter formed mostly by neutrons with some exceptions of protons and electrons [9]. One another say, it consists of strange-degenerate matter including strange quarks in addition to up and down quarks [10]. While, some theorize it consists of high energy pions and kaons in addition to neutrons [6] or ultra-dense quark-degenerate matter [11]. What actually exists inside core is not yet known absolutely we are having only some models based on whom density of core of neutron star is calculated. The actual density will remain uncertain as long as we are not certain about its composition. In fact, this is one of the prominent reasons why the picture about the limiting mass of neutron star remained unclear from more than seventy years even after the pioneering attempt taken by Oppenheimer and Volkoff [12].

Between the neutron stars and black holes it has been theorized that there remains one more type of neutron star commonly called as quark star. The mass of quark star falls between mass limit of neutron star and the minimum mass required for the star to become black hole. The whole idea of quark star is hypothetical and no strong confirmation has been observed about the presence of such stars, but the observations taken on stars like RX JI856.5-3754 and 3C58 suggest these stars to be quark stars due to available density above the density prescribed for neutron stars [13]. There are also some suggestions about the presence of quark star like the stars PSR B0943 + 10, SN2006gy, SN2005gj, SN2005ap and SN1987 [14]. Quark stars get formed when the mass of star is sufficient to create gravitational pressure above the degeneracy pressure caused by neutrons inside the star. Under such conditions neutrons can break into their constituent particles, that is, quarks create quark matter [15]. The matter will consist of up and down quarks—symbolized as u for up quark and d for down quark. However, the quark matter with some composition of strange s quarks, or of other different compositions is also possible.

It is theorized that neutron stars consisting core of ordinary quark matter, that is, u and d quark matter is stable, under extreme temperatures or pressures. However, quark stars consisting entirely of this ordinary quark matter are highly unstable and therefore dissolve spontaneously in another kind of quark matter commonly called as strange quark matter [13] [16]. Such type of stars possessing this strange quark matter is called as strange quark star [16]. Strange quark or s quark is the 3rd lightest quark in the quark family. It is mostly found in hadrons like kaons (k), sigma baryons (Σ), strange D mesons (D) and some other strange particles like the white dwarfs' and neutron stars. The quark star's gravitational collapse is controlled by quark degeneracy pressure since the quarks belong to fermions family like electrons and neutrons and thus obey Pauli's exclusion principle. In the paper we will use the fact what is possible when the gravitational pressure will exceed the Pauli's exclusion principle. The idea that they will turn into the densest packing system is used here in the paper.

Now let us move to another important aspect of the paper, that is, sphere packing in crystallography. Sphere packing is a possible regular or irregular arrangement of similar or dissimilar spheres in any volume. The spherical packing in 3-dimensional Euclidean space is very common. Basically there are two known arrangements of spheres known these are regular or lattice arrangement and irregular or random arrangement. FCC (face centered cubic) or CCP (cubic close packing) and HCP (hexagonal close packing) are the two unit structures known possible in the close packing of spherically symmetrical and mono sized particles. These two structures have the highest packing density as proved by Gauss in 1831 and originally proposed by Johannes Kepler in 1611 in his so called Kepler's conjecture theorem [16] [17]. The highest average atomic packing function (APF) Φ possible due to Keplers' Conjecture theorem is 0.74048048 [4] [18]. However, this value validates only to mono sized sphere packing. This conjecture was further proved by Thomas Callister Hales, following the suggested ideas of Laszlo Fejes Toth. In both the FCC and HCP packing each sphere has twelve neighbors. There is one gap surrounded by six spheres (octahedral) and two smaller gaps surrounded by four spheres (tetrahedral).

For close packing of two or more different sized spheres, to calculate the Φ becomes much difficult. Now the APF Φ becomes function of radius ratio between the spheres and sphere concentration over other sized concentrations, like for binary sphere packing the small sphere concentration in the mixture is calculated. Different methods like Monte Carlo [19]-[21], genetic algorithms [22] and TJ sphere packing algorithm [23], have been applied to attempt to find binary sphere packaging over certain ranges of functions α and χ [19]. There is no clear conjecture as to which structures are the densest at different radius ratios. Structures are known which exceed the close packing density for radius ratios up to 0.659786 only [19]. For what structures are possible with radius ratios above 0.659786 we don't have any idea about it, no work has been done about it. In binary sphere packing of mono sized spheres it is possible to get the packing density more than Kepler's Conjecture density, that is, 0.74048, if small spheres up to 0.29099 of the radius of the larger sphere are inserted into the vacant voids between the larger spheres [18].

In the paper we are using the same (mono) sized sphere packing and binary sphere packing concepts to to solve the problem to some extent. A one more packing of three sized spheres which we can call ternary sphere packing has been included about which no work has been yet done. This paper is the preceding paper of previous paper of author Jahangir Ahmad Dar which came in IJAA read reference No. 4 regarding the technique used in the paper to derive value of density. In the previous paper neutrons where taken as spheres and here quarks. We can say this is the second work done on the technique, that is, use of sphere packing in astronomy to derive density of stars.

2. Calculations of Density of Quarks Core of Neutron Star

Different models of neutron star matter suggest different forms and types of quark matter inside the core of the neutron star. However, there is not any form/state of any type of quark matter known of which the exact mass density ρ has been calculated. So the density of core of neutron star has remained a mystery. Here we will use the packing fraction concept well known in crystallography, to discuss and calculate the possible packing and thus the mass density ρ of quark matter inside the core of neutron star. We will try to solve the density by considering three possible solid quark matter types that are possible to exist in the core of the star.

2.1. Binary Sphere Packing to Sought Out Density

It is well known that neutron stars are balanced by neutron degeneracy pressure [3] [24] [25]. And at some depth particularly at the core of the neutron star, the gravitational inward pressure is so high such that neutrons will disintegrate in to their constituents u and d quarks. The formed quark matter will consist of u and d quarks by considering the mass of gluons that held the quarks together inside proton or neutron as zero [26].

If n is the number of neutrons that has crushed down to form quark matter occurring volume V then the total number of quarks available in the volume V is 3n, since each neutron contributes 3 quarks-1 u quark and 2 d quarks to the quark matter. The ratio between total number of u quarks and d quarks in volume V will be equal to the ratio between u quarks and d quarks occupying a single neutron, *i.e.*,

$$\frac{\text{No. of u quarks in V}}{\text{No. of d quarks in V}} = \frac{\text{No. of u quarks in a neutron}}{\text{No. of d quarks in a neutron}} = \frac{1}{2} \tag{1}$$

Similarly, in any structure or unit cell created by this binary quark matter, the ratio between u and d quarks occupying the cell must be same $\frac{1}{2}$, *i.e.*,

$$\frac{\text{No. of u quarks in a unit cell}}{\text{No. of d quarks in a unit cell}} = \frac{1}{2} \tag{2}$$

The packing density of this quark matter depends on functions α and χ, *i.e.*, radius ratio between the quarks and the sphere concentration [27]. For any packing of spheres not necessarily identical in size, packing fraction is given by

$$\Phi = \frac{\sum_{i=1}^{N} v_i(r_i)}{V_u} \qquad (3)$$

where V_u is the volume occupied by the cell, $v_i(r_i)$ is the volume of the ith quark with radius (r_i) and is given as

$$v_1(r) = \pi^{x/2} r^x / \Gamma(1 + x/2) \qquad (4)$$

where Γ is Euclid's function.

To calculate the χ we must calculate the d quark concentration in the matter as they are lesser in size than u quark. If k is the number of u quarks in the cell therefore as per u and d number ratio indicated in the Equation (2), the number of d quarks will be $2k$. Thus we have for small sphere concentration in a case as

$$\chi = \frac{2k}{k + 2k} = \frac{2}{3} = 0.666 \qquad (5)$$

on the basis of mass difference between u and d quarks we have assumed that they possess different sizes therefore have different radii. However, at this present stage we don't have any data nor have been done any work regarding sizes of quarks. So it is impossible to know anything about the value of function α and the biggest problem is that we can't even assume or guess it, so that to use it in deriving the partition function of binary quark matter. Therefore by considering the densest solid core of neutron stars as the quark matter possessing u and d quark flavors with different sizes it is impossible to calculate the density of the core or matter using any crystallography technique. The problem will be same for any other flavored quark matter as per no size of any type of quark is known yet.

2.2. Ternary Sphere Packing to Sought out Density

In standard model of the particles, neutron consist of three quarks binding together by the exchange of force carrying particles called as gluons symbolized by g. Gluons are the force carrying particles which carry simultaneously color and anticolor and couple to the color charge. In theoretical predictions the mass of gluon, m_g is taken as zero [28], therefore in the above condition we have assumed the core consisting of quark matter only. However by some work referenced as 29, 30, 31 it is predicted that gluons indeed have some mass limits. Some of m_g predictions are 1 MeV, 20 MeV, 500 MeV, 370 MeV, etc. [29]-[31]. Since it is clear from different predicted values of m_g the actual value of m_g is still a mystery but it is clear that mesons have some mass and its contribution to any matter can't be neglected like in the case for the crushed and deformed neutrons under the effect of gravitational pressure inside the neutron star to form quark matter. The matter will associate mesons as third kind of solid spheres possessing some value of m_g therefore possess some volume. The matter therefore will be quark-gluon matter. If the state of matter is taken solid [32] [33] as taken in the previous case and close packing of quark-gluon is considered as is possible due high gravitational pressure. The packing problem will be now ternary sphere packing. For ternary case we don't have any solution and procedure to estimate the packing density therefore to calculate the mass density of quark-gluon matter. The main problem in crystallography to sought out before using the packing fraction of matter to calculate mass density of densest quark-gluon matter are:

1. What is the impact on function radius ratio α when there are more than two different sized spheres used for packing or how will α behave for ternary sphere packing.
2. We don't know radius of gluon even the gluon mass is not known well.
3. We can calculate any sized sphere concentration χ in a any mixture by expanding Equation (5) as

$$\chi = \frac{\sum_{i=1}^{N} k_i}{\sum_{N=1}^{M} \sum_{i=1}^{N} k_i +} \qquad (6)$$

but the effect by other sized spheres on the structure of the cell is not known.

4. No structure is known consisting any ternary sphere packing.

A new method is required to invent for this discussed ternary sphere packing or to solve any ternary sphere packing in any branch of science. In a nutshell by considering gluon as a massive particle and its significant contribution in creating quark-gluon mass for any solid and close packing possible inside any star like neutron star, at this era it is impossible to calculate its packing density and therefore to calculate mass bound inside the region consisting quark-gluon matter, that is the core of neutron star. At this time we can suggest only that the density packing fraction for quark-gluon matter inside the star will be higher than Kepler's limit or any greatest known packing density in binary sphere packing if the radius of gluon and the number of gluons will be sufficient to fill every vacant void formed between the packing of u and d quarks as shown in **Figure 1** below, by supposing that the radius or size of u quark, d quark and g gluon varies as they vary in masses.

In the figure, the small spheres are considered as gluons g, big sized spheres as u quarks and the remaining sized as d quarks. In the image where the gluon concentration is higher than other regions indicates the void vacancy created by packing of u and d quarks is higher than other voids.

2.3. Mono Sized Sphere Packing to Sought out Density

Since we have not been able to calculate the density by considering the quark matter inside the core of the neutron star as a binary sphere packing and ternary sphere packing, we can consider the problem as mono sized spherical quark problem which is the only option for us at this time to imagine the density of the matter in the core of neutron star.

Now the value of α in this case becomes 1 called as Kepler's limit [27]. Due to the gravitational collapse the quarks will take highest possible packing fraction, which according to Kepler's Conjecture Theorem is $\Phi = 0.740480$ [17] [27]. So the possible structures constituting the said Kepler's limits will be FCC and CCP. The basic structure of FCC and HCP is shown in **Figure 2** as below.

The images above has been taken from the previous work done by the author Jahangir A.D. on neutron stars in the paper ''Mass limit of neutron star'' see reference no. 4.

In both structures coordinate number is 12 and both have the same packing fraction [17] [27], therefore we can take any structure to calculate the mass density obtained by the quark matter. Take HCP as a unit cell. The

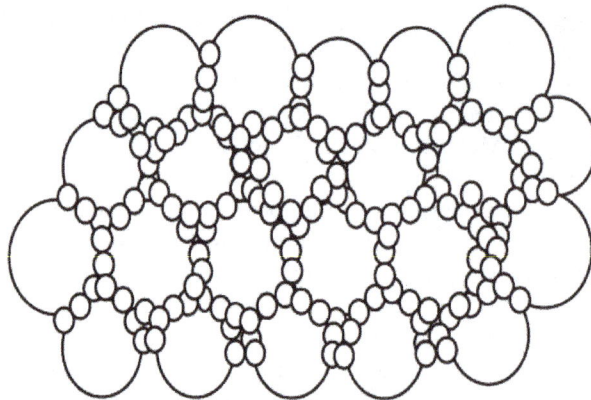

Figure 1. Schematic of quarks-gluons packing.

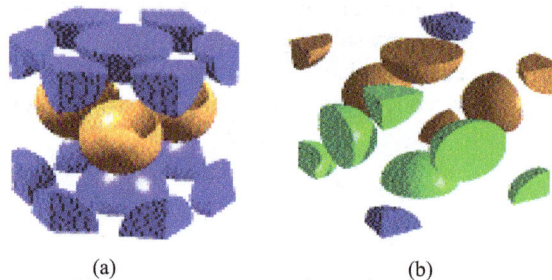

(a)   (b)

Figure 2. (a) HCP unit cell; (b) FCC unit cell.

mass of this unit cell indicated as M is the total mass of the quarks available in the cell, that is,

$$M = \sum_{i=1}^{N} m_i (r_i) \tag{7}$$

where, N is the number of quarks which is 6 and m is the mass of the individual quark.

The densest available packing fraction for HCP and FCC as per Kepler's Conjecture Theorem is 0.74048...

Rearranging Equation (3), we get

$$V_u = \frac{\sum_{i=1}^{N} v_i (r_i)}{\Phi} \tag{8}$$

But $V_u = M/\rho$.

Putting the values of V_u and M in Equation (8), and rearranging the relation and by taking ρ as the density of the unit cell or in other words density of the quark matter and ρ' as the density of the quark particle, we get

$$\rho = \frac{\sum_{i=1}^{N} m_i (r_i)}{\sum_{i=1}^{N} v_i (r_i)} \Phi = \rho' \Phi \tag{9}$$

where $\Phi = 0.74048$. By taking value of $\times$ as 3 in Equation (4), that is, considering whole case as a 3-dimensional, we have

$$\rho' = \frac{m}{\pi^{x/2} r^x / \Gamma (1 + x/2)} = \frac{3m}{4\pi r^3} \tag{10}$$

Therefore, we have

$$\rho = \frac{3m}{4\pi r^3} \Phi \tag{11}$$

It is clear that the density of quark matter or the core of neutron star depend on two functions m and r, that is, $\rho \rightarrow f(m,r)$. But question arises what values of r and m we should take for calculation? We have some predictions about the masses of quarks in standard model but no radius or in other words density of the quarks, therefore here we can use the radius limits for quarks predicted by some well known experiments. We have upper limits in the value for r as 0.85×10^{-16} cm predicted by ZEUS Collaboration [34], 1.0×10^{-16} cm predicted by HI Collaboration, 0.79×10^{-16} cm predicted by CDF Collaboration and 0.42×10^{-16} cm predicted by L3 Collaboration [35] [36]. From the standard model of particles [37] we have six flavors of quarks (q)—up (u), down (d), strange (s), charm (c), bottom (b) and top (t) [28] with masses as; $m_u = 1.7 - 3.1$ MeV/c^2, $m_d = 4.1 - 5.7$ MeV/c^2, $m_c = 1.25 - 1.30$ GeV/c^2, $m_b = 4.145 - 4.21$ GeV/c^2, $m_s = 90 - 100$ MeV/c^2, $m_t = 172.31 - 174.94$ GeV/c^2.

Using the above mentioned values in the Equation (11), we have calculated different possible values about the density of mono sized quark-matter and are given in the following **Table 1**.

In the table m_u^l and m_u^u are the lower and upper mass limits of u quark, m_d^l and m_d^u are the lower and upper mass limits of d quark, m_s^l and m_s^u are the lower and upper mass limits of s quark, m_c^l and m_c^u are the lower and upper mass limits of c quark, m_b^l and m_b^u are the lower and upper mass limits of b quark, m_t^l and m_t^u are the lower and upper mass limits of t quark. The calculated limits in ρ has been written clearly in the proceeding **Table 2**.

The density of mono-sized densest packing quark matter will be somewhere between the ranges predicted in the tables as per the limits in radii of quark particles predicted by some successful experiments in the world. The values predicted in the tables are the only values we can consider about the density of densest mono-sized quark matter that is believed to be is available inside the core of neutron stars. In the strange stars like quark stars it is possible at different layers these all densities are possible

3. Conclusion

We have used packing fraction concept to derive the density of quark matter inside neutron stars. We studied the three cases possible for quark type concentration—mono-sized single flavor packing of quark matter, binary sized u and d flavor packing of quark matter and u and d quark with gluon as ternary sphere packing. Among the all three cases we have concluded that only density of mono-sized case is possible to calculate density ranges

Table 1. Calculated density of mono sized quark-matter.

Flavor Of Quark Matter↓	At Mass of Quark ↓	Using L3 Collaboration Value of $r = 0.42 \times 10^{-16}$ cm $\rho = ... \times 10^{48}$ ↓	Using CDF Collaboration Value of $r = 0.79 \times 10^{16}$ cm $\rho = ... \times 10^{48}$ ↓	Using ZEUS Collaboration Value of $r = 0.85 \times 10^{-16}$ cm $\rho = ... \times 10^{48}$ ↓	Using HI Collaboration Value of $r = 1.0 \times 10^{-16}$ cm $\rho = ... \times 10^{48}$ ↓
u	$m_u^l = 1.7$ MeV/c^2	4.0587	0.6098	0.48959	0.30667
	$m_u^u = 3.1$ MeV/c^2	7.40038	1.11204	0.89278	0.54828
D	$m_d^l = 4.1$ MeV/c^2	9.78769	1.47077	1.18078	0.72515
	$m_d^u = 5.7$ MeV/c^2	13.6072	2.04472	1.64157	1.00813
S	$m_s^l = 90$ MeV/c^2	214.85179	32.28535	25.91970	15.91794
	$m_s^u = 100$ MeV/c^2	238.724	35.87261	28.79967	17.6866
C	$m_c^l = 1.25$ GeV/c^2	2.98405	0.44840	0.35999	0.22108
	$m_c^u = 1.30$ GeV/c^2	3.10341	0.46634	0.37439	0.22992
B	$m_b^l = 4.14$ GeV/c^2	9.89511	1.48692	1.19374	0.73310
	$m_b^u = 4.21$ GeV/c^2	10.05028	1.51023	1.21246	0.74460
T	$m_t^l = 172.31$ GeV/c^2	411.34570	61.81210	49.62471	30.47578
	$m_t^u = 174.31$ GeV/c^2	417.62415	62.75555	50.38215	30.94093

Table 2. Range of mass density of monosized quark-matter.

Quark Matter	ρ Range Using L3 Collaboration Results	ρ Range Using CDF Collaboration Results	ρ Range Using ZEUS Collaboration Results	ρ Range Using HI Collaboration Results
U	4.0587×10^{48} - 7.40038×10^{48} MeV/c^2 cm^3	0.6098×10^{48} - 1.11204×10^{48} MeV/c^2 cm^3	0.48959×10^{48} - 0.89278×10^{48} MeV/c^2 cm^3	0.30667×10^{48} - 0.54828×10^{48} MeV/c^2 cm^3
D	9.78769×10^{48} - 13.6072×10^{48} MeV/c^2 cm^3	1.47077×10^{48} - 2.04472×10^{48} MeV/c^2 cm^3	1.18078×10^{48} - 1.64157×10^{48} MeV/c^2 cm^3	0.72515×10^{48} - 1.00813×10^{48} MeV/c^2 cm^3
S	214.85179×10^{48} - 238.724×10^{48} MeV/c^2 cm^3	32.28535×10^{48} - 35.87261×10^{48} MeV/c^2 cm^3	25.91970×10^{48} - 28.79967×10^{48} MeV/c^2 cm^3	15.91794×10^{48} - 17.6866×10^{48} MeV/c^2 cm^3
C	2.98405×10^{48} - 3.10341×10^{48} GeV/c^2 cm^3	0.44840×10^{48} - 0.46634×10^{48} GeV/c^2 cm^3	0.35999×10^{48} - 0.37439×10^{48} GeV/c^2 cm^3	0.22108×10^{48} - 0.22992×10^{48} GeV/c^2 cm^3
B	9.89511×10^{48} - 10.05028×10^{48} GeV/c^2 cm^3	1.48692×10^{48} - 1.51023×10^{48} GeV/c^2 cm^3	1.19374×10^{48} - 1.21246×10^{48} GeV/c^2 cm^3	0.73310×10^{48} - 0.74460×10^{48} GeV/c^2 cm^3
T	411.34570×10^{48} - 417.62415×10^{48} GeV/c^2 cm^3	61.81210×10^{48} - 62.75555×10^{48} GeV/c^2 cm^3	49.62471×10^{48} - 50.38215×10^{48} GeV/c^2 cm^3	30.47578×10^{48} - 30.94093×10^{48} GeV/c^2 cm^3

between two values for every experimental value about the radius of quark predicted by the work of four known Collaborations—HI, ZEUS, L3 and CDF.

References

[1]   Chandrashekhar, S. (1931) The Maximum Mass of Ideal White Dwarfs. *Astrophysical Journal*, **74**, 81.

[2]   Potehin, A.Y. (2011) The Physics of Neutron Stars. http://arxiv.org/pdf/1102.5735.pdf

[3]   Srinivasan, G. (2002) The Maximum Mass of Neutron Stars. *The Astronomy and Astrophysics Review*, **11**, 67-96. http://dx.doi.org/10.1007/s001590200016

[4]   Jahangir, A.D. (2014) Mass Limit of Neutron Star. *International Journal of Astronomy and Astrophysics*, **4**, 414-418. http://dx.doi.org/10.4236/ijaa.2014.42036

[5]   Wald, R.M. (1997) Gravitational Collapse and Cosmic Censorship. http://arxiv.org/abs/gr-qc/9710068

[6]   Backer, D.C. (1976) Pulsar Average Wave Forms and Hollow-Cone Beam Models. *Astrophysical Journal*, **209**, 895-907. http://dx.doi.org/10.1086/154788

[7]   Schafner, J. and Mishustin, I.N. (1996) Hyperon-Rich Matter in Neutron Stars. *Physical Review C*, **53**, 1416.

[8]   Miller, M.C. (2004) Introduction to Neutron Stars. University of Maryland, College Park.

[9]   Villian, L. and Haensel, P. (2008) *Astron. and Astrophys*, **444**, 539.

[10]   Farhi, E. and Jaffe, R.L. (1984) Strange Matter. *Physical Review D*, **30**, 2379-2390. http://dx.doi.org/10.1103/PhysRevD.30.2379

[11]   Haensel, P., Potekhin, A.Y. and Yakovlev, D.G. (2007) Neutron Stars. Springer, Berlin. http://dx.doi.org/10.1007/978-0-387-47301-7

[12]   Oppenheimer, J. R. and Volkoff, G. M. (1939) On Massive Neutron Cores. *APS Journals*, **55**, 374.

[13]   Drae, J.J., Marshall, H.L., Dreizler, S., *et al.* (2002) Is RX J185635-375 a Quark Star? *Astrophysical Journal*, **572**, 996-1001. http://arxiv.org/abs/astro-ph/0204159 http://dx.doi.org/10.1086/340368

[14]   Chan, T.C., Cheng, K.S., Haro, T., Lau, H.K., Lin, L.M., Suen, W.M. and Tian, X.L. (2009) Could the Compact Remnant of SN 1987A Be a Quark Star? *Astrophysical Journal*, **695**, 732-746. http://dx.doi.org/10.1088/0004-637X/695/1/732

[15]   O'Toole, P.I. and Hudson, T.S. (2011) New High-Density Packings of Similarly Sized Binary Spheres. *The Journal of Physical Chemistry C*, **115**, 19037-19040. http://dx.doi.org/10.1021/jp206115p

[16]   Weber, F., *et al.* (1994) Strange-Matter Stars. In: *Proceedings: Strangeness and Quark Matter*, World Scientific, Singapore, 87.

[17]   Hales, T.C. (2005) A Proof of the Kepler Conjecture. *Annals of Mathematics*, **162**, 1065-1185. http://dx.doi.org/10.1088/0954-3899/33/1/001

[18]   Zong, C. (2002) From Deep Holes to Free Planes. *Bulletin of the American Mathematical Society*, **39**, 533-555. http://dx.doi.org/10.1090/S0273-0979-02-00950-3

[19]   Shapiro, S.L. and Teukolsky, S.A. (2008) Black Holes, White Dwarfts and Neutron Stars: The Physics of Compact Objects. Wiley, New York.

[20]   Filion, L., Marechal, M., van Oorschot, B., Pelt, D., Smallenburg, F. and Dijkstra, M. (2009) Efficient Method for Predicting Crystal Structures at Finite Temperature: Variable Box Shape Simulations. *Physical Review Letters*, **103**, Article ID: 188302.

[21]   Kummerfeld, J.K., Hudson, T.S. and Harrowell, P. (2008) The Densest Packing of AB Binary Hard-Sphere Homogeneous Compounds across All Size Ratios. *Journal of Physical Chemistry B*, **112**, 10773-10776. http://dx.doi.org/10.1021/jp804953r

[22]   Filion, L. and Dijkstra, M. (2009) Prediction of Binary Hard-Sphere Crystal Structures. *Physical Review E*, **79**, Article ID: 046714. http://dx.doi.org/10.1103/physreve.79.046714

[23]   Torquato, S. and Jiao, Y. (2010) Robust Algorithm to Generate a Diverse Class of Dense Disordered and Ordered Sphere Packings via Linear Programming. *Physical Review E*, **82**, Article ID: 061302. http://dx.doi.org/10.1103/physreve.82.061302

[24]   Lattimer, J.M. and Prakesh, M. (2004) The Physics of Neutron Stars. *Science*, **304**, 536-542. http://arxiv.org/abs/astro-ph/0405262 http://dx.doi.org/10.1126/science.1090720

[25]   Baym, G. and Pethick, C. (1979) Physics of Neutron Stars. *Annual Review of Astronomy and Astrophysics*, **17**, 415-443. http://dx.doi.org/10.1146/annurev.aa.17.090179.002215

[26]   Yao, W.M., *et al.* (2006) Review of Particle Physics: Neutrino Mass, Mixing, and Flavor Change. *Journal of Physics G*, **33**, 1-1232. http://dx.doi.org/10.1088/0954-3899/33/1/001

[27]   Hopkins, A.B. and Stillinger, F.H. (2012) Densest Binary Sphere Packings. *Physical Review E*, **85**, Article ID: 021130. http://dx.doi.org/10.1103/physreve.85.021130

[28]   Nave, R. (2008) Quarks. HyperPhysics. Georgia State University, Department of Physics and Astronomy, Atlanta.

[29]   Halzen, F., Krein, G. and Natale, A.A. (1993) Relating the QCD Pomeron to an Effective Gluon Mass. *Physical Review D*, **47**, 295. http://dx.doi.org/10.1103/PhysRevD.47.295

[30] Cornwall, J.M. and Soni, A. (1983) Glueballs as Bound States of Massive Gluons. *Physics Letters B*, **120**, 431-435. http://dx.doi.org/10.1016/0370-2693(83)90481-1

[31] Yndurain, F. (1995) Limits on the Mass of the Gluon. *Physics Letters B*, **345**, 524-526. http://dx.doi.org/10.1016/0370-2693(94)01677-5

[32] Yu, M. and Xu, R.X. (2011) Toward an Understanding of Thermal X-Ray Emission of Pulsars. *Astroparticle Physics*, **34**, 493-502. http://dx.doi.org/10.1016/j.astropartphys.2010.10.017

[33] Dai, S. and Xu, R.X. (2012) Quark-Cluster Stars: Hints from the Surface. http://arxiv.org/pdf/1201.3759.pdf

[34] ZEUS Collaboration (2004) Search for Contact Interactions, Large Extra Dimensions and Finite Quark Radius in ep Collisions at HERA. *Physics Letters B*, **591**, 23-41. http://dx.doi.org/10.1016/j.physletb.2004.03.081

[35] Truemper, J.E., Burwitz, V., Haberl, F. and Zavlin, V.E. (2004) The Puzzles of RX J1856.5-3754: Neutron Star or Quark Star? *Nuclear Physics B Proceedings Supplements*, **132**, 560-565. http://dx.doi.org/10.1016/j.nuclphysbps.2004.04.094

[36] Chekanov, S., Derrick, M. and Krakauer, D. (2003) Argonne National Laboratory, Chicago (And Others); Deutsches Elektronen-Synchrotron (DESY), Hamburg; ZEUS Collaboration, IAEA, INIS, 35032556.

[37] Oerter, R. (2006) The Theory of Almost Everything: The Standard Model. Penguin Group, London.

16

Existence of Black Neutron Star

Trivedi Rajesh

Caterpillar Electric Pvt Limited, Delhi, India
Email: advocate.dma@gmail.com

Abstract

A sufficiently large star will collapse to form a Black Hole Singularity due to Gravitational Pressure beyond Neutron Degeneracy. A Black Hole exhibits extremely strong Gravitational attraction that no particle or electromagnetic radiation can escape from it. The boundary of the region from which no escape is possible is called Event Horizon. In this work it is proposed that there exists a Neutron star smaller than Event Horizon, which is termed as Black Neutron Star. Furthermore an alternative method is proposed to ascertain the maximum permissible mass limit of the Neutron Star and the minimum mass limit of the naturally occurring gravitationally collapsed Black hole.

Keywords

Neutron Star, Black Hole, Gravitational

1. Introduction

The primary source of energy of a Main Sequence Star is nuclear fusion at its central core, this nuclear fusion releases energy which exerts an outward radiation pressure countering the gravitational collapse and maintaining the hydrostatic equilibrium. The onset of fusion process in significant mass star is always with Hydrogen fusion, first to deuterium and then to Helium. Eventually the supply of Hydrogen gets exhausted and the Gravitational Pressure forces the core to get contracted.

Depending on the core mass this Gravitational contraction is countered either by Electron degeneracy or by further initiation of fusion of Helium present in the core. In 1935, Chandrasekhar [1] proposed an upper limit of formation of white dwarf through Electron Degeneracy which is known as Chandrasekhar Limit (1.44 Solar Mass).

If the mass of the core is more than the Chandrasekhar Limit, then Gravitational Collapse causes sufficient increase in temperature to enable Helium fusion to higher atoms to finally Fe^{56}. Once the iron is formed in the fusion process, no longer sustainable fusion process can take place as beyond this, the fusion process is endo-thermic.

Once this stage is achieved, the star starts collapsing as now no longer counter balancing fusion energy is produced. Since the core mass is higher than the Chandrasekhar Limit and hence even the Electron degeneracy can no longer provide sufficient counter balancing pressure to hold on to Gravitational Collapse. These electrons get captured by protons and in the process Neutrons are left in abundance, the continued inward pressure forces Neutrons to go to their lowest Energy States and developed Neutron Degeneracy Pressure counters the Gravitational Pressure and a Hydrostatic Equilibrium is established thus forming a Neutron Star [2] [3].

In 1939, J R Oppenheimer and G Volkoff using the work of R C Tolman proposed the limiting mass of the Neutron Star at 0.7 Solar Mass, this is called as TOV limit [4]. This limit is quite low, in fact lower than the Chandrasekhar Limit for Electron Degeneracy which is not realistic and hence modern estimates put this limit to 1.4 to 3.0 Solar Masses.

The Neutron Degeneracy Pressure calculations define the relationship between mass of the Neutron Star core and its radius as:

$$R(n) = k(M)^{-1/3}$$

If the mass of the core of the collapsing star is more than the TOV limit then even the Neutron Degeneracy cannot provide sufficient counter balancing pressure against Gravitational Collapse and a Black hole is formed. A Black Hole exhibits extremely strong Gravitational attraction that no particle or electromagnetic radiation can escape from it. The boundary of the region from which no escape is possible is called Event Horizon. This is mathematically defined as Schwarzschild radius; therefore if an object of mass M is smaller than the size of Schwarzschild radius then it is termed as Black Hole.

For an object of mass M the Schwarzschild radius $R(s)$ is

$$R(s) = 2GM/c^2$$

In this work the formation of Neutron Star is understood from a perspective of the Spherical Packing aspect, and it is observed that there are three distinct core mass ranges, one from 1.4 - 2.7 Solar Mass, second from 2.7 to 3.24 Solar Mass and third from 3.24 solar Mass and higher. The first range can form a Visible Neutron Star, the second range star is also a visible star but it will have to have certain exotic matter beyond Neutrons and third range will form a Neutron Star of the radius smaller than Schwarzschild radius. This star can be termed as Black neutron Star as no light can escape out of it.

1.1. Revised Pauli's Exclusion Principle and Neutron Degeneracy

As referred above once the Hydrostatic equilibrium is disturbed due to exhaustion of Hydrogen required for fusion, the star collapses towards Neutron Degeneracy if the mass of the core of the star is higher than Chandrasekhar Limit. J R Oppenheimer and G Volkoff carried out certain computations by assuming this compressed Neutron matter as Cold Fermi Gas and arrived at TOV Limit.

In this work an alternative method is proposed to arrive at Neutron Star Mass limit based on Spherical Packing aspect. It is proposed that this Gravitational Pressure will create extra ordinary situation in which Neutrons are pressed like any other spherical object. The fermions cannot stick to Pauli's Exclusion Principle under such extreme pressure. This is no violation of Pauli's Exclusion Principle; it is just that the same gets redefined a bit which paves way for bosonic compactness as...,

"Two identical fermions (particles with half-integer spin) cannot occupy the same quantum state simultaneously, but they can be forced to occupy the same quantum state under extreme external pressure as the Δx is made to approach zero, where Δx is the uncertainty in position".

1.2. Spherical Packing of Neutrons

Under such extreme gravitational pressure, it is most likely that the Neutrons will pack themselves completely densely in the given core space. This dense packing can be understood from the Kepler Conjecture [5] [6]. So if there are n Neutrons each of radius r, then these Neutrons will just fit inside a spherical space of radius $R(p)$ where we call this as packing radius.

$$R(p) = \left(\frac{n}{0.74}\right)^{1/3} r$$

2. Discussions and Black Neutron Star

A comparative analysis of various radii is carried out to establish that even inside the Event Horizon a neutron Star may exist which is termed as Black Neutron Star.

Table 1 represents the typical values of these radii in Kms for various solar masses, for computation purpose it is taken that one Solar Mass has 1.18×10^{57} Neutrons and the realistic core radius of neutron is 0.55 fm, the value of radius of Neutron has substantial impact on tabulated values.

It is shown that the Schwarzschild radius up to 3.24 solar mass is less than the packing radius for the available neutrons. This is a very crucial result simply because for any object smaller than 3.24 solar mass, the collapse to Event Horizon is not possible without actually pushing the neutrons beyond their extreme possible packing arrangement thus defining the lower limit on the mass of Invisible Star (Black hole) as 3.24 Solar Mass.

Similarly if the star is more than 3.24 solar mass, then there is still room for neutron compactness beyond Event Horizon as $R(p) < R(s)$. The conclusion is that we can have a Black Neutron star. With this conclusion author proposes that a physical Neutron Star exists which is smaller than Event Horizon in size.

Another interesting observation is that the packing radius up to 2.65 solar mass is less than the neutron star radius, signifying that a core of more than 2.65 solar mass will have to press the neutrons beyond their best possible packing arrangement, if the neutron star can at all be formed for a core of more than 2.65 solar mass. The inescapable conclusion is that we cannot have a visible neutron star of more than 2.65 Solar Mass. This observation is categorized as per mass range as follows:

2.1. Mass of Star Core between 1.4 to 2.65 Solar Mass

As per **Table 1** above for this range

$$R(n) > R(p) > R(s)$$

so technically from the pure dynamics point of view, there is ample space for Neutrons to move around even after neutron degeneracy pressure has countered the gravity, it cannot be stated that due to Heisenberg uncertainty principle they have become completely relativistic (speed at par with speed of light). So a stable visible neutron Star can be formed for such mass range without causing any ambiguity.

2.2. Mass of Star Core between 2.65 to 3.24 Solar Mass

A very critical mass which is neither conducive for neutron Star and nor for black Hole.

In this case the neutron star radius required to sustain the Gravitational Pressure becomes less than the Packing radius, that implies that something must happen to such densely packed neutrons like either neutrons get converted into quarks or convert into energy, but then it no longer can be stated that this is a Neutron Star, so a visible neutron star is possible only up to around 2.65 Solar Mass. This also explains and predicts that in general visible neutron Stars will be in existence between 1.4 to 2.65 Solar Mass and also of radius around 11 - 9 Kms.

This inflection point of 2.65 Solar Mass defines the upper limit of a visible Neutron Star Mass. The significance of this point is that above this mass the required Neutron Star radius is less than packing radius, thus the same is not possible with all Neutron in the core.

It is also useful to note that even for this mass the Schwarzschild radius is still lower than these two radii, thus effectively light can still escape and star will be visible.

2.3. Mass of Star Greater than 3.24 Solar Mass

This is the case where

Table 1. The comparative analysis of various radii of a collapsing star core.

Solar mass	1	1.4	2	2.65	3.00	3.24	5	10
R(s)	2.9	4.1	5.8	7.7	8.8	9.5	14.7	29.3
R(p)	6.4	7.2	8.1	8.8	9.2	9.5	11.0	13.8
R(n)	12	11	9.7	8.8	8.5	-	-	-

$$R(s) > R(p) > R(n)$$

That means a visible Neutron star of 3.24 solar mass and above is completely ruled out, now it is the question if the same gets converted into Black Hole Singularity or adopts some other mechanism.

Since the star is large enough the additional Gravitational pressure will squeeze the Neutron core to an extent that it shrinks beyond the Event Horizon but still larger than packing radius. In fact depending on the star mass the internal neutron degeneracy pressure will sustain the gravitational pressure even beyond the Event Horizon since the position of Neutrons keeps getting extremely rigid and fixed with reducing positional uncertainty and Heisenberg uncertainty principle would require the Neutrons to become increasingly relativistic, this will ensure certain stability of star even if its radius is smaller than Event Horizon. So it is quite likely that we get stellar objects of larger mass than 3.24 solar mass but inside the   Event Horizon, which are perfectly definable structures (without singularity) and they can be termed as Invisible Neutron Stars as their radii is less than Schwarzschild radius and more than packing radius.

3. Conclusion

It is proposed that an invisible Neutron Star may exist wherein $R(s) > R(p)$, this will enable us to look for physical presence of a star beyond Event Horizon, which so far being treated as singularity thus out of bounds of physical understanding. An upper limit is also evaluated for the mass of the visible Neutron star and the lower mass limit for the naturally occurring gravitationally collapsed Black hole is also computed. These parameters are Neutron radius dependent.

References

[1]   Chandrasekhar, S. (1935) The Highly Collapsed Configurations of a Stellar Mass. *Monthly Notices of the Royal Astronomical Society*, **95**, 207-225. http://dx.doi.org/10.1093/mnras/95.3.207

[2]   Bombaci, I. (1996) The Maximum Mass of a Neutron Star. *Astronomy and Astrophysics*, **305**, 871-877.

[3]   Carroll, B.W. and Ostlie, D.A. (2006) §16.3. The Physics of Degenerate Matter. In: *An Introduction to Modern Astrophysics*, 2nd Edition, Addison-Wesley, Boston.

[4]   Oppenheimer, J.R. and Volkoff, G.M. (1939) On Massive Neutron Cores. *Physical Review Letters*, **55**, 374. http://dx.doi.org/10.1103/PhysRev.55.374

[5]   Sloane, N.J.A. (1998) The Sphere-Packing Problem. *Documenta Mathematika*, **3**, 387-396.

[6]   Hales, T.C. (1998) The Kepler Conjecture. http://front.math.ucdavis.edu/math.MG/9811078

17

Diurnal Variations and Spikes by the Torsind Registered and Their Impact on the Accuracy of G Measurement

A. F. Pugach

Main Astronomical Observatory of the NASU, Kiev, Ukraine
Email: pugach@yandex.ru, pugach@mao.kiev.ua

Abstract

The article reports on the results of an analysis of the torsind behavior long-term observations. The torsind is a species of ultralight disc torsion balance. The data analysis showed that the signal recorded contains the 24-hour periodic component presumably associated with the Sun. Moreover, unpredictable strong impacts, forcing torsind disk to rotate in one or another direction, were revealed. Presumably the reason of these effects is the Sun. This indicates the existence of an unknown radiation that bears a torque which may impact on the mechanical systems dynamics. This fact leads to the need to measure the gravitational constant G overnight and during periods of minimum of the solar activity, provided that the G measurements are carried out using a torsion balance.

Keywords

Torsion Indicator, Torsind Measurements, Gravitational Constant G

1. Introduction

American scientists E. Saxl and M. Allen were observing diurnal variations of the period of a massive torsion pendulum over the 16-day interval starting on September 20, 1979 and terminating on October 5, 1979 [1]. The pendulum regularly speeded up by approximately 1 part in 6000 shortly before or after local noon. That process was quite regular, implying a connection with the circadian rhythm of the Sun.

Since the observation lasted 15 days, the length of a resulting series of the measurements was large enough to get a statistically significant conclusion about the reality of the periodic variations. Minimum of the readings always come near to noon, from which it was concluded that the 24-hour variability of the signal exists.

This finding has long remained unknown to the general physical community, despite the fact that the measurements were done by well-known experts, and the measurements themselves were carried out methodically correctly (for details, see [2] pp. 127-137).

It is fair to say that E. Saxl and M. Allen were not the first in this issue. Similar diurnal variations were studied in the first half of the twentieth century yet. The French astronomer of Russian origin N.M. Stoyko tracked diurnal oscillations of a vertical pendulum and published his results in [3].

However, even these studies can not be considered as pioneering ones, since long before that the daily fluctuations of a torsion pendulum were first reported by N.P. Myshkin—a professor of the Novo Alexandria Institute. For many years he had been studying an effect of heat and light on rotation of a horizontally suspended mica disk. In 1898 he came to the conclusion that "the fluctuations are in some connection with the position of the Sun relative to horizon" [4]. After that he was trying to find out "whether these vibrations of the system have not only daily, but the annual periodicity?" [4].

Thus, the history of study of the diurnal variations of a torsion pendulum has more than a century. However, the very small amplitude of these oscillations and significant impact on the behavior of pendulums both the researcher and environmental conditions made it difficult to study. Therefore, these works had more reconnaissance rather than deep exploratory mode.

New possibilities for the study of this delicate phenomenon appeared recently in connection with the use of the torsind [5] and appropriate methods of analysis.

2. Brief Description of the Torsind and Methods of Observation

The torsind—**tors**ion **ind**icator—is a specific type of torsion balance that uses a very light metal or paper disc instead of the linear beam of a classical unit, suspended from a monofilament made from natural silk, instead of quartz or a rigid suspension [5]. A specific feature of this thread allows the device disk to make a lot of revolutions on its axis without accumulation of reverse torque in the thread itself. In other words, if the torsind disc, being influenced with an external torque, slowly makes several revolutions then after the termination of the torque it retains at the last position having no backspin. This surprising property of a silk thread is due to its specific molecular structure. The basis of a thread are fibroin protein molecules, substances even stronger than Kevlar. Repetitive amino acid sequences of this protein form antiparallel pleated β-layers which are connected to each other by hydrogen bonds. These bonds are not very strong and allow for moderate mechanical offsetting of the layers. This explains the fact that under slow (necessarily slow!) rotation the layers can slide relative to each other. Moreover, it does not reduce the mechanical strength of the thread. The housing of the torsind is made from a quartz cylinder.

A web-camera monitors the disc rotation. The webcam is mounted above the upper face of the cylinder and is connected to a computer.

The design of such a balance makes it insensitive to variations in gravitational potential and ensures that it is unaffected by gravitational (tidal) influences from any direction

The instrument can not measure a force, but indicates the direction of the applied torque only. The low weight of the suspended unit (the aluminum disk mass does not exceed 100 mg) provide high sensitivity of the device. The device is clearly responds to a torque:

$$6.47 \times 10^{-12} \ \text{N} \cdot \text{m}$$

It should be added that the torsind is equipped with an automatic registration system and its continuous operation does not require the presence of an operator. Since mid-2009 the torsind has been sited at the Main Astronomical Observatory (Kiev, Ukraine) in an isolated, shaded room with tightly closed doors and windows, the entrance being barred to outsiders.

A detailed description of the instrument and its properties, identified for several years of our examination, is published in [5].

From this description it follows that the torsind **does not respond to** changes in:

- gravitation potential;
- temperature of the device itself;
- meteorological parameters (pressure, humidity);
- static electric field during a thunderstorm;

- moderate magnetic field;
- degree of ionization of the ionosphere;
- operation of electromechanical units;
- change in the load on the torsind thread;
- possible microvibration of the floor;
- sound waves of moderate intensity;
- radio waves and cell phones.

Possible weak convective air movement inside the torsind housing can also not be a source of significant interference.

On the other hand, torsind responds to some astronomical phenomena under such unusual circumstances, when other astrophysical instruments are powerless. For example, torsind "sees" a solar eclipse, which occurs on the opposite side of the globe [6] [7]. Moreover, torsind reacts to some syzygy effects, being underground at a depth of 40 meters [8].

The torsind ability to respond to some planetary configuration is truly surprising. For example, in June 2012, there was the transit of the planet Venus across the sun's disk. This phenomenon has been recorded with two devices shown qualitatively similar picture (see [5], Figure 10). It should be noted here that at the time of the transit the Venus was at a distance of 42 million km from the Earth. This fact has been reliably recorded, but it lacks any solid explanation. Thus, the torsind disk rotation seems to have a much deeper cause than the ostensible reasons (changes in temperature, pressure, etc.).

The measurement accuracy defined 13.02.2013 under a stable space weather is characterized by the standard error (se) of a single measurement of:

$$se = \pm 0.157°.$$

Thus, the torsind as a unique instrument simultaneously combines two qualitatively opposite features. On the one hand, the torsind operation is not influenced by many external circumstances which seem to be strong noise-producing factors for many other scientific instruments. On the other hand, torsind can register some events, not be observed with other devices.

Using this specific torsind's features allowed us for several years to observe periodic changes in signals registered that we previously called "*the diurnal variations*".

3. Observational Material

We did not plan to observe the diurnal variations as an independent scientific task. It is a specific "by product" of our long-term monitoring of the torsind reactions. All discussed observations were performed using the torsind WEB_1. These observations contained some fragments where repeatable mode of records was present.

One of the most spectacular manifestation of the diurnal variations was registered in June, 2013 (see **Figure 1**). That short fragment clearly shows the daily periodicity.

The followings are the observations when stable periodic oscillations were present in their "pure" form without significant interference for a long time. Thus, **Figure 2** shows a record of relatively stable oscillations, when the torsind counts at night time during 8 days almost did not change, however during daytime counts increased by about 14 - 25 degrees.

Figure 3 shows the diurnal changes which are superimposed on variable background substrate. Moreover, the amplitude of the variations varies by almost an order. These data prompted us to more thoroughly analyze all the available observations in order to confirm the reality of diurnal variations and determine their exact period. The autocorrelation analysis method was used as an adequate one for this task.

4. Determination of the Period of the Diurnal Variation

We have analyzed the least noisy series of observations where the varying signal was more regular in appearance, including the short sequences shown on the above figures. That is, the most plain observations were chosen, including no other processes with large amplitudes.

That fact that torsind registers diurnal fluctuations is not in doubt. The question remains open: is it solar or sidereal day. The autocorrelation analysis (ACA) method was used to accurately determine the period of the diurnal variation. Several observations which were made at the moments of quiet space weather, when there

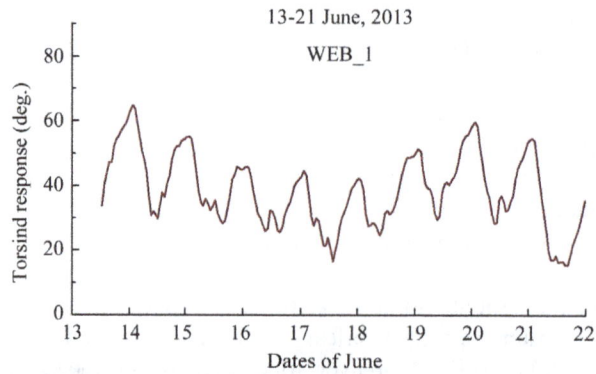

Figure 1. One of the results with the records of the diurnal variations.

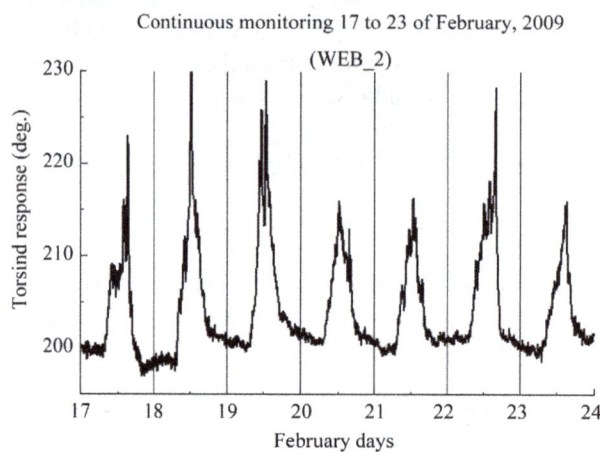

Figure 2. Sustained diurnal variations in February 2009.

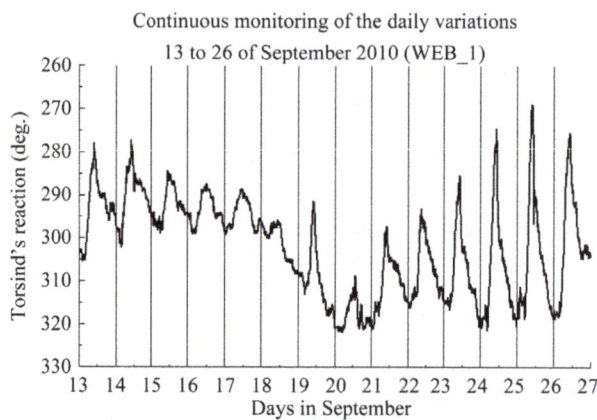

Figure 3. Periodic variations of alternating amplitude with a marked trend in 2010.

were no spikes, significant astro-space phenomena or other strong influences, were chosen for the analysis. In these periods of the observation the most significant signals were the same daily fluctuations. All autocorrelation functions obtained for these series have a form similar to that shown in **Figure 4**.

The average values of the periods were drawn from the first six partitive maxima for each individual observation series. The resulting values of the periods (in minutes) are presented in the last column of **Table 1**. For all of these data the general average value of the period 1440.24 ± 2.69 minutes was obtained.

Autocorrelation function for continuous measurements
in the range of June 13-21, 2013

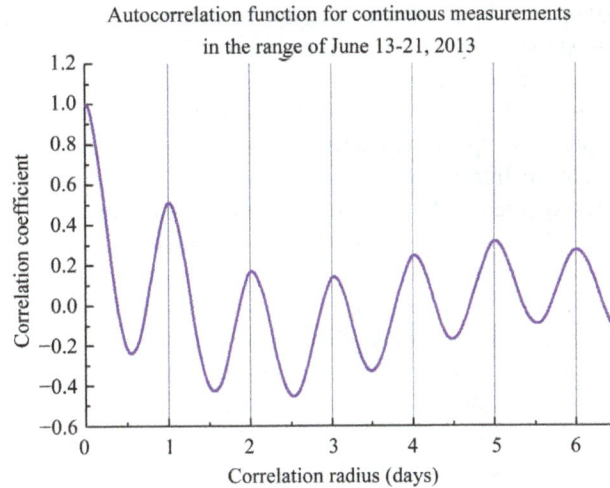

Figure 4. The partitive maxima of the autocorrelation function follow by exactly 1 day.

Table 1. Observational series used and result.

Measurement interval	N	Period (min) ± SE
17-23 February, 2009	10150	1448.31 ± 1.87
26 Sep-06 Oct, 2009	14396	1431.22 ± 10.51
04-25 June, 2010	31200	1435.54 ± 33.2
13-27 September, 2010	20160	1438.70 ± 3.31
10 May-05 Jun, 2011	36840	1444.03 ± 3.46
11-25 May, 2011	21600	1434.38 ± 2.88
13-21 June , 2013	12300	1449.50 ± 2.59
Mean		**1440.24 ± 2.69**

N—quantity of individual counts in a series.

Thus, the results of ACA clearly indicate that the daily variations are modulated by the Sun, rather than the other sources (the galaxy core, the Moon, star sources, planets, etc.).

5. The So-Called "T-Momentum"

As a result of the analysis performed, it was found that in almost all cases there is the periodic signal which is close to 24 hours. This conclusion is valid in cases when it was possible to carry out the long-term continuous monitoring in quiet space weather conditions. What does "non-quiet" space weather mean will be mentioned further down.

On the basis of our visual assessment and taking into account the results of the preliminary analysis, we estimate the amplitude of the periodic component in range of 20 - 50 degrees. It is important to note that the 24-hour periodic component manifests itself in different seasons of the year, which is likely to indicate its constant presence.

We do not consider a reason which forces the torsind disk to rotate. At the present stage of research this problem seems to be premature, since a carrier that transmits torque on the unit is not known yet. A purpose of this publication is to draw the attention of researchers on the fact of the periodic rotation of the torsind disk in different directions. This fact indicates the presence of an unknown torque, affecting the state of the mechanical system. Let's call it conditionally "Torsind-momentum" or "T-momentum".

The detection of periodic torsind oscillations proposes a certain scientific interest because it indicates a new, unexplained phenomenon. But that's not all. We are clear aware that intermittent heteropolar torque can affect the readings of torsion balances, which are still used in the determination of the Newtonian gravitational constant G. According to small statistics given by S. Shlamminger [9]-[11] about two-thirds of all works on measuring of G are based on the use of torsion balances. This means that two-thirds of all measurement results have been in some extend distorted by the presence of the unaccounted T-momentum.

This circumstance may perhaps explain why universal constant G is measured so far with a rather large error, and the accuracy of its determination is by several orders of magnitude lower than the accuracy of the determination of other universal constants. At the moment there is no possibility to calculate the impact of the T-momentum very accurate. However, if we look at **Figure 2**, **Figure 3** and **Figure 4** it becomes clear that extreme perturbing effect manifests itself in the middle of the day, close to local noon.

In reading the recent history of the G measurement the apparent contradiction between the increased measurement accuracy and the decrease in the accuracy of the final result becomes obvious. For example we refer to recent measurements of G performed by two teams: the Franco-British [12] and Chinese [13]. Having reached the accuracy of the G determination in each of its methods at 26 - 27 ppm, they have, nevertheless, got results that differed by an amount that was almost an order of magnitude higher than the standard error of each method. We can offer an explanation for this discrepancy based on our long-term torsind observations.

6. Spikes

The fact that the torsind responds well to syzygies [7], solar and lunar eclipses [6] [8] [14], Venus' transit [5] and other astronomical phenomena has been defined reliably. However besides the torsind very often registers strong bursts, responding to some unknown astro-space phenomena. At these moments the torsind disk can make 5, 10, 20 or more revolutions in a row. Such a sharp reaction of the torsind we will conventionally call **SPIKE**.

Figure 5 shows one spike recorded by the torsind on 2-3 of May, 2013. Within 18 hours the torsind disk made 6 revolutions clockwise, and then it began to rotate in the opposite direction, making 8 full revolutions more.

Apparently, at this time a strong rotational momentum has sprung up, carrying a right- and then a left-screw torques. Evidently, a G measurement, being performed elsewhere in any other laboratory with a torsion balance at the same time, could be distorted by the additional torque. And besides, the sign of the perturbation would depend on what a branch of the spike (ascending or descending) was active during the measurement process. Of course, this statement is true under the assumption that the sphere of the spike action covers the entire Earth. The fact that torsind feels syzygy effects, even when deep underground, is the basis for this assumption [7] [8].

A double spike—is a spike when the disc rotation in one direction is immediately replaced by rotation in the opposite direction. They are relatively rare. More often there are cases when the torsind disc turns round several hundreds or thousands of degrees then ceases its rotation and stops.

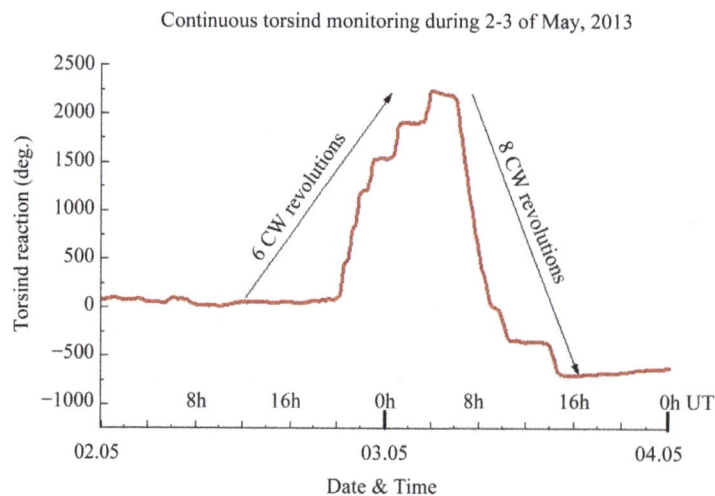

Figure 5. The double strong spike recorded on 2-3 May 2013.

Figure 6 shows the torsind reaction in the summer of 2010. Here there are weak double negative spike recorded on July 15 and a small single positive spike at night on July 2-3, and a small single negative spike on July 27. It should be noted that in this case the amplitudes of the spikes were rather small, so the diurnal variations of low amplitude are readily traced throughout the curve.

The nature and the mode of the spike appearances were not understood yet, but it was noted that the spikes began to appear more and more frequently in recent years. This was established as follows.

The regular torsind observations were conducting over the last several years. Each series of daily measurements was considered as a random sequence, and the standard deviations SD were calculated for it. **Figure 7** illustrates how the value of SD changed over time.

We believe that the large SD values are conditioned namely by the spikes so far as the diurnal variations and the syzygy effects have not large amplitudes.

It should be noted that the **Figure 7** shows the statistical parameter D only, rather than the amplitudes of the spike. But even this parameter as we see clearly reflects the increased spike activity.

7. Discussion

Based on the foregoing, it follows that the processing of the G measurement results must take into account the

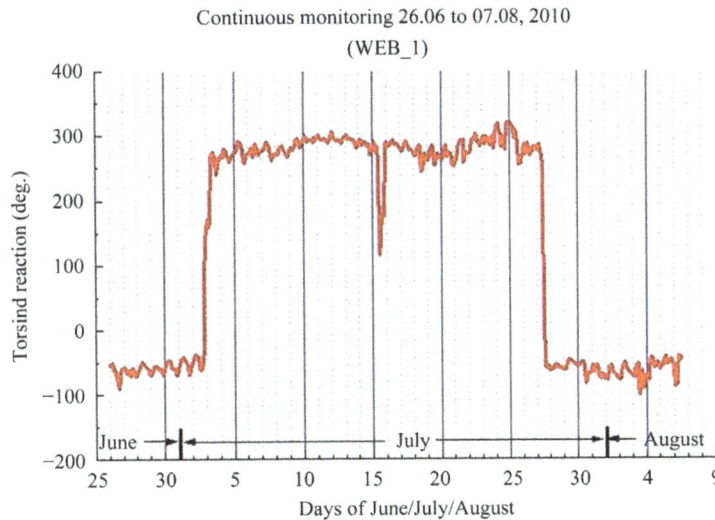

Figure 6. The result of the continuous observations in June-August 2010.

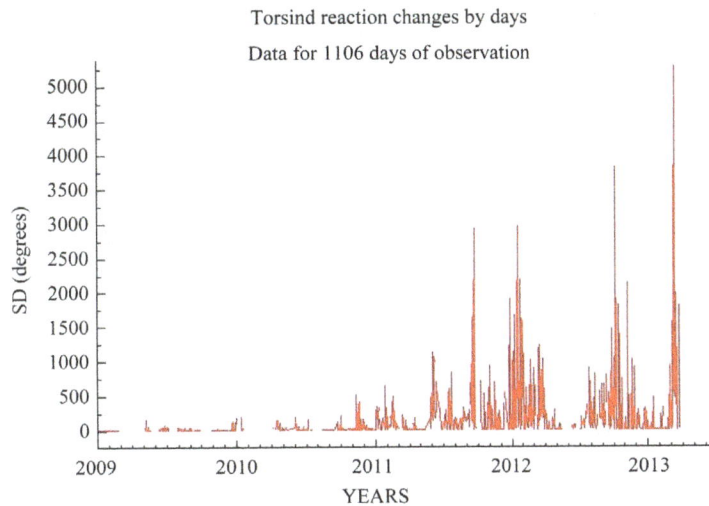

Figure 7. Change in the parameter SD over time.

possible influence of the diurnal variations, and especially of the spikes.

More than 20 years ago in the publication [15] dedicated to finding the cause of significant variation of the measured values of G, the following was stated.

"*Analysis of measurement results of the gravitational constant shows that they are associated with a variety of space and geophysical phenomena. It is reasonable to assume that this analysis reveals some unrecorded by researchers factors that directly or indirectly affect the measurement results rather than the change in the physical parameter—the gravitational constant G. The perennial search for these factors ... has not been successful*".

We believe that we have managed to identify one of the possible factors that is periodically or randomly affecting the beam position of a torsion balance, which is used for the measuring of G.

It is effect of the above mentioned T-momentum. The nature of its effect is not known, but it is obvious that it can be considered as an unknown field or pulse carrying a torque that can influence the dynamic of a mechanical system. The torque of such exposure is extremely small.

The torsind disc, with a sensitivity at least of 6.47×10^{-12} N·m rotates with an angular velocity of no more than $0.01°$ per second in the most cases. But even such a low impact must be considered. Especially it is necessary to take into account the effect of strong spikes when the angular velocity of the disk rotation reaches $1°$/s. According to our preliminary estimates the T-momentum imparts to the torsind disk the left-screw rotation in the most cases.

We propose an idea of the key experiment to verify our hypothesis. If the suspected T-momentum does have an impact on the behavior of the torsion balance, then instrumentation counts should deviate compared to the true (unperturbed) readings during spike-activity. Sign of the deviation depends on the direction of torque (left- or right-handed).

In **Table 2**, we present the complete list of all spikes recorded in early 2012, when the spike activity was very high (see **Figure 7**). The table shows the precise times of start and end of the spikes, their amplitude and direction of the T-momentum—a right-handed (+) or left-handed (−).

If measurements of G have been somewhere conducted in January 2012, the staff of that laboratory may use the data in **Table 2** for adequate analysis in order to assess whether there were large deviations in the results of their measurements. This key experiment should clearly answer the question whether there is the influence of spikes on the measurement result.

For the full understanding we give a graph of the continuous torsind record made in January 2012 (**Figure 8**). The graph helps to understand at what time and in what direction the T-momentum was. The spikes listed in **Table 2** can be easily detected in this graph.

Table 2. Spike activity within January, 2012.

Date	UT (h)		Amplitude, deg.	
	Start	End	+	−
11	08.33	10.45	2890	-
14	07.22	8.78	-	380
15	11.28	12.30	-	1080
16	12.63	15.55	6580	-
16	15.55	19.86	-	7560
20	3.28	7.40	-	730
23	7.78	10.64	2120	-
26	8.92	19.00	1910	-
27	11.52	13.56	1160	-
28	01.57	14.95	4120	-
30	08.23	10.95	-	2560

Continuous data for 8 to 31 of January, 2012

(WEB_1)

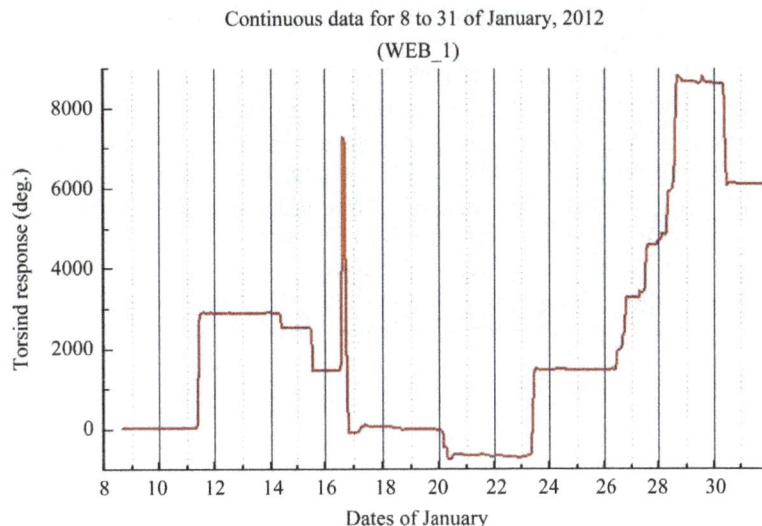

Figure 8. Continuous recording in January, 2012.

8. Conclusions

We hope that clarification of the G value will be significantly improved if the gravimetric measurements should be conducted at night time provided there is no spike activity. Since the spike activity increases with the increasing of the solar activity [10], then for the reliable measurements of G, some quiet periods at minimum 11-year solar cycle should be chosen.

And the most important, the gravilaboratories where the measurements of G are carrying out must be equipped with sensitive instruments like the torsion indicator.

Such instruments should allow conducting continuous monitoring of the environment situation for the selection of suitable working hours, when there is no spike activity.

References

[1] Saxl, E.J. and Allen, M. (1971) 1970 Solar Eclipse as "Seen" by a Torsion Pendulum. *Physical Review D*, **3**, 823-825. http://dx.doi.org/10.1103/PhysRevD.3.823

[2] Munera, H.A. (Ed.) (2011) Should the Laws of Gravitation Be Reconsidered? Apeiron, Montreal, 448 p.

[3] Stoyko, N.M. (1947) Sur la variation jouraliere de la marche des pendules et de la deviation de la verticale. *Acad. Sci. Comptes, Rendus.*, Paris, **224**, 1440-1441.

[4] Myshkin, N.P. (1906) The Movement of a Body in the Flow of Radiant Energy. *Russian Journal of Physical and Chemical Society*, **3**, 151-184. (In Russian)

[5] Pugach, A.F. (2013) The Torsind—A Device Based on a New Principle for Non-Conventional Astronomical Observations. *International Journal of Astronomy and Astrophysics*, **3**, 33-38. http://dx.doi.org/10.4236/ijaa.2013.32A006

[6] Pugach, A.F. (2009) Observations of the Astronomical Phenomena by Torsion Balances. *Physics of Consciousness and Life, Cosmology and Astrophysics*, **9**, 30-51.

[7] Olenici, D., Pugach, A.F., Cosovanu, I., Lesanu1, C., Deloly, J.-B., Vorobyov, D., Delets, A. and Olenici-Craciunescu, S.-B. (2014) Syzygy Effects Studies Performed Simultaneously with Foucault Pendulums and Torsinds during the Solar Eclipses of 13 November 2012 and 10 May 2013. *International Journal of Astronomy and Astrophysics*, **4**, 39-53. http://dx.doi.org/10.4236/ijaa.2014.41006

[8] Olenici, D. and Pugach, A.F. (2012) Precise Underground Observations of the Partial Solar Eclipse of 1 June 2011 Using a Foucault Pendulum and a Very Light Torsion Balance. *International Journal of Astronomy and Astrophysics*, **2**, 204-209. http://dx.doi.org/10.4236/ijaa.2012.24026

[9] Pugach, A.F. (2011) Is the Maurice Allais's Effect Exclusively Gravitational in Nature? In: Munera, H., Ed., *Should the Laws of Gravitation Be Reconsidered*? Apeiron, Montreal, 257-264.

[10] Pugach, A.F. (2013) Torsind as a Recorder of a Possibly New Energy. *Thermal Energy and Power Engineering*, **2**, 129-133. www.vkingpub.com/tepe

[11] Schlamminger, S. (2013) The Measurement of Newton's Constant of Gravitation.
      http://www.schlammi.com/ri_g.html

[12] Quinn, T., Parks, H, Speake, C. and Davis, R. (2013) Improved Determination of G Using Two Methods. *Physical Review Letters*, **111**, Article ID: 101102. http://dx.doi.org/10.1103/PhysRevLett.111.101102

[13] Luo, J., Liu, Q., Tu, L.-C., *et al.* (2009) Determination of the Newtonian Gravitational Constant G with Time-of-Swing Method. *Physical Review Letters*, **102**, Article ID: 240801. http://dx.doi.org/10.1103/PhysRevLett.102.240801

[14] Pugach, A.F. and Olenici, D. (2012) Observations of Correlated Behavior of Two Light Torsion Balances and a Paraconical Pendulum in Separate Locations during the Solar Eclipse of January 26th, 2009. *Advances in Astronomy*, **2012**, Article ID: 263818. http://dx.doi.org/10.1155/2012/263818

[15] Izmailov, V.P., Karagioz, O.V., Kuznetsov, V.A., *et al.* (1993) Temporal and Spatial Variations of the Measured Values of the Gravitational Constant. *Measurement Techniques*, **36**, 1065-1069. http://dx.doi.org/10.1007/BF00979446

18

On Kantowski-Sachs Viscous Fluid Model in Bimetric Relativity

R. C. Sahu[1], S. P. Misra[2], B. Behera[3]

[1]Department of Mathematics, K.S.U.B. College, Bhanjanagar, India
[2]Department of Mathematics, Sri Jagannath Mahavidyalaya, Rambha, India
[3]Department of Mathematics, U.N. College, Soro, India
Email: rcsahu2@rediffmail.com, rcsahu1958@gmail.com, sibaprasada_misra@rediffmail.com, benudharbhr@gmail.com

Abstract

Kantowski-Sachs plane symmetric models are investigated in bimetric theory of gravitation proposed by Rosen [1] in the context of bulk viscous fluid. Taking conservation law and the equation of state, two different models of the universe are obtained. It is observed that Kantowski-Sachs vacuum model obtained in first case and bulk viscous fluid model obtained in second case. It is also observed that the bulk viscous cosmological model always represents an accelerated universe and consistent with the recent observations of type-1a supernovae. Some physical and geometrical features of the viscous fluid model are studied.

Keywords

Bimetric Theory, Viscous Fluid, Kantowski-Sachs

1. Introduction

General relativity established by Einstein serves as a basis for constructing mathematical models of the universe. This theory has some controversies and lapses for which various alternative and modified theories of it have been proposed by authors from time to time to unify gravitation and matter fields in various forms. Most of the cosmological models based on general relativity and its modified theories such as Barber's second self creation theory, Einstein-Cartan, Gauge theory gravity, Brans-Dicke theory, Scalar-tensor theories, Scalar theories contain an initial singularity (the big-bang) from which the universe expands. Thus to get rid of the singularities that occur in general relativity and other theories, Rosen [1] proposed his bimetric theory of relativity. Other bimetric theories of gravitation are Born-Infeld (1934) bimetric theory (according to Moffat); J Moffat's non-symmetric gravitation theory(1979-1995); J Bekenstein's (1992) treatment of gravitational lensing and MOND; Clayton-Moffat (1998-2003) scalar-vector-tensor theory. Rosen's bimetric theory of relativity consists of two metric tensors at each point of the space time whose role is to determine physical situation. The first Riemannian metric tensor g_{ij}, which describes gravitation and the back ground metric tensor γ_{ij}, which enters into the field equations and interacts with g_{ij} but does not interact directly with matter. One can regards γ_{ij} as giving the geometry that would exist if there were no matter. Accordingly, at each point of the space-time one has two line elements

$$ds^2 = g_{ij}dx^i dx^j \tag{1}$$

and

$$d\sigma^2 = \gamma_{ij}dx^i dx^j \tag{2}$$

where ds is the interval between two neighbouring events as measured by a clock and a measuring rod. The interval $d\sigma$ is an abstract or a geometrical quantity which is not directly measurable. One can regard it as describing the geometry that exists if no matter was present. Moreover, this theory also satisfies the covariant and equivalence principles and agrees with the theory of general relativity up to the accuracy of observations made till the date.

As in general relativity, the variation principle also leads to the conservation law

$$T^{ij}_{;j} \equiv 0 \tag{3}$$

where (;) denotes covariant differentiation with respect to g_{ij}. Accordingly the geodesic equation of a rest particle is the same as that of general relativity.

The field equations of Rosen's bimetric theory of gravitation are

$$N^i_j - \frac{1}{2}N\delta^i_j = -8\pi k T^i_j \tag{4}$$

where

$$N^i_j = \frac{1}{2}\gamma^{ab}\left(g^{hi}g_{hj|a}\right)_{|b}, \quad N = N^i_j, \quad (i,j = 1,2,3,4)$$

and $k = \sqrt{\dfrac{g}{\gamma}} = \sqrt{\dfrac{-A^2 B^4 \sin^2\theta}{-1}} = AB^2 \sin\theta$ together with g = determinant of g_{ij} and γ = determinant of γ_{ij}.

Here the vertical bar (|) denotes the covariant differentiation with respect to γ_{ij} and T^i_j is the energy momentum tensor of the matter.

Usually the investigation of relativistic models has the energy momentum tensor of matter and generated by a perfect fluid. But to obtain more realistic models, one must consider the viscosity mechanism because the effect of bulk viscosity exhibits essential influence on the characteristic of the solution. The viscosity mechanism in cosmology has attracted the attention of many researchers as it can account for high entropy of the present universe (Weinberg [2] [3]). High entropy per baryon and the remarkable degree of isotropy of the cosmic microwave background radiation suggests that one should analyze dissipative effects in cosmology. Moreover, there are several processes which are expected to give rise to viscous effects. These are the decoupling of neutrinos during the radiation era and the decomposition of matter and radiation during the recombination era (Kolb and Turner [4]), decay of massive superstring models into massless models (Myung and Cho [5]), gravitational string production (Turok [6] and Barrow [7]) and particle creation effect in the grand unification era. Murphy [8] shows that introduction of bulk viscosity can avoid the big bang singularity. Hence one should consider the presence of material distribution other than the perfect fluid to get realistic cosmological models (see Gron [9] for a review on cosmological models with bulk viscosity) of the universe. If the present entropy is not due to bulk viscosity then perhaps it is produced by the effects of shear viscosity or heat conduction in an initially anisotropic or inhomogeneous expansion. Indeed, it may be just these dissipative processes that are responsible for smoothing out initial anisotropies and hence producing the high degree of isotropy observed in the cosmic microwave radiation background.

In general relativity, the relativists are generally using various symmetries to get physically viable information from the complicated structure of the field equations. The field equations of general relativity are non-linear in nature with ten unknowns $\left(g_{ij}\right)$. So it is very difficult to determine the exact solutions of the field equations.

The involvement of symmetry *i.e.* spherical or cylindrical or plane reduces the number of gravitational potentials g_{ij} and thus helps one in simplifying the field equation to some extent. A space-time that admits the three-parameter group of motions of Euclidian plane is said to possess plane symmetry and is called plane symmetric space-time. The origin of structure in the universe is one of the greatest mysteries even today. The present day observations indicate that the universe at large scale is homogeneous and isotropic and it is witnessing an accelerating phase as reported recently by Gasperini and Veneziano [10]. It is well known that the exact solution of general theory of relativity for homogeneous space-time belongs to either Bianchi types or Kantowski-Sachs [11].

Rosen [1] [12] [13], Yilmaz [14], Karade and Dhoble [15], Karade [16], Israelit [17]-[19], Liebscher [20], Reddy and Venkateswaralu [21], Deo and Thengane [22], Sahoo [23], Mohanty, Sahoo and Mishra [24] are some of the eminent authors, who have studied various aspects of bimetric theory.

Sahoo [23] has studied Kantowski-Sachs model in presence of cosmic cloud strings coupled with electromagnetic field in bimetric theory. He has shown that there is no contribution from Maxwell's field but established the geometric string model and vacuum model of the universe. Sahoo [25] has also studied Spherically symmetric Kantowski-Sachs space-time in bimetric theory of gravitation, considering the source of gravitation perfect fluid coupled with scalar meson field and has shown that the macro cosmological model-represented by perfect fluid does not exist, where as the micro cosmological model represented by scalar meson field exists. Sahu, Nayak and Behera [26] have found that Bianchi type-I cosmological models do not exist in bimetric theory of gravitation in presence of viscous fluid or mesonic viscous fluid with or without a mass parameter in general. Further, Kantowski-Sachs cosmological models are also studied by different authors like Tiwari and Dwibedi [27], Rahaman, Chakraborty, Bera and Das [28], Chaubey [29], Rao and Neelima [30], Adhav, Dawande and Raut [31], Hector Martinez and Carlos Peralta [32] in different angles.

To the best of our knowledge no author has studied Kantowski-Sachs plane symmetric model in the context of bimetric theory of relativity, when source of the gravitational field is governed by bulk viscous fluid. Therefore, in this paper we are interested to study this problem for two different cases. The work reported in first case concludes that Kantowski-Sachs plane symmetric model does not accommodate bulk viscous fluid in bimetric theory of relativity. However, Kantowski-Sachs bulk viscous fluid model obtained in second case.

2. Field Equations

Consider the Kantowski-Sachs [33] metric in the form

$$\mathrm{d}s^2 = \mathrm{d}t^2 - A^2\mathrm{d}r^2 - B^2\left(\mathrm{d}\theta^2 + \sin^2\theta\mathrm{d}\varnothing^2\right) \tag{5}$$

where the metric potentials A and B are functions of cosmic time "t" only.

The background flat space-time metric is

$$\mathrm{d}\sigma^2 = \mathrm{d}t^2 - \mathrm{d}r^2 - \left(\mathrm{d}\theta^2 + \sin^2\theta\mathrm{d}\varnothing^2\right). \tag{6}$$

The energy momentum tensor for bulk viscous fluid distribution is given by

$$T_{ij}^v = \left(\rho + \bar{p}\right)u_i u_j - \bar{p}g_{ij}; \tag{7}$$

with

$$\bar{p} = p - \eta u_{;i}^i, \tag{8}$$

where p is the proper pressure, ρ is the energy density, $\bar{p}$ is the effective pressure, u_i is the four velocity vector of the fluid and η is the bulk viscous coefficient of the fluid.

Since the bulk viscous pressure represents only a small correction to the thermo dynamical pressure, it is reasonable assumption that the inclusion of viscous term in the energy momentum tensor does not change fundamentally the dynamics of the cosmic evolution. For the specification of η, we assume that the fluid obeys an equation of state of the form

$$p = \lambda\rho, \quad 0 \leq \lambda \leq 1. \tag{9}$$

Here λ is called the adiabatic parameter.

Using comoving co-ordinate system, the field Equation (4) for the metrics (5) and (6) corresponding to the energy momentum tensor (7) can be written as

$$\left(\frac{A_4}{A}\right)_4 - 2\left(\frac{B_4}{B}\right)_4 = 16\pi k\overline{p}, \tag{10}$$

$$\left(\frac{A_4}{A}\right)_4 = -16\pi k\overline{p} \tag{11}$$

And

$$\left(\frac{A_4}{A}\right)_4 + 2\left(\frac{B_4}{B}\right)_4 = 16\pi k\rho. \tag{12}$$

Equation (8) can be expressed as

$$\overline{p} = p - \eta\left(\frac{A_4}{A} + 2\frac{B_4}{B}\right). \tag{13}$$

Here and afterwards the suffix "4" after a field variable represents ordinary differentiation with respect to time "t" only.

3. Solution of the Field Equations

Equations (10), (11) and (12) yield

$$\left(\frac{A_4}{A}\right)_4 = \left(\frac{B_4}{B}\right)_4 = -16\pi k\overline{p} = \frac{16}{3}\pi k\rho. \tag{14}$$

Taking last two terms of Equation (14), we get

$$3\overline{p} + \rho = 0. \tag{15}$$

Equations (15) and (13) yield

$$(3p + \rho) - 3\eta\left(\frac{A_4}{A} + 2\frac{B_4}{B}\right) = 0. \tag{16}$$

Case-1: From the reality conditions, we have

$$\overline{p} \geq 0 \text{ and } \rho \geq 0. \tag{17}$$

So from Equation (15), we find

$$\overline{p} = \rho = 0. \tag{18}$$

Use of (18) in Equation (9), we obtain

$$p = 0. \tag{19}$$

By help of Equation (18), equations (10) and (11) yield

$$\left(\frac{A_4}{A}\right)_4 = \left(\frac{B_4}{B}\right)_4 = 0. \tag{20}$$

On integration, (20) yields

$$A = B = e^{a_1 t + a_2} \tag{21}$$

where a_1 and a_2 are constants of integration.

Putting the values of A and B from (21), and use of Equations (18) and (19) in Equation (16), we have

$$\eta = 0. \tag{22}$$

Thus the metric (5) corresponding to Equations (21) & (22) takes the form

$$ds^2 = dt^2 - e^{2(a_1 t + a_2)}\left(dr^2 + d\theta^2 + \sin^2\theta d\varnothing^2\right). \tag{23}$$

With proper choice of co-ordinates Equation (23) can be transformed to

$$ds^2 = dT^2 - e^{2a_1 T} \left(dr^2 + d\theta^2 + \sin^2 \theta d\varnothing^2 \right). \tag{24}$$

As $\eta = 0$, so Kantowski-Sachs viscous fluid model does not survive in bimetric theory but vacuum model of the universe only exists.

It is observed from (18), (19) and (22) that

$$\eta = \rho = p = 0. \tag{25}$$

Thus the above result reduces to that of result already obtained by Sahoo [25].

Case-2:

With help of the conservation property (3), metric (5) takes the form

$$\rho_4 + \left(\overline{p} + \rho \right) \left(\frac{A_4}{A} + 2\frac{B_4}{B} \right) = 0. \tag{26}$$

By the help of (15) Equation (26) yields

$$\rho_4 + \frac{2\rho}{3} \left(\frac{A_4}{A} + 2\frac{B_4}{B} \right) = 0. \tag{27}$$

To avoid complexity in the problem substituting the relation $\dfrac{A_4}{A} = \dfrac{B_4}{B}$ from Equation (14) in Equation (27), we have

$$\rho = \frac{C_1}{B^2}, \tag{28}$$

where C_1 is the constant of integration.

Use of (28) and value of "k" from (4) in Equation (14), we get

$$\left(\frac{A_4}{A} \right)_4 = \frac{16}{3} \pi k \rho = \left(\frac{16}{3} \pi C_1 \sin \theta \right) \cdot (A) = \alpha A \ (\text{say}), \tag{29}$$

where $\alpha = \dfrac{16}{3} \pi C_1 \sin \theta$ (constant).

Now Equation (29) can be expressed as

$$\frac{1}{A} \cdot \left(\frac{A_4}{A} \right)_4 = \alpha. \tag{30}$$

Integrating (30), one can obtain

$$A = \frac{4}{\left(\sqrt{2\alpha} t + C_2 \right)^2}, \tag{31}$$

where C_2 is the constant of integration.

As we have consider the relation $\dfrac{A_4}{A} = \dfrac{B_4}{B}$, so we can find

$$A = \frac{4}{\left(\sqrt{2\alpha} t + C_2 \right)^2} = B. \tag{32}$$

Thus (28) with the help of (32) yields

$$\rho = \frac{C_1}{16} \left(\sqrt{2\alpha} t + C_2 \right)^4. \tag{33}$$

Now use of Equation (33) in Equation (15), we get

$$\bar{p} = -\frac{\rho}{3} = -\frac{C_1}{48}\left(\sqrt{2\alpha}t + C_2\right)^4. \tag{34}$$

Putting the value of ρ from (33) in Equation (9) and assigning different values to λ, we get

$$p = \rho = \frac{C_1}{16}\left(\sqrt{2\alpha}t + C_2\right)^4; \text{ when } \lambda = 1 \tag{35}$$

$$p = 0; \text{ when } \lambda = 0 \tag{36}$$

and

$$p = \frac{\rho}{3} = \frac{C_1}{48}\left(\sqrt{2\alpha}t + C_2\right)^4, \text{ when } \lambda = \frac{1}{3} \tag{37}$$

Using (32), Equation (13) yields

$$\eta = \frac{p - \bar{p}}{3\dfrac{A_4}{A}} = \frac{p - \bar{p}}{\dfrac{-6\sqrt{2\alpha}}{\left(\sqrt{2\alpha}t + C_2\right)}} \tag{38}$$

By use of (35), (36), (37) separately in (38) and then using (34) in each case, we get

$$\eta = -\frac{C_1\left(\sqrt{2\alpha}t + C_2\right)^5}{72\sqrt{2\alpha}}, \text{ for the case of stiff fluid} \tag{39}$$

$$\eta = -\frac{C_1\left(\sqrt{2\alpha}t + C_2\right)^5}{288\sqrt{2\alpha}}, \text{ for the case of dust fluid} \tag{40}$$

$$\eta = -\frac{C_1\left(\sqrt{2\alpha}t + C_2\right)^5}{144\sqrt{2\alpha}}, \text{ for the case of radiating fluid} \tag{41}$$

Therefore in view of (32), the line element (5) can be written in the form

$$ds^2 = dt^2 - 16\left(\sqrt{2\alpha}t + C_2\right)^{-4}\left(dr^2 + d\theta^2 + \sin^2\theta d\varnothing^2\right). \tag{42}$$

The above model of the universe can be transformed through a proper choice of coordinates to the form

$$dS^2 = dT^2 - 16\left(T\right)^{-4}\left(dr^2 + d\theta^2 + \sin^2\theta d\varnothing^2\right). \tag{43}$$

4. Physical and Geometrical Properties of the Model (43)

i. The Spatial Volume V of the Universe:
The spatial volume V of the universe is found to be

$$V = \sqrt{-g} = \frac{64\sin\theta}{\left[\sqrt{2\alpha}t + C_2\right]^6}.$$

Now $V \rightarrow$ constant as $t \rightarrow 0$ and $V \rightarrow 0$ as $t \rightarrow \infty$.

Thus we inferred from the results obtained above that the universe starts from a constant volume and collapse at infinite future.

ii. The Expansion Scalar θ:
The Expansion Scalar "θ" in the model is found to be

$$\theta = u^i_{;i} = \frac{A_4}{A} + 2\frac{B_4}{B} = -\frac{6\sqrt{2\alpha}}{\left(\sqrt{2\alpha}t + C_2\right)}.$$

Hence as $t \rightarrow 0$, $\theta \rightarrow$ constant and as $t \rightarrow \infty$, $\theta \rightarrow 0$.

This result shows that the model has the constant rate of expansion at initial time but as time increases the rate of expansion becomes slow and there will be no expansion at infinite future.

iii. Anisotropy of the Universe:

The shear scalar σ (Ray Choudhuri [34]), defined by

$$\sigma^2 = \frac{1}{12}\left\{\left[\frac{g_{11,4}}{g_{11}} - \frac{g_{22,4}}{g_{22}}\right]^2 + \left[\frac{g_{22,4}}{g_{22}} - \frac{g_{33,4}}{g_{33}}\right]^2 + \left[\frac{g_{33,4}}{g_{33}} - \frac{g_{11,4}}{g_{11}}\right]^2\right\},$$

for the model yield $\sigma^2 = \dfrac{32\alpha}{3\left[\sqrt{2\alpha}t + C_2\right]^2}$.

Therefore $\sigma^2 \to a$ constant as $t \to 0$ and $\sigma^2 \to 0$ as $t \to \infty$. Thus it is inferred that the model is anisotropic at initial time but gradually approaches to isotropic as time increases. It is interesting that at infinite future the universe may turns to isotropic state. Since the universe in a smaller case is neither homogeneous, so the transition from anisotropic to isotropic state might have happened in the early universe which is not supported by any observed or experimental data. However there are theoretical arguments that sustain the existence of an anisotropic phase that approaches an isotropic case (Misner, [35] Chaotic Cosmology). The early universe could also be characterized by an irregular expansion mechanism. Thus it would be useful to explore models in which anisotropies existing at an early stage of expansion, are damped out in the course of evolution and such models have received some attention (Hu & Parker, [36]). As $\mathrm{Lim}_{t\to\infty}\left(\dfrac{\sigma}{\theta}\right) = \pm\dfrac{2}{3\sqrt{3}}$, the model is anisotropic at the initial time and continues throughout the evolution.

iv. Hubble parameter:

The Hubble parameter H in the model is found to be $H = \dfrac{-2\sqrt{2\alpha}}{\left[\sqrt{2\alpha}t + C_2\right]^2}$. As H is a function of time so the model is not a steady state model.

v. Scale factor:

The scale factor S^3 in the model is found to be $S^3 = \dfrac{64\sin\theta}{\left[\sqrt{2\alpha}t + C_2\right]^2}$. Thus "S" decreases as time increases.

vi. The deceleration parameter:

The deceleration parameter 'q' in the models defined by $q = -\dfrac{VV_{44}}{V_4^2} = -\dfrac{7}{6}$, which has $-$ve sign. Hence the model of the universe corresponds to an inflationary model. The model represents an accelerating universe in bimetric theory of gravitation and also consistent with the recent observations of type-Ia supernovae.

vii. Energy conditions for viscous fluid:

The strong, weak and dominant energy conditions *i.e.* $\rho + 3\bar{p} \geq 0, \rho > 0$ and $\rho + \bar{p} \geq 0$ are given by in the model as

$$\rho + 3\bar{p} = 0, \quad \rho = \frac{C_1}{16}\left(\sqrt{2\alpha}t + C_2\right)^4 \quad \text{and} \quad \rho + \bar{p} = \frac{C_1}{24}\left(\sqrt{2\alpha}t + C_2\right)^4.$$

It is observed from above data that the strong energy conditions is satisfied in the model. The weak and dominant energy conditions are also satisfied when $t > 0$. Again as $t \to 0$, $\rho \to a$ constant and as $t \to \infty$, $\rho \to \infty$. Thus we inferred that the model has no singularity at $t = 0$ and the space time reduces to flat space time.

viii. Bulk viscous coefficient:

The bulk viscous coefficient η are found in the model as given below:

$$\eta = -\frac{C_1\left(\sqrt{2\alpha}t + C_2\right)^5}{72\sqrt{2\alpha}}, \quad \text{for the case of stiff fluid}$$

$$\eta = -\frac{C_1 \left(\sqrt{2\alpha}t + C_2 \right)^5}{288\sqrt{2\alpha}}, \ \text{for the case of dust fluid}$$

$$\text{and} \ \ \eta = -\frac{C_1 \left(\sqrt{2\alpha}t + C_2 \right)^5}{144\sqrt{2\alpha}}, \ \text{for the case of radiating fluid}$$

In all the cases it is observed that as $t \to 0$, $\eta \to -ve$ constant and as $t \to \infty$, $\eta \to -\infty$. So it is evident from the above result that the solutions leads to unphysical situations and hence there is no singularity involved in the model.

5. Conclusion

In this paper, Kantowski-Sachs models are constructed in Rosen's bimetric theory of gravitation when the energy momentum tensor is bulk viscous fluid. Applying the conservation equation and also the equation of state, two different models of the Kantowski-Sachs universe are obtained *i.e.* vacuum model and bulk viscous fluid model. It is observed that the bulk viscous cosmological model always represents an accelerated universe and also is consistent with the recent observations of type-1a supernovae. The model obtained is not of a steady state model and has no singularity. Also the model is anisotropic at initial time but approaches to isotropy at infinite future. As there is one way to avoid singularity is energy density ρ to vanish, so Rosen's model in the context of bulk viscous fluid is only valid when the energy density ρ is not zero.

Acknowledgements

The authors thank the reverend referee for his constructive comments to bring the paper in improvement form.

References

[1]   Rosen, N. (1973) A Bi-Metric Theory of Gravitation. *General Relativity and Gravitation*, **4**, 435-447. http://dx.doi.org/10.1007/BF01215403

[2]   Weinberg, S. (1971) Entropy Generation and the Survival of Protogalaxies in an Expanding Universe. *Astrophysical Journal*, **168**, 175. http://dx.doi.org/10.1086/151073

[3]   Weinberg, S. (1972) Gravitational and Cosmology. Wiley and Sons, New York.

[4]   Kolb, E.W. and Turner, M.S. (1990) The Early Universe. Additson-Wesley, Boston.

[5]   Myung, S. and Cho, B.M. (1986) Entropy Production in a Hot Heterotic String. *Modern Physics Letters A*, **1**, 37. http://dx.doi.org/10.1142/S0217732386000075

[6]   Turok, N. (1988) String-Driven Inflation. *Physical Review Letters*, **60**, 549. http://dx.doi.org/10.1103/PhysRevLett.60.549

[7]   Barrow, J.D. (1988) String-Driven Inflationary and Deflationary Cosmological Models. *Nuclear Physics B*, **310**, 743-763. http://dx.doi.org/10.1016/0550-3213(88)90101-0

[8]   Murohy, G.L. (1973) Big-Bang Model without Singularities. *Physical Review D*, **8**, 4231. http://dx.doi.org/10.1103/PhysRevD.8.4231

[9]   Grøn, Ø. (1990) Viscous Inflationary Universe Models. *Astrophysics and Space Science*, **173**, 191-225. http://dx.doi.org/10.1007/BF00643930

[10] Gasperini, M. and Veneziano, G. (2003) The Pre-Big Bang Scenario in String Cosmology. *Physics Reports*, **373**, 1. http://dx.doi.org/10.1016/S0370-1573(02)00389-7

[11] Roy Choudhuary, A.K. (1979) Theoretical Cosmology. Clarendon, Oxford.

[12] Rosen, N. (1975) A Bi-Metric Theory of Gravitation. II. *General Relativity and Gravitation*, **6**, 259-268. http://dx.doi.org/10.1007/BF00751570

[13] Rosen, N. (1978) Bimetric Gravitation Theory on a Cosmological Basis. *General Relativity and Gravitation*, **9**, 339-351. http://dx.doi.org/10.1007/BF00760426

[14] Yilmaz, H. (1975) On Rosen's Bi-Metric Theory of Gravitation. *General Relativity and Gravitation*, **6**, 269-276. http://dx.doi.org/10.1007/BF00751571

[15] Karade, T.M. and Dhoble, Y.S. (1980) Axially Symmetric Vacuum Solutions of the Bimetric Relativity Theory. *Let-

tere al Nuovo Cimento, **29**, 390-392. http://dx.doi.org/10.1007/BF02743238

[16] Karade, T.M. (1980) *Indian Journal of Pure and Applied Mathematics*, **11**, 1202.

[17] Israelit, M. (1976) Equations of Motion in Rosen's Bimetric Theory of Gravitation. *General Relativity and Gravitation*, **7**, 623-641. http://dx.doi.org/10.1007/BF00770720

[18] Israelit, M. (1979) Background Killing Vectors and Conservation Laws in Rosen's Bimetric Theories of Gravitation. *General Relativity and Gravitation*, **11**, 25-36. http://dx.doi.org/10.1007/BF00756669

[19] Israelit, M. (1981) Spherically Symmetric Fields in Rosen's Bimetric Theories of Gravitation. *General Relativity and Gravitation*, **13**, 681-688. http://dx.doi.org/10.1007/BF00759411

[20] Liebscher, D.E. (1975) Bi-Metric Theories of Gravitation. *General Relativity and Gravitation*, **6**, 277-280. http://dx.doi.org/10.1007/BF00751572

[21] Reddy, D.R.K. and Venkateswaralu, R. (1989) Non-Existence of Biachi Type-1 Perfect Fluid Cosmological Models in a Bi-Metric Theory of Gravitation. *Astrophysics and Space Science*, **158**, 169-171. http://dx.doi.org/10.1007/BF00637454

[22] Deo, S. and Thengane, K.D. (2002) *FIZIKABII*, **3**, 155.

[23] Sahoo, P.K. (2009) On Kantowski-Sachs Cosmic Strings Coupled with Maxwell Fields in Bimetric Relativity. *International Journal of Theoretical Physics*, **49**, 25-30.

[24] Mohanty, G., Sahoo, P.K. and Mishra, B. (2002) On Bianchi Type-I Mesonic Cosmological Model in Bimetric Theory. *Astrophysics and Space Science*, **281**, 609-612. http://dx.doi.org/10.1023/A:1015868106122

[25] Sahoo, P.K. (2005) Spherically Symmetric Cosmological Models in Bimetric Theory. *Bulgarian Journal of Physics*, **32**, 175-180.

[26] Sahu, R.C., Nayak, B. and Behera, B. (2013) Viscous and Mesonic Viscous Fluid Models in Bimetric Theory of Gravitation. *Bulletin of The Allahabad Mathematical Society*, **28**, 1-17.

[27] Tiwari, R.K. and Dwivedi, U. (2010) Kantowski-Sachs Cosmological Models with Time-Varying G and K. *FIZIKA B*(Zegreb), **19**, 1-8.

[28] Rahaman, F., Chakraborty, N., Bera, J. and Das, S. (2002) Homogeneous Kantowski-Sachs Model in Lyra Geometry. *Bulgarian Journal of Physics*, **29**, 91-96.

[29] Chaubey, R. (2011) Bianchi Type-III and Kantowski-Sachs Universes with Wet Dark Fluid. *International Journal of Astronomy and Astrophysics*, **1**, 25-38.

[30] Rao, V.U.M. and Neelima, D. (2013) Kantowski-Sachs String Cosmological Model with Bulk Viscosity in General Scalar Tensor Theory of Gravitation. *ISRN Mathematical Physics*, **2013**, Article ID: 759274.

[31] Adhav, K.S., Dawande, M.V. and Raut, R.B. (2012) *G.J.P &A.Sc. and Tech.*, **2**, 65-77.

[32] Martinez, H. and Peralta, C. (2005) *Rev.Mex.Fis.*, **51**, 22-26.

[33] Kantowski, R. and Sachs, R.K. (1966) Some Spatially Homogeneous Anisotropic Relativistic Cosmological Models. *Journal of Mathematical Physics*, **7**, 433. http://dx.doi.org/10.1063/1.1704952

[34] Roy Chaudhury, A.K. (1955) *Physical Review*, **98**, 1123.

[35] Misner, C.W. (1968) The Isotropy of the Universe. *Astrophysical Journal*, **151**, 431. http://dx.doi.org/10.1086/149448

[36] Hu, B.L. and Parker, L. (1978) Anisotropy Damping through Quantum Effects in the Early Universe. *Physical Review D*, **17**, 933. http://dx.doi.org/10.1103/PhysRevD.17.933

19

Restricted Three Body Problem with Stokes Drag Effect

Mamta Jain[1], Rajiv Aggarwal[2]

[1]Department of Mathematics, Shri Venkateshwara University, Gajraula, India
[2]Department of Mathematics, Sri Aurobindo College, University of Delhi, Delhi, India
Email: mamtag27@gmail.com, rajiv_agg1973@yahoo.com

Abstract

The existence and stability of stationary solutions of the restricted three body problem under the effect of the dissipative force, Stokes drag, are investigated. It is observed that there exist two non collinear stationary solutions. Further, it is also found that these stationary solutions are unstable for all values of the parameters.

Keywords

Restricted Three Body Problem, Libration Points, Linear Stability, Dissipative Forces, Stokes Drag

1. Introduction

Two finite masses, called primaries, are moving in circular orbits around their common centre of mass, and an infinitesimal mass is moving in the plane of motion of the primaries. To study the motion of the infinitesimal mass is called the restricted three body problem. [1] proved that there existed five points of equilibrium, or points of libration (often denoted by L_1, ... L_5), which were the stationary solutions of the restricted problem. Out of them, three are collinear and two are non collinear. The collinear libration points are unstable for all values of mass parameter μ and the triangular libration points are stable for $0 < \mu < \mu_c$, where $\mu_c = 0.03852 \cdots$ is a critical value of mass parameter [2].

As we know, dissipative forces are those where there is a loss of energy such as friction and one of the most important mechanisms of dissipation is the Stokes drag which is a force experienced by a particle moving in a gas, due to the collisions of the particle with the molecules of the gas.

[3] has determined some results on the global dynamics of the regularized restricted three body problem with dissipative forces. Their investigations have motivated us to study the motion of the restricted three body problem under dissipative forces such as Stokes drag. In the synodic frame, Stokes drag force is defined by [4]:

$$\left(F_x, F_y\right) = -k\left(\dot{x} - y + \alpha\,\Omega_y,\ \dot{y} + x - \alpha\,\Omega_x\right)$$

where $k \in (0,1)$ is the dissipative constant, depending on several physical parameters like the viscosity of the gas, the radius and mass of the particle. Here

$$\Omega = \Omega(r) \equiv r^{-\frac{3}{2}}$$

is the keplerian angular velocity at distance $r = \sqrt{x^2 + y^2}$ from the origin of the synodic frame and $\alpha \in (0,1)$ is the ratio between the gas and keplerian velocities.

A number of authors have investigated the location and stability of the equilibrium point in the presence of specific dissipative forces. [5] has used the Jacobi constant to investigate the effect of an external drag force proportional to the velocity in the rotating frame and has concluded that L_4 and L_5 are unstable to this type of drag force. In their studies of the motion of dust particles in the vicinity of the Earth, [6] has analyzed the stability of the equilibrium points in the presence of radiation pressure which includes the Poynting Robertson drag terms. They have shown that the libration points are unstable to such a drag force. The effects of radiation pressure and Poynting Robertson light drag on the classical equilibrium points are analyzed by [7] and [8]. [9] has systematically discussed the dynamical effect of general drag in the planar circular restricted three body problem and has found that L_4 and L_5 are asymptotically stable with this kind of dissipation. It has been shown by [10], [11], [12] and [13] that, in the case of Stokes drags, exterior resonances may compensate the decrease of the semi major axis and that stationary solutions still exist. A numerical analysis of 1:1 resonance, taking into account the effect of the inclination and the eccentricity, has been studied by [14]. An analytical study of the linearised stability of L_4 and L_5 is provided in [4].

Furthermore, [15] has examined the linear stability of triangular equilibrium points in the generalized photo gravitational restricted three body problem with Poynting Robertson drag. They have considered the smaller primary as an oblate body and bigger one as radiating and they have concluded that the triangular equilibrium points are unstable in linear sense. [16] has discussed the nonlinear stability in the generalized restricted three body problem with Poynting Robertson drag considering smaller primary as an oblate body and bigger one radiating. They have proved that the triangular points are stable in nonlinear sense. [17] has discussed the stability of triangular equilibrium points in photo gravitational circular restricted three body problem with Poynting Robertson drag and a smaller triaxial primary. They proved that the parameters involved in the problem (radiation pressure, oblateness and Poynting Robertson drag) influenced the position and linear stability of triangular points. In the presence of Poynting Robertson drag, triangular points are unstable, and in the absence of Poynting Robertson drag, these points are conditionally stable. In a series of papers, [18] has performed an analysis in the restricted three body problem with Poynting Robertson drag effect. They found that there existed two noncollinear stationary solutions which were linearly unstable.

In the present paper, we study the same problem but with the effects of stokes drag instead Poynting Robertson drag on noncollinear libration points L_4 and L_5 in the restricted three body problem.

2. Equations of Motion

Suppose m_1 and m_2 are the primaries revolving with angular velocity n in circular orbits about their centre of mass O, an infinitesimal mass m_3 is moving in the plane of motion of m_1 and m_2. The line joining m_1 and m_2 is taken as X-axis and "O" their center of mass as origin and the line passing through O and perpendicular to OX and lying in the plane of motion of m_1 and m_2 is the Y-axis. We consider a synodic system of coordinates O (xyz); initially coincident with the inertial system O (XYZ), rotating with the angular velocity n about Z-axis; (the z-axis is coincident with Z-axis) (**Figure 1**).

In the synodic axes the equation of motion of m_3 in the restricted three body problem with Stokes drag S is

$$m_3\left(\frac{\partial^2 r}{\partial t^2} + 2\omega \times \frac{\partial r}{\partial t} + \frac{\partial \omega}{\partial t} \times r + \omega \times (r \times \omega)\right) = F \tag{1}$$

where

$$F = F_1 + F_2 + S,$$

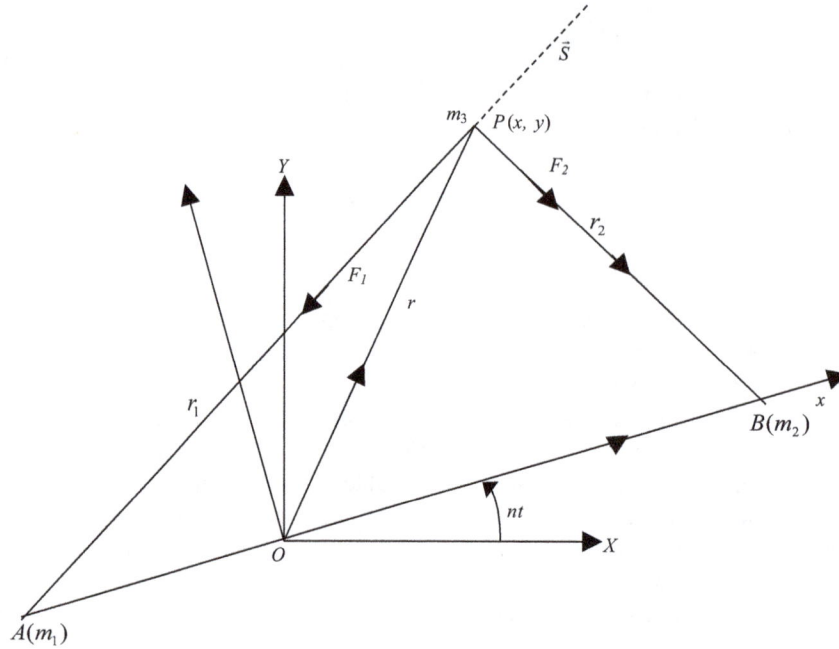

Figure 1. Configuration of the restricted three body problem with Stokes drag S.

F_1 = Gravitational Force acting on m_3 due to $m_1 = G\dfrac{m_3 m_1}{r_1^2}\hat{r}_1$,

F_2 = Gravitational Force acting on m_3 due to $m_2 = G\dfrac{m_3 m_2}{r_2^2}\hat{r}_2$,

S = Stokes drags Force acting on m_3 due to m_1 along $\overline{AP}$.

Its components along the synodic axes (x, y) are $S_x = k(\dot{x} - y) + \alpha S_y'$ and $S_y = k(\dot{y} + x) - \alpha S_x'$.
where

$$S' = S'(r) = r^{\frac{-3}{2}},$$

$$r = \overline{OP} = xi + yj,$$

$$\boldsymbol{\omega} = n\text{K} = \text{ Angular velocity of the axes } O(xy) = \text{const.}$$

The equations of motion of m_3 in Cartesian coordinates (x, y) are

$$\ddot{x} - 2n\dot{y} - n^2 x = -Gm_1 \frac{(x - x_1)}{r_1^3} - Gm_2 \frac{(x - x_2)}{r_2^3} - Gk(\dot{x} - y + \alpha S_y'),$$

$$\ddot{y} + 2n\dot{x} - n^2 y = -Gm_1 \frac{y}{r_1^3} - Gm_2 \frac{y}{r_2^3} - Gk(\dot{y} + x - \alpha S_x'),$$

where

n = Mean motion, G = Gravitational constant,

$(x_1, 0)$ and $(x_2, 0)$ = coordinates of A and B in the synodic system.

Using [2] terminology, the distance between primaries is unchanged and same is taken equal to one; the sum of the masses of the primaries is also taken as one. The unit of time is chosen so as to make the gravitational constant unity. The equations of motion of the infinitesimal mass m_3 in the synodic coordinate system (x, y) and dimensionless variables are

$$\ddot{x} - 2\dot{y} = \Omega_x - k(\dot{x} - y + \alpha S_y'), \tag{2}$$

$$\ddot{y} + 2\dot{x} = \Omega_y - k\left(\dot{y} + x - \alpha\, S_x'\right) \tag{3}$$

where

$$\Omega = \frac{1}{2}\left(x^2 + y^2\right) + \frac{(1-\mu)}{r_1} + \frac{\mu}{r_2},$$

$$r_1^2 = \left(x + \mu\right)^2 + y^2, \tag{4}$$

$$r_2^2 = \left(x + \mu - 1\right)^2 + y^2, \tag{5}$$

$$\mu = \frac{m_2}{m_1 + m_2} \le \frac{1}{2} \Rightarrow m_1 = 1 - \mu;\ m_2 = \mu.$$

The Stokes drag effect is of the order of $k = 10^{-5}, \alpha = 0.05$ (generally $k \in (0,1)$ and $\alpha \in (0,1)$ as stated in the introduction).

3. Stationary Solutions (Libration Points)

The solutions (x, y) of Equations (2) and (3) with $\ddot{x} = 0, \ddot{y} = 0, \dot{x} = 0, \dot{y} = 0$ are given by

$$x - (1-\mu)\frac{(x+\mu)}{r_1^3} - \mu\frac{(x+\mu-1)}{r_2^3} + k\left(y + \alpha\, S_y'\right) = 0, \tag{6}$$

and

$$y\left(1 - \frac{(1-\mu)}{r_1^3} - \frac{\mu}{r_2^3}\right) - k\left(x - \alpha\, S_x'\right) = 0. \tag{7}$$

Here, if we take $k = 0$, then it will be the classical case of the restricted three body problem and the solutions of these equations are just the five classical Lagrangian equilibrium points L_i (i = 1, 2, 3, 4, 5). The L_i (i = 1, 2, 3) are three collinear libration points which lie along the x-axis and L_i (i = 4, 5) are the two non collinear libration points which make the equilateral triangles with the primaries. Due to the presence of the Stokes drag force, it is clear from Equations (6) and (7) that collinear equilibrium solution does not exist. Since there is a possibility of non collinear libration points under the effect of drag forces, so we restrict our analysis to these points. Their locations when $k = 0$, are (see, e.g., [19])

$$L_{4,5}\left[x_0 = \frac{1}{2} - \mu,\quad y_0 = \pm\frac{\sqrt{3}}{2}\right].$$

Now, we suppose that the solution of Equations (6) and (7) when $k \ne 0$ and $y \ne 0$ are given by

$$\bar{x} = x_0 + \pi_1,\quad \bar{y} = y_0 + \pi_2 \qquad \pi_1, \pi_2 \ll 1$$

Making the above substitutions in Equations (6) and (7), and applying Taylors series expansion around the libration points by using that (x_0, y_0) is a solution of these equations when $k = 0$, we can get a linear set of equations.

$$\pi_1\left[1 + (1-\mu)\frac{3(x_0+\mu)^2}{\left\{(x_0+\mu)^2 + y_0^2\right\}^{\frac{5}{2}}} - \frac{1}{\left\{(x_0+\mu)^2 + y_0^2\right\}^{\frac{3}{2}}} + \mu\frac{3(x_0+\mu-1)^2}{\left\{(x_0+\mu-1)^2 + y_0^2\right\}^{\frac{5}{2}}} - \frac{1}{\left\{(x_0+\mu-1)^2 + y_0^2\right\}^{\frac{3}{2}}}\right]$$

$$+ \pi_2\left[(1-\mu)\frac{3(x_0+\mu)y_0}{\left\{(x_0+\mu)^2 + y_0^2\right\}^{\frac{5}{2}}} + \mu\frac{3(x_0+\mu-1)y_0}{\left\{(x_0+\mu-1)^2 + y_0^2\right\}^{\frac{5}{2}}}\right] + k\left[y_0 + \frac{3}{2}\alpha\left(x_0^2 + y_0^2\right)^{\frac{-7}{4}}y_0\right] = 0 \tag{8}$$

and

$$\pi_2 \left[1 + (1-\mu) \frac{3y_0^2}{\left\{ (x_0+\mu)^2 + y_0^2 \right\}^{\frac{5}{2}}} - \frac{1}{\left\{ (x_0+\mu)^2 + y_0^2 \right\}^{\frac{3}{2}}} + \mu \frac{3y_0^2}{\left\{ (x_0+\mu-1)^2 + y_0^2 \right\}^{\frac{5}{2}}} - \frac{1}{\left\{ (x_0+\mu-1)^2 + y_0^2 \right\}^{\frac{3}{2}}} \right]$$

$$+ \pi_1 \left[(1-\mu) \frac{3(x_0+\mu)y_0}{\left\{ (x_0+\mu)^2 + y_0^2 \right\}^{\frac{5}{2}}} + \mu \frac{3(x_0+\mu-1)y_0}{\left\{ (x_0+\mu-1)^2 + y_0^2 \right\}^{\frac{5}{2}}} \right] - k \left[x_0 + \frac{3}{2} \alpha \left(x_0^2 + y_0^2 \right)^{\frac{-7}{4}} x_0 \right] = 0 \tag{9}$$

After substituting the values of the constants x_0 and y_0 in the above equations and rejecting the second and higher order terms in π_1 and π_2, we get the values of π_1 and π_2 as

$$\pi_1 = -\frac{1}{2\sqrt{3}} (2+3\alpha) \mu k,$$

$$\pi_2 = \frac{5}{18} (2+3\alpha) \mu k.$$

Hence, putting the values of π_1 and π_2, the displaced equilibrium points are given by

$$L_{4,5} \left[\overline{x} = \frac{1}{2} - \mu - \frac{1}{2\sqrt{3}} (2+3\alpha) \mu k, \overline{y} = \pm \frac{\sqrt{3}}{2} + \frac{5}{18} (2+3\alpha) \mu k \right] \tag{10}$$

Here, the shifts in L_4 and L_5 are of $O(k/\mu)$. If we calculate $(\overline{x}, \overline{y})$ numerically, taking $k = 10^{-5}$, $\alpha = 0.05$ for different values of μ, we find that while using Stokes drag, as far as the values of μ increase corresponding $\overline{x}$ values decrease and the $\overline{y}$ values increase.

4. Stability of $L_{4,5}$

We write the variational equations by putting $x = \overline{x} + \xi$ and $y = \overline{y} + \eta$, ξ, $\eta \ll 1$, in the equations of motion (2) and (3), where $(\overline{x}, \overline{y})$ are the coordinates of the libration point. Therefore, expanding $f(\overline{x}, \overline{y})$ and $g(\overline{x}, \overline{y})$ by Taylors Theorem, we get

$$\ddot{\xi} - 2\dot{\eta}$$

$$= \Omega_x (\overline{x}, \overline{y}) + \xi \left[1 - \frac{\mu}{(\overline{r}_2)^3} + \frac{3\mu(\overline{x}+\mu-1)^2}{(\overline{r}_2)^5} + \frac{3(1-\mu)(\overline{x}+\mu)^2}{(\overline{r}_1)^5} - \frac{(1-\mu)}{(\overline{r}_1)^3} - k - \frac{21\overline{x}\,\overline{y}\alpha}{4} \left(\overline{x}^2 + \overline{y}^2 \right)^{\frac{-11}{4}} k \right] \tag{11}$$

$$+ \eta \left[\frac{3\overline{y}\mu(\overline{x}+\mu-1)}{(\overline{r}_2)^5} + \frac{3\overline{y}(1-\mu)(\overline{x}+\mu)}{(\overline{r}_1)^5} + k + \frac{3}{2}\alpha \left(\overline{x}^2 + \overline{y}^2 \right)^{\frac{-7}{4}} k - \frac{21}{4} \overline{y}^2 \alpha \left(\overline{x}^2 + \overline{y}^2 \right)^{\frac{-11}{4}} k \right],$$

$$\ddot{\eta} + 2\dot{\xi}$$

$$= \Omega_y (\overline{x}, \overline{y}) + \xi \left[\frac{3\overline{y}\mu(\overline{x}+\mu-1)}{(\overline{r}_2)^5} + \frac{3\overline{y}(1-\mu)(\overline{x}+\mu)}{(\overline{r}_1)^5} - k - \frac{3}{2}\alpha \left(\overline{x}^2 + \overline{y}^2 \right)^{\frac{-7}{4}} k + \frac{21}{4}\overline{x}^2\alpha \left(\overline{x}^2 + \overline{y}^2 \right)^{\frac{-11}{4}} k \right] \tag{12}$$

$$+ \eta \left[1 + \frac{3\overline{y}^2\mu}{(\overline{r}_2)^5} - \frac{\mu}{(\overline{r}_2)^2} + \frac{3(1-\mu)\overline{y}^2}{(\overline{r}_1)^5} - \frac{(1-\mu)}{(\overline{r}_1)^3} - k + \frac{21\overline{x}\,\overline{y}\alpha}{4} \left(\overline{x}^2 + \overline{y}^2 \right)^{\frac{-11}{4}} k \right].$$

Let us consider the trial solution of Equations (11) and (12),

$$\xi = \xi_0 e^{\lambda t}, \quad \eta = \eta_0 e^{\lambda t}$$

where ξ_0 and η_0 are constants and λ is a complex constant. Then we have

$$\lambda^2 \xi_0 e^{\lambda t} - 2\lambda \eta_0 e^{\lambda t}$$

$$= \xi_0 e^{\lambda t}\left[1 - \frac{\mu}{(\bar{r}_2)^3} + \frac{3\mu(\bar{x}+\mu-1)^2}{(\bar{r}_2)^5} + \frac{3(1-\mu)(\bar{x}+\mu)^2}{(\bar{r}_1)^5} - \frac{(1-\mu)}{(\bar{r}_1)^3} + \lambda k - \frac{21\bar{x}\,\bar{y}\,\alpha}{4}\left(\bar{x}^2+\bar{y}^2\right)^{\frac{-11}{4}}k\right] \tag{13}$$

$$+ \eta_0 e^{\lambda t}\left[\frac{3\bar{y}\,\mu(\bar{x}+\mu-1)}{(\bar{r}_2)^5} + \frac{3\bar{y}(1-\mu)(\bar{x}+\mu)}{(\bar{r}_1)^5} + \lambda k + \frac{3}{2}\alpha\left(\bar{x}^2+\bar{y}^2\right)^{\frac{-7}{4}}k - \frac{21}{4}\bar{y}^2\alpha\left(\bar{x}^2+\bar{y}^2\right)^{\frac{-11}{4}}k\right],$$

$$\lambda^2 \eta_0 e^{\lambda t} + 2\lambda \xi_0 e^{\lambda t}$$

$$= \xi_0 e^{\lambda t}\left[\frac{3\bar{y}\,\mu(\bar{x}+\mu-1)}{(\bar{r}_2)^5} + \frac{3\bar{y}(1-\mu)(\bar{x}+\mu)}{(\bar{r}_1)^5} - \lambda k - \frac{3}{2}\alpha\left(\bar{x}^2+\bar{y}^2\right)^{\frac{-7}{4}}k + \frac{21}{4}\bar{x}^2\alpha\left(\bar{x}^2+\bar{y}^2\right)^{\frac{-11}{4}}k\right] \tag{14}$$

$$+ \eta_0 e^{\lambda t}\left[1 + \frac{3\bar{y}^2\mu}{(\bar{r}_2)^5} - \frac{\mu}{(\bar{r}_2)^2} + \frac{3(1-\mu)\bar{y}^2}{(\bar{r}_1)^5} - \frac{(1-\mu)}{(\bar{r}_1)^3} - \lambda k + \frac{21\bar{x}\,\bar{y}\,\alpha}{4}\left(\bar{x}^2+\bar{y}^2\right)^{\frac{-11}{4}}k\right].$$

Now, from Equations (13) and (14), we derive the following simultaneous linear equations

$$\xi\left\{\lambda^2 + \frac{1-\mu}{(\bar{r}_1)^3}\left(1 - \frac{3(\bar{x}+\mu)^2}{(\bar{r}_1)^2}\right) + \frac{\mu}{(\bar{r}_2)^3}\left(1 - \frac{3(\bar{x}+\mu-1)^2}{(\bar{r}_2)^2}\right) - 1 - \lambda k + \frac{21\bar{x}\,\bar{y}\,\alpha}{4}\left(\bar{x}^2+\bar{y}^2\right)^{\frac{-11}{4}}k\right\}$$

$$+ \eta\left\{-2\lambda - \frac{3\,\bar{y}\,\mu(\bar{x}+\mu-1)}{(\bar{r}_2)^5} - \frac{3\bar{y}(1-\mu)(\bar{x}+\mu)}{(\bar{r}_1)^5} - \lambda k - \frac{3}{2}\alpha\left(\bar{x}^2+\bar{y}^2\right)^{\frac{-7}{4}}k - \frac{21}{4}\bar{y}^2\alpha\left(\bar{x}^2+\bar{y}^2\right)^{\frac{-11}{4}}k\right\} = 0 \tag{15}$$

and

$$\xi\left\{2\lambda - \frac{3\,\bar{y}\,\mu(\bar{x}+\mu-1)}{(\bar{r}_2)^5} - \frac{3\bar{y}(1-\mu)(\bar{x}+\mu)}{(\bar{r}_1)^5} + \lambda k + \frac{3}{2}\alpha\left(\bar{x}^2+\bar{y}^2\right)^{\frac{-7}{4}}k - \frac{21}{4}\bar{x}^2\alpha\left(\bar{x}^2+\bar{y}^2\right)^{\frac{-11}{4}}k\right\}$$

$$+ \eta\left\{\lambda^2 + \frac{1-\mu}{(\bar{r}_1)^3}\left(1 - \frac{3\bar{y}^2}{(\bar{r}_1)^2}\right) + \frac{\mu}{(\bar{r}_2)^3}\left(1 - \frac{3\bar{y}^2}{(\bar{r}_1)^2}\right) - 1 - \lambda k - \frac{21\bar{x}\,\bar{y}\,\alpha}{4}\left(\bar{x}^2+\bar{y}^2\right)^{\frac{-11}{4}}k\right\} = 0 \tag{16}$$

The simultaneous linear Equations (15) and (16) can be written as

$$\xi\left(\lambda^2 + e - h - 1 - \lambda k_{x,\dot{x}} - k_{x,x}\right) + \eta\left(-2\lambda - g - k_{x,y}\right) = 0 \tag{17}$$

$$\xi\left(2\lambda - g + k_{y,x}\right) + \eta\left(\lambda^2 + e - f - 1 - \lambda k_{y,\dot{y}} - k_{y,y}\right) = 0 \tag{18}$$

where

$$e = \frac{1-\mu}{(\bar{r}_1)^3} + \frac{\mu}{(\bar{r}_2)^3}, \tag{19}$$

$$f = 3\left[\frac{1-\mu}{(\bar{r}_1)^5} + \frac{\mu}{(\bar{r}_2)^5}\right]\bar{y}^2, \tag{20}$$

$$g = 3\left[\frac{(1-\mu)(\bar{x}+\mu)}{(\bar{r}_1)^5} + \frac{\mu(\bar{x}+\mu-1)}{(\bar{r}_2)^5}\right]\bar{y}, \tag{21}$$

$$h = 3\left[\frac{(1-\mu)(\bar{x}+\mu)^2}{(\bar{r}_1)^5} + \frac{\mu(\bar{x}+\mu-1)^2}{(\bar{r}_2)^5}\right]. \tag{22}$$

and

$$k_{x,x} = \left(\frac{\partial S_x}{\partial x}\right)_- = \frac{21}{4}\alpha\left(\bar{x}^2 + \bar{y}^2\right)^{\frac{-11}{4}}\bar{x}\,\bar{y}\,k, \quad k_{x,\dot{x}} = \left(\frac{S_x}{\partial \dot{x}}\right)_- = k,$$

$$k_{x,y} = \left(\frac{\partial S_x}{\partial y}\right)_- = -k + \frac{21}{4}\bar{y}^2\alpha\left(\bar{x}^2 + \bar{y}^2\right)^{\frac{-11}{4}}k,$$

$$k_{x,\dot{y}} = \left(\frac{\partial S_x}{\partial \dot{y}}\right)_- = 0, \quad k_{y,x} = \left(\frac{\partial S_y}{\partial x}\right)_- = k + \frac{21}{4}\bar{x}^2\alpha\left(\bar{x}^2 + \bar{y}^2\right)^{\frac{-11}{4}}k, \quad k_{y,\dot{x}} = \left(\frac{\partial S_y}{\partial \dot{x}}\right)_- = 0,$$

$$k_{y,y} = \left(\frac{\partial S_y}{\partial y}\right)_- = \frac{21}{4}\alpha\left(\bar{x}^2 + \bar{y}^2\right)^{\frac{-11}{4}}\bar{x}\,\bar{y}\,k, \quad k_{y,\dot{y}} = \left(\frac{\partial S_y}{\partial \dot{y}}\right)_- = k.$$

(23)

Neglecting terms of $O\left(k^2\right)$, the condition for the determinant of the linear equations defined by the Equations (17) and (18) to be zero is

$$\lambda^4 - \left(k_{x,\dot{x}} + k_{y,\dot{y}}\right)\lambda^3 + \left[2\left(1+e\right) - f - h - k_{x,x} + 2\left(k_{x,\dot{y}} - k_{y,\dot{x}}\right) - k_{y,y}\right]\lambda^2$$

$$+ \left[\left(1-e+f\right)k_{x,\dot{x}} + \left(1-e+h\right)k_{y,\dot{y}} + 2\left(k_{x,y} - k_{y,x}\right) - g\left(k_{x,\dot{y}} + k_{y,\dot{x}}\right)\right]\lambda$$

$$+ \left[\left(e-h-1\right)\left(e-f-1\right) - \left(g\right)^2 + \left(1-e+f\right)k_{x,x} + \left(1-e+h\right)k_{y,y} - g\left(k_{x,y} + k_{y,x}\right)\right] = 0$$

(24)

This quadratic Equation (24) has the general form

$$\lambda^4 + \sigma_3\lambda^3 + \left(\sigma_{20} + \sigma_2\right)\lambda^2 + \sigma_1\lambda + \left(\sigma_{00} + \sigma_0\right) = 0$$

(25)

where

$$\sigma_0 = \left(1-e+f\right)k_{x,x} + \left(1-e+h\right)k_{y,y} - f\left(k_{x,y} + k_{y,x}\right),$$

$$\sigma_1 = \left(1-e+f\right)k_{x,\dot{x}} + \left(1-e+h\right)k_{y,\dot{y}} + 2\left(k_{x,y} - k_{y,x}\right),$$

$$\sigma_2 = -k_{y,y} - k_{x,x},$$

$$\sigma_3 = -k_{x,\dot{x}} - k_{y,\dot{y}},$$

$$\sigma_{20} = 2\left(1+e\right) - f - h,$$

$$\sigma_{00} = \left(e-f-1\right)\left(e-f-1\right) - g^2.$$

Here σ_{00}, σ_{20} and $\sigma_i\left(i=0,1,2,3\right)$ can be derived by evaluating e, f, g and h defined earlier. The value of the coefficient in the zero drag case is denoted by adding additional subscript 0. If we neglect product of powers of μ with any of the constants defined in Equation (23), we obtain

$$\sigma_{00} = \frac{27}{4}\mu,$$

$$\sigma_{20} = 1,$$

$$\sigma_0 = \frac{-189\bar{x}\,\bar{y}\,\alpha}{16}\left(\bar{x}^2 + \bar{y}^2\right)^{\frac{-11}{4}}k - \frac{63\bar{x}\,\bar{y}\,\alpha}{16}\left(\bar{x}^2 + \bar{y}^2\right)^{\frac{-11}{4}}k + \left[\frac{63\sqrt{3}\,\alpha}{16}\left(\bar{x}^2 + \bar{y}^2\right)^{\frac{-7}{4}} - \frac{63\sqrt{3}\,\alpha\,\mu}{8}\left(\bar{x}^2 + \bar{y}^2\right)^{\frac{-7}{4}}\right]k,$$

$$\sigma_1 = \left[-1 + \frac{21}{2}\bar{x}^2\alpha\left(\bar{x}^2 + \bar{y}^2\right)^{\frac{-11}{4}} - \frac{21}{2}\alpha\bar{y}^2\left(\bar{x}^2 + \bar{y}^2\right)^{\frac{-11}{4}}\right]k,$$

(26)

$$\sigma_2 = -\frac{21}{2}\alpha\left(\bar{x}^2 + \bar{y}^2\right)^{\frac{-11}{4}}\bar{x}\,\bar{y}\,k,$$

$$\sigma_3 = -2k.$$

By assuming σ_i to be small, we investigate the stability of the non zero drag case. We can use the classical solutions of the zero drag case (i.e. when $k=0$). Equation (25) reduces to

$$\lambda^4 + \sigma_{20}\lambda^2 + \sigma_{00} = 0 \tag{27}$$

The four classical solutions for L_4 and L_5 to $O(\mu)$ are given by the pair of values

$$L_{4,5} : \lambda_{1,2} = \pm\sqrt{-1 + \frac{27}{4}\mu}$$

$$\lambda_{3,4} = \pm\sqrt{-\frac{27}{4}\mu} \tag{28}$$

Since we are primarily interested in the stability of L_4 and L_5 under the effects of a drag force, we restrict our analysis to these points. The four roots of the classical characteristic equation can be written as

$$\lambda_n = \pm T i \quad (n = 1, \cdots, 4) \tag{29}$$

where

$$T = \sqrt{\frac{\sigma_{20\pm}\sqrt{\sigma_{20}^2 - 4\sigma_{00}}}{2}} \tag{30}$$

is a real quantity for L_4 and L_5. Using the values of σ_{00} and σ_{20} given in Equations (26) we have

$$T^2 = 1 - \frac{27}{4}\mu \quad \text{or} \quad T^2 = \frac{27}{4}\mu \tag{31}$$

With the introduction of drag we assume a solution of the form

$$\lambda = \lambda_n(1 + \rho + \upsilon i) = \left[\mp\upsilon \pm (1+\rho)i\right]T \tag{32}$$

where ρ and υ are small real quantities. To lowest order we have

$$\lambda^2 = \left[-(1+2\rho) - 2\upsilon i\right]T^2 \tag{33}$$

$$\lambda^3 = \left[\pm3\upsilon \mp (1+3\rho)i\right]T^3 \tag{34}$$

$$\lambda^4 = \left[(1+4\rho) + 4\upsilon i\right]T^4 \tag{35}$$

Substituting these in Equation (25), and neglecting products of ρ or υ with σ_i, and solving the real and imaginary parts of the resulting simultaneous equations for ρ or υ we get

$$\upsilon = \frac{\pm\sigma_3 T^2 \mp \sigma_1}{2T(2T^2 - \sigma_{20})}, \tag{36}$$

$$\rho = \frac{(\sigma_{00} + \sigma_0) - (\sigma_{20} + \sigma_2)T^2 + T^4}{2T^2(\sigma_{20} - 2T^2)}. \tag{37}$$

(i) The stability of L_4
For L_4, we have

$$\upsilon = \frac{\sigma_3 T^2 - \sigma_1}{2T(2T^2 - \sigma_{20})}, \tag{38}$$

$$\rho = \frac{(\sigma_{00} + \sigma_0) - (\sigma_{20} + \sigma_2)T^2 + T^4}{2T^2(\sigma_{20} - 2T^2)}. \tag{39}$$

On putting the values of σ_i, in Equations (38) and (39) from Equation (26) and also taking, $T^2 = \frac{27}{4}\mu$, we have

$$\upsilon = \frac{k}{3\sqrt{3\mu}}, \quad \rho = \frac{27\mu}{8 - 108\mu}.$$

Now, putting these values of ρ and υ in Equation (35), and neglecting the terms of $O(k\mu)$, we get the characteristic equation as

$$\lambda^4 - \frac{729\mu^2}{16 - 216\mu} = 0$$

whose roots are

$$\lambda_1 = -\frac{3\sqrt{3\mu}}{2^{\frac{3}{4}}(2 - 27\mu)^{\frac{1}{4}}}, \qquad \lambda_2 = -\frac{3i\sqrt{3\mu}}{2^{\frac{3}{4}}(2 - 27\mu)^{\frac{1}{4}}},$$

$$\lambda_3 = \frac{3i\sqrt{3\mu}}{2^{\frac{3}{4}}(2 - 27\mu)^{\frac{1}{4}}}, \qquad \lambda_4 = \frac{3\sqrt{3\mu}}{2^{\frac{3}{4}}(2 - 27\mu)^{\frac{1}{4}}}.$$

Also on taking $T^2 = 1 - \frac{27}{4}\mu$ in Equations (38) and (39) from Equation (26), we get the characteristic equation as

$$\lambda^4 + \frac{(-4 + 27\mu)(-4 + 81\mu)}{8(-2 + 27\mu)} - ik = 0$$

whose roots are

$$\lambda_1 = -\frac{(-16 + 432\mu) - 16ik}{2^{\frac{3}{4}}(-2 + 27\mu)^{\frac{1}{4}}}, \qquad \lambda_2 = -\frac{i(-16 + 432\mu) + 16k}{2^{\frac{3}{4}}(-2 + 27\mu)^{\frac{1}{4}}},$$

$$\lambda_3 = \frac{i(-16 + 432\mu) + 16k}{2^{\frac{3}{4}}(-2 + 27\mu)^{\frac{1}{4}}}, \qquad \lambda_4 = \frac{(-16 + 432\mu^2) - 16ik}{2^{\frac{3}{4}}(-2 + 27\mu)^{\frac{1}{4}}}.$$

If $\upsilon \neq 0$,

According to [9], the resulting motion of a particle is asymptotically stable only when all the real parts of λ are negative and the condition for asymptotically stable under the arbitrary drag force is given by

$$0 < \sigma_1 < \sigma_3 \tag{40}$$

where σ_1 and σ_3 are defined in Equation (26). But we see that the linear stability of triangular equilibrium points does not depend on the value of $k_{x,x}$ and $k_{y,y}$. Therefore the condition $\sigma_3 > 0$ can only be satisfied when k is positive and the drag force is a function of $\dot{x}$ and $\dot{y}$.

But here in our case of Stokes drag $\sigma_1 = -k, \sigma_3 = -2k$ and therefore $\sigma_1 > \sigma_3$ and hence L_4 is not asymptotically stable. Further one of the roots of λ i.e. λ_4 has positive real root. Therefore L_4 is not stable. Thus we conclude that L_4 is neither stable nor asymptotically stable and hence linearly unstable.

Similarly, we conclude that L_5 is neither stable nor asymptotically stable and hence linearly unstable.

5. Conclusions

We have studied the existence of the triangular libration points and their linear stability by using Stokes drag. We have shown that there exist two noncollinear stationary points $L_4(\bar{x}, \bar{y})$ and $L_5(\bar{x}, -\bar{y})$ (Equation (10)). If we put $k = 0$, these results agree with the classical restricted three body problem.

In the classical case i.e. when $k = 0$, we observe that as the value of μ increases, the abscissa $\bar{x}$ of L_4 decreases and the ordinate $\bar{y}$ of L_4 remains constant, while in our case (i.e. Stokes drag), when $k = 10^{-5}$, we observe that the abscissa $\bar{x}$ of L_4 decreases and the ordinate $\bar{y}$ of L_4 changes slightly. In our previous paper ([18]) i.e. in the case of Poynting Robertson drag, the abscissa and the ordinate decrease with μ. As regards, the stability of L_4 in both the cases (Poynting Robertson drag and Stokes drag) is always unstable for all values of μ. The result of stability is quite different when we compare with the classical case. In the classical case, L_4 is stable for μ, whereas in the case of drag forces, motion is unstable for all values of μ.

In the case of Stokes drag, we have derived a set of linear equations in terms of ξ and η (Equations (17) and (18)), which involve the components of the Stokes drag force evaluated at the libration points (Equations (19)-(23)). From these, we derive a characteristic equation having the general form (Equation (25)).

Further, we have derived the approximate expressions for $\sigma_0, \sigma_1, \sigma_2, \sigma_3, \sigma_{00}$ and σ_{20} occurring in the above characteristic equation. These expressions are given in terms of the partial derivatives of the Stokes drag, evaluated at the libration points.

Using the [9] terminology, in the case of drag force, we assume a solution of the form (Equation (32)), where υ and ρ are small real quantities and

$$\lambda_n = \pm \mathrm{T}i \quad (n = 1, \cdots, 4)$$

is a real quantity for L_4 and L_5 in the classical case. After substituting the values of λ, λ^2, λ^3 and λ^4 in the characteristic equation, we get the values of υ and ρ (Equations (36) and (37)).

Further to investigate the stability of the shifted points, by using [9] terminology, the resulting motion of a particle is asymptotically stable only when all the real parts of λ are negative. Also, the condition for asymptotical stability under the drag force is given by Equation (40).

The condition $\sigma_3 > 0$ can only be satisfied when $k > 0$. In the case of Stokes drag $\sigma_1 = -k$ and $\sigma_3 = -2k$, Equation (40) is not satisfied. Therefore, L_4 and L_5 are not asymptotically stable. Further, we have seen that one of the roots of λ i.e. λ_4 has positive real root; thus, L_4 and L_5 are not stable. Hence, due to Stokes drag, L_4 and L_5 are neither stable nor asymptotically stable but unstable whereas in the classical case L_4 and L_5 are stable for the mass ratio $\mu < 0.03852$ [19].

References

[1] Euler, L. (1772) Theoria Motuum Lunae, Typis Academiar Imperialis Seientiarum, Petropoli. Reprinted in Opera Omnia, Series 2, Courvoisier, L., Ed., Vol. 22, Orell, Fussli Turici, Lausansje, 1958.

[2] Szebehely, V. (1967) Theory of Orbits, the Restricted Problem of Three Bodies. Academic Press, New York and London

[3] Celletti, A., Stefanelli, L., Lega, E. and Froeschlé, C. (2011) Some Results on the Global Dynamics of the Regularized Restricted Three-Body Problem with Dissipation. *Celestial Mechanics and Dynamical Astronomy*, **109**, 265-284. http://dx.doi.org/10.1007/s10569-010-9326-y

[4] Murray, C.D. and Dermott, S.F. (1999) Solar System Dynamics. Cambridge University Press, Cambridge.

[5] Jeffreys, H. (1929) The Earth. 2nd Edition, Cambridge University Press, Cambridge.

[6] Colombo, G.D., Lautman, I. and Shapiru, I. (1966) The Earth's Dust Belt: Fact or Fiction? Gravitational Focusing and Jacobi Capture. *Journal of Geophysical Research*, **71**, 5705-5717. http://dx.doi.org/10.1029/JZ071i023p05705

[7] Schuerman, D. (1980) The Restricted Three-Body Problem Including Radiation Pressure. *Astrophysical Journal*, **238**, 337-342. http://dx.doi.org/10.1086/157989

[8] Simmons, J.F.L., Mcdonald, A.J.C. and Brown, J.C. (1985) The Restricted Three-Body Problem with Radiation Pressure. *Celestial Mechanics*, **35**, 145-187. http://dx.doi.org/10.1007/BF01227667

[9] Murray, C.D. (1994) Dynamical Effects of Drag in the Circular Restricted Three Body Problems: 1. Location and Stability of the Lagrangian Equilibrium Points. *Icarus*, **112**, 465-184. http://dx.doi.org/10.1006/icar.1994.1198

[10] Beauge, C. and Ferraz-Mello, S. (1993) Resonance Trapping in the Primordial Solar Nebula: The Case of a Stokes Drag Dissipation. *Icarus*, **103**, 301-318. http://dx.doi.org/10.1006/icar.1993.1072

[11] Beauge, C. and Ferraz-Mello, S. (1994) Capture in Exterior Mean Motion Resonance Due to Poynting Robertson Drag. *Icarus*, **110**, 239-260. http://dx.doi.org/10.1006/icar.1994.1119

[12] Beauge, C., Aarseth, S.J. and Ferraz-Mello, S. (1994) Resonance Capture and the Formation of the Outer Planets. *Monthly Notices of the Royal Astronomical Society*, **270**, 21-34. http://dx.doi.org/10.1093/mnras/270.1.21

[13] Sicardy, B., Beauze, C., Ferraz-Mello, S., Lazzaro, D. and Roques, F. (1993) Capture of Grains into Resonances through Poynting-Robertson Drag. *Celestial Mechanics & Dynamical Astronomy*, **57**, 373-390. http://dx.doi.org/10.1007/BF00692487

[14] Liou, J.C., Zook, H.A. and Jackson, A.A. (1995) Radiation Pressure, Poynting Robertson Drag and Solar Wind Drag in the Restricted Three Body Problem. *Icarus*, **116**, 186-201. http://dx.doi.org/10.1006/icar.1995.1120

[15] Ishwar, B. and Kushvah, B.S. (2006) Linear Stability of Triangular Equilibrium Points in the Generalized Restricted Three Body Problem with Poynting Robertson Drag. *Journal of Dynamical Systems and Geometric Theories*, **4**, 79-86.

http://dx.doi.org/10.1080/1726037X.2006.10698504

[16] Kushvah, B.S., Sharma, J.P. and Ishwar, B. (2007) Nonlinear Stability in the Generalised Photogravitational Restricted Three Body Problem with Poynting-Robertson Drag. *Astrophysics and Space Science*, **312**, 279-293. http://dx.doi.org/10.1007/s10509-007-9688-0

[17] Singh, J. and Emmanuel, A.B. (2014) Stability of Triangular Points in the Photogravitational CR3BP with Poynting Robertson Drag and a Smaller Triaxial Primary. *Astrophysics and Space Science*, **353**, 97-103. http://dx.doi.org/10.1007/s10509-014-2023-7

[18] Jain, M. and Aggarwal, R. (2015) Existence and Stability of Non-Collinear Librations Points in the Restricted Problem with Poynting Robertson Light Drag Effect. *International Journal of Mathematics Trends and Technology*, **19**, 20-33.

[19] Brouwer, D. and Clemence, G.M. (1961) Methods of Celestial Mechanics. Academic Press, New York.

20

Numerical Investigation of a Shock Accelerated Heavy Gas Cylinder in the Self-Similar Regime

Bing Wang[1], Jing-Song Bai[1,2], Tao Wang[1]

[1]Institute of Fluid Physics, China Academy of Engineering Physics, Mianyang, China
[2]National Key Laboratory of Shock Wave and Detonation Physics, Institute of Fluid Physics, China Academy of Engineering Physics, Mianyang, China
Email: bjsong@foxmail.com

Abstract

A detailed numerical simulation of a shock accelerated heavy gas (SF_6) cylinder surrounded by air gas is presented. It is a simplified configuration of the more general shock-accelerated inhomogeneous flows which occur in a wide variety of astrophysical systems. From the snapshots of the time evolution of the gas cylinder, we find that the evolution of the shock accelerated gas cylinder is in some ways similar to the roll-ups of a vortex sheet for both roll up into a spiral and fall into a self-similar behavior. The systemic and meaningful analyses of the negative circulation, the center of vorticity and the vortex spacing are in a good agreement with results obtained from the prediction of vorticity dynamics. Unlike the mixing zone width in single-mode or multi-mode Richtmyer-Meshkov instability which doesn't exist, a single power law of time owing to the bubble and spike fronts follow a power law of t^θ with different power exponents, the normalized length of the shock accelerated gas cylinder follows a single power law with $\theta = 0.43$ in its self-similar regime obtained from the numerical results.

Keywords

Interficial Instabilities, Vortex Dynamics, Power Laws

1. Introduction

When an impulsive acceleration impinges on the corrugated interface between two fluids of different densities, the instability at the interface will arise due to the deposited vorticity induced by the baroclinic torque production term $\nabla\rho\times\nabla p/\rho^2$ (where p is the pressure and ρ is the density). The class of problems is generally referred to as the Richtmyer-Meshkov (RM) instability [1] [2]. One of the eventual goals in investigating RM instability is to shed light on the resultant mixing. Mixing is of contemporary interest in many fields of research, among which are the inertial confinement fusion [3], the fuel mixing in a Scramjet [4], and the explosion of supernovas [5].

In the RM instability researches, an interesting configuration is a shock wave interacting with a cylindrical interface (circular interface in two dimensions) between two fluids. When a planar shock wave impacts on a heavy gas (e.g., SF$_6$) cylinder around by an ambient gas (e.g., air), a shock wave is reflected and a refracted shock wave transmits into the heavy gas cylinder. Because the heavy gas acoustic impedance exceeds that of the ambient gas, the refracted shock is slower than the incident shock wave, and a convergent shock refraction pattern occurs. Because of this and the curvature of the cylinder, the transmitted shock focuses at the downstream vertex. This focusing will induce a pressure rising that eventually leads to a cusp-like protrusion [6]. The baroclinically deposited vorticity due to the shock-cylinder interaction stretches and distorts the interface and rolls up into a counter-rotating vortex pair. Then, the evolution of the interface will be dominated by the vortex pair and falls into a self-similar regime, which is in some ways similar to the roll-ups of a vortex sheet [7]-[9].

The propagation of a shock wave in an inhomogeneous medium, and the response of the medium to impulsive acceleration are of fundamental interest in astrophysical systems. The evolution of the interstellar medium in spiral galaxies is significantly influenced by the strong shock waves generated by supernovae explosion [10]. Moreover, it is well-known that the generated shock waves remarkably alter the morphology of the cloud (a region of higher density). The bright eastern knot of the Puppis A supernova remnant results in a distorted shock front due to a cloud-shock interaction as seen in images from the Chandra X-ray telescope [11]. In this paper, the shock-cylinder interaction is a particularly simple configuration to investigate the problem of shock accelerated inhomogeneous flows.

There are many experimental and numerical researches of the shock-cylinder interactions which concentrate on different subjects. Haas and Sturtevant [12], Jacobs [13] and Tomkins et al. [14] studied this problem experimentally for exploring the wave patterns and the distortion of volume in the RM instability, studying the effects of centrifugal force and viscosity, and investing the mixing mechanisms in a shock-accelerated flow, respectively. Based on the experiments performed by Haas and Sturtevant, Picone and Boris [15] studied the early and the late time phenomena of these experiments numerically. Additionally, Quirk and Karni [16] also simulated the same experiments with the concentration on the early stages of the shock-cylinder interaction. Recently, in light of the experiment performed by Tomkins et al., Weirs et al. [17] investigated the three dimensional effects and Shankar et al. [18] studied the effect of the tracer particle in the shock cylinder interactions.

As be mentioned just, there were some studies on the shock accelerated heavy gas cylinder, however, in the aforementioned papers no attention has been paid on a perhaps existing scaling law in contrast with the single mode or multi mode RM instability [19]-[21]. Moreover, most numerical studies were performed for validating codes with comparing to the experimental data. In this work, we systemically and meaningfully analyze the obtained numerical results. Comparing with the results from the prediction of vorticity dynamics, we want to present the behavior of the characterized variables in the evolution of the shock accelerated cylinder. Through studying the evolution of the integral length of the shock-accelerated cylinder, we shed light on the scaling law of the growth of the normalized length follows in the self-similar regime.

2. Numerical Methods and Initialization

This paper applies our large eddy simulation code MVFT (multi-viscous flow and turbulence) [22] to simulate the multi-viscosity-fluid and turbulence. The code MVFT is used for the compressible large eddy simulation that was developed by Institute of Fluid Physics at the China Academy of Engineering Physics. MVFT can be used to simulate multi-component flows, and compute shocks, contact discontinuities and material interfaces at high accuracy. It splits the flow into an inviscid flow and a viscous flow by using an operator splitting technique, where the former is computed by employing the piecewise parabolic method with a third-order Godunov scheme and the latter is calculated by utilizing a central difference scheme in conjunction with a second-order Runge-Kutta method for the time integration. MVFT applies based on the piecewise parabolic method [23] to interpolate physical quantities, the Vreman [24] subgrid scale eddy viscosity model to conduct large eddy simulation, and to solve Navier-Stokes equations.

The initial conditions are significantly important in numerical simulations, especially for the membrane less RM instability researches. Initially, a sharp interface [25] [26] was used but leaded to an ill-posed simulation compared with the experiments. This is because the interspecies diffusion would not be deniable in the membrane less technique [27] which is used to form the interface between two fluids in experiments. More concentrations should be required on the interfacial diffusion. Consequently, an interfacial transition layer with finite thickness [28] was introduced to characterize the diffusive interface. Recently, a well-characterized initial concentration profile [17] was specified with experimentally measured data. And the contour map of scalar mass fraction and dissipation rate obtained from numerical calculations with the profile were in a good agreement with experimental data qualitatively. For we will simulate the same experiment, the profile is chosen as our initial concentration profile.

In the present simulations, the initial conditions were adapted to the Mach 1.2 shock tube experiment of Tomkins *et al.* [14] which was performed at Los Alamos National Laboratory (LANL). **Figure 1** shows a schematic of the computational domain. To match the shock tube test section dimensions, the computational domain has (x, y) dimension $[0, L_x] \times [0, L_y]$ with $L_x = 14.4$ cm, $L_y = 3.75$ cm. The inflow/outflow boundary condition is used on the left and right x boundaries. At the upper y boundary, the reflecting boundary condition is used and the symmetry condition is enforced at the lower y boundary. Three different grid resolutions with $\Delta x = \Delta y = 50$ μm, 30 μm and 15 μm are used for grid convergence study. In the simulations, the initial concentration profile of the cylinder in [17] is used. The simulations are run at a CFL number of 0.2. The main gas parameters are presented in **Table 1**.

3. Results

Figure 2 depicts the snapshots of the time evolution of the cylinder, which shows volume fraction maps that are corresponding to the densities for seven times after shock passage on three different grid resolutions. Counter rotating vortex pair formed at the interface owing to the baroclinic vorticity deposited stretches and rolls the interface in ward the vortex cores after the shock collides. At about 220 μs, the vortex pair starts to roll up into the vortex core, and then falls into a self similar state. In **Figure 2** one also can see a cusp-like protrusion [6] produced by shock refraction in the simulations with finer grid. This demonstrates that finer grid is needed in the shock-cylinder simulations for better capturing more fine-scale structures.

Figure 3 presents the negative circulation evolution over time of the flow field. Once a straight vortex sheet rolls up, the circulation of each branch is either positive or negative. For the roll-ups of the straight vortex sheet, we only consider the negative circulation here. The amount of circulation deposited during the interaction of the

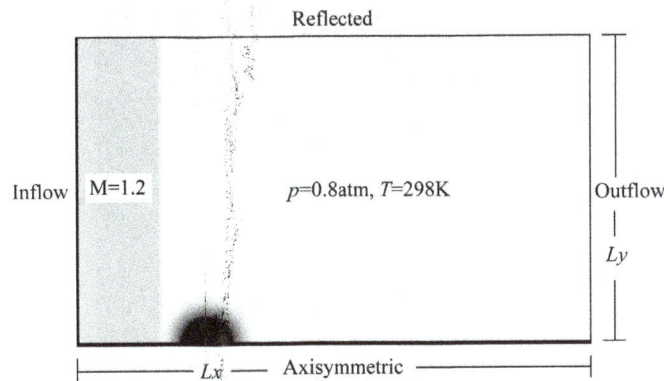

Figure 1. Schematic diagram of the computational domain.

Table 1. Properties of air and SF$_6$ gases.

Gases	Density (kg/m³)	Specific Heat Ratio	Kinematic Viscosity (10^{-6} m²/s)	Prandtl Number	Diffusion Coefficient in Air (cm²/s)
Air	0.95	1.40	15.7	0.71	0.204
SF$_6$	4.85	1.09	2.47	0.90	0.097

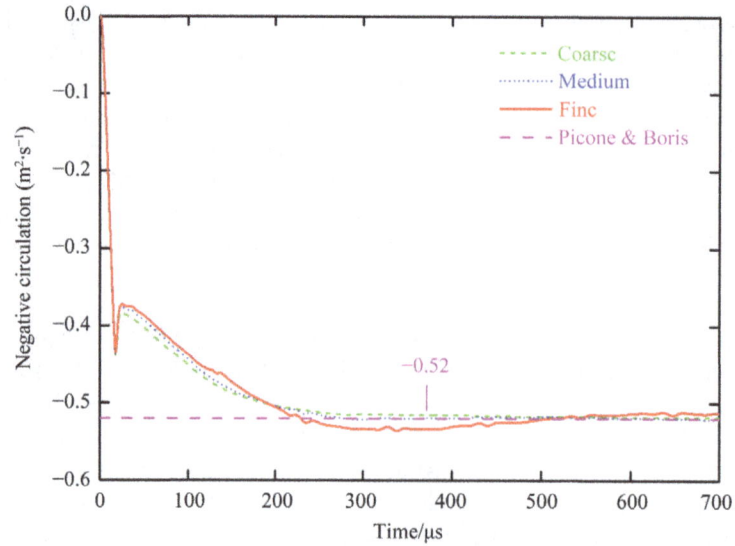

Figure 2. Image sequence of SF$_6$ volume fraction at t = 130, 220, 310, 400, 560, and 650 μs after shock impingement on the coarse (left column), medium (middle column), and fine (right column) mesh resolutions. The shock has traversed the cylinder from top to bottom.

Figure 3. Negative circulation as a function of time on three different grid resolutions. The short dashed, short dotted and solid lines correspond to the results computed on the coarse, medium, and fine grid resolutions respectively. The dashed line is the theoretical prediction for the negative circulation in [29].

shock wave with the interface is calculated using a path integration of velocity

$$\Gamma \equiv \oint V \cdot ds, \tag{1}$$

where V is the velocity vector and s is the path. From Stokes theorem, the circulation is a measure of the vorticity over an area A. In the two dimensional flow, the circulation is

$$\Gamma = \int_A \omega_z (x, y, t) \, dA, \tag{2}$$

with

$$\omega_z (x, y, t) = \partial v / \partial x - \partial u / \partial y. \tag{3}$$

where ω_z is the vorticity component which is perpendicular to the x-y plane, u and v are x and y components of the velocity respectively.

The vorticity generated by a shock wave propagating through a circular cross-section has been studied by Picon and Boris [29]. They gave the magnitude of the vortex strength or circulation,

$$\Gamma \approx 2V_2 \left(1 - \frac{V_2}{2W} \right) R_0 \ln \left(\frac{\rho_\infty}{\rho_c} \right). \tag{4}$$

where V_2 is the flow velocity behind the shock in the laboratory frame, W is the shock velocity, R_0 is the radius of the cross-section of the cylinder, ρ_∞ is the ambient density, and ρ_c is the density of the gas in the cylinder. For an initial temperature of 298 K and pressure of 0.8 atm, the resulting 1D gas dynamic velocities are the following, $V_2 = 105.6$ m/s, and $W = 414.75$ m/s. In the simulations, the effective radius R_0 is 2.57 mm, the ambient density ρ_∞ is the density of air, 0.95 kg/m^3, and ρ_c is the density of the cylinder 2.85 kg/m^3. From these data, one can get the amount of circulation is -0.52 m^2/s. From **Figure 3**, one can see that the simulation data is shown a good agreement with the value.

Additionally, one can calculate the vortex strength by applying the approximate model of the compact vortex. When a shock impacts on a heavy gas cylinder, the cylinder will stand here relative to the ambient gas because of its inertia. Then the cylinder will have a velocity V_2 with the opposite shock direction relative to the background. There is a discontinuity of the tangential velocity at both edge of the cylinder, so the cylinder could be considered as a two dimensional vortex sheet. From the vorticity dynamics, one can calculate the magnitude of the circulation of the compact vortex rolled up from the vortex sheet $\Gamma = 2V_2 R_0$ [30]. One can obtain a value of -0.54 m^2/s which is also in a good agreement with the numerical results.

In two dimensional case, as analogous to center of mass, one can define the coordinate of center of vorticity,

$$x_{cv} = \frac{\int x \omega_z \, dA}{\int \omega_z \, dA}, \quad y_{cv} = \frac{\int y \omega_z \, dA}{\int \omega_z \, dA}. \tag{5}$$

The time evolution of the center of vorticity is presented in the **Figure 4**. One can see that the x-component velocity of center of voriticity is 85.6 m/s. In [15], Picone and Boris gave an equation to compute the perpendicular distance d of the vortex core to the y axis or the half vortex spacing,

$$d \approx \frac{\Gamma}{4\pi (V_{cm} - V_2)}. \tag{6}$$

Here, V_{cm} is the velocity of center of vorticity in x direction. Then, one can get the value of d, 2.3 mm. From **Figure 4**, the value is in a good agreement with the center of vorticity in y direction.

The vorticity distribution along y direction computed on the fine mesh resolution at seven different times is shown in **Figure 5**. The vorticity trends to get together inward and extend outward as time goes by. Because of roll-ups, there are some troughs appeared in the vorticity distribution. After about 400 µs, the vorticity is almost fixed at the outer edge as time elapses, although it also moves inward to the y axis. The behavior is consistent with the evolution of the center of vorticity in y direction which decreases after about 400 µs shown in **Figure 4**.

Considering the mole fraction $X(x, y, t)$, the spatial maximum of the mole fraction in the stream wise direction gives

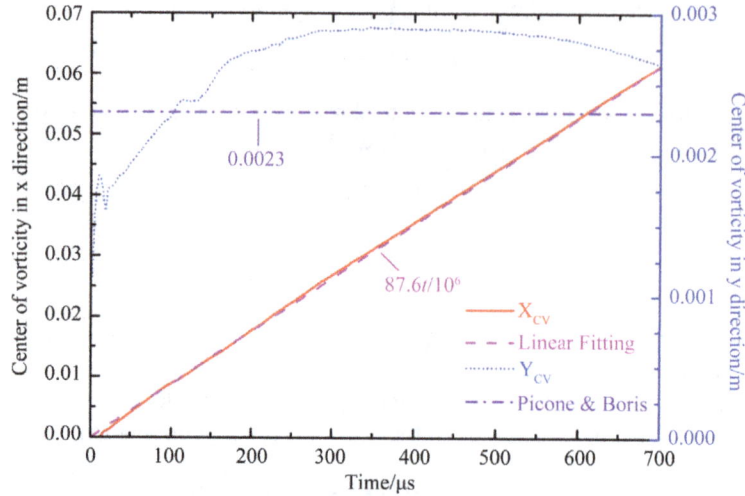

Figure 4. Center of vorticity in x direction (solid line) and y direction (shot dot) as a function of time computed on the fine grid resolution. The dashed line is obtained from linear fitting. The dash dot line corresponds to the theoretical prediction for the half of vortex spacing in [29].

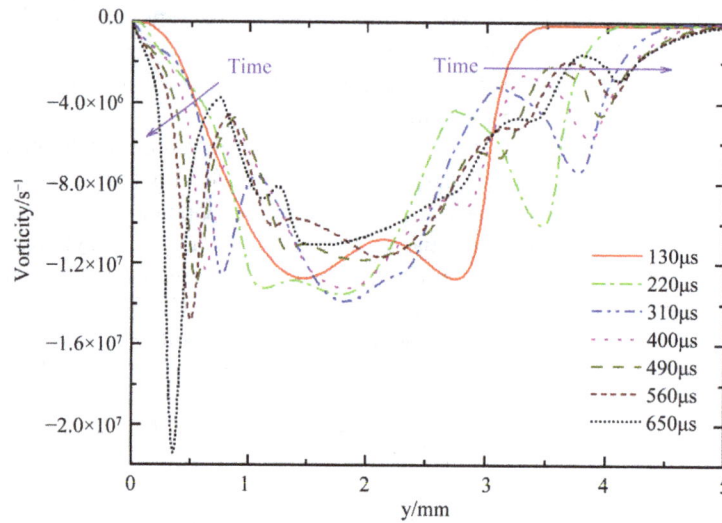

Figure 5. Vorticity distribution along y direction at seven different times on the resolution of $\Delta x = \Delta y = 15\ \mu m$.

$$X_{max}(x,t) = \max(X(x,y,t)), \quad \text{for all } y. \tag{7}$$

The left and right edge locations of the cylinder, $x_l(t)$ and $x_r(t)$, are defined as the x position $X_{max} \le \varepsilon$, with $\varepsilon = 0.05$ in the present simulations. The length is given by $h(t) = x_r(t) - x_l(t)$; which divided by the diameter of the cylinder $2R_0$, then one can get the normalized length $\eta(t) = h(t)/2R_0$. In **Figure 6**, the log-log plot of normalized length vs. time shows that from about 200 μs to 650 μs, there is a linear relationship. Then, a power law of $t^{0.43}$ for the normalized length is obtained.

4. Discussions and Conclusions

To the authors' knowledge, there hasn't been an investigation of the power law of the length for a shock accelerated gas cylinder, although the power law followed by the mixing zone width has been investigated in the single mode or multi-mode Richtmyer-Meshkov instability. Dimonte and Schneider [19] reported a power law t^θ

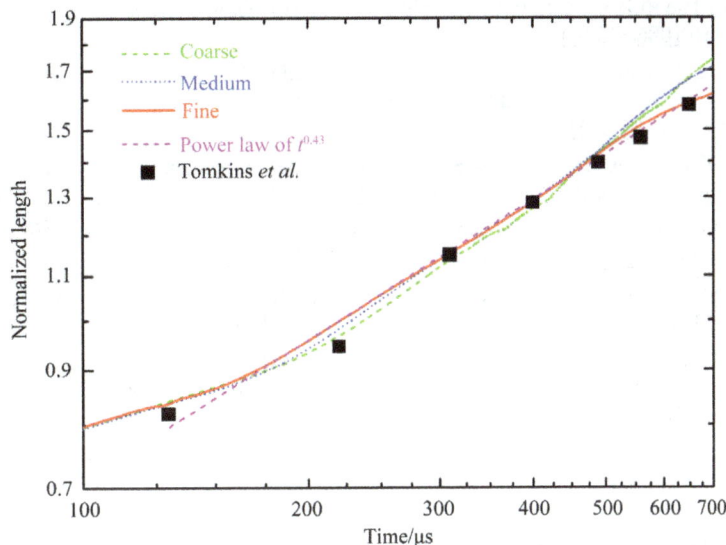

Figure 6. Log-plot of normalized length vs. time computed on the coarse (shot dashed line), medium (short dotted line) and fine (solid line) mesh resolutions. The dashed line corresponds to the power law of $t^{0.43}$. The scattered data correspond to the experimental data obtained from [14].

with $\theta \sim 0.2$ - 0.32 for $0.1 < A < 0.5$ with linear electric motor experiments in three dimensions. Using full numerical simulations, Oron *et al.* [20] gave the power exponent $\theta = 0.4 \pm 0.02$ for the bubbles. Recently, Sohn [21] predicted that the bubble fronts follow a power law with θ in the range of $(0.3$ - $0.35) \pm 0.02$ by applying a quantitative model of bubble completion. For the roll-ups of vortex sheet, the power law of $r \sim t^{\theta}$ with $\theta = 2/3$ has been gained [7], where r is the polar radius of the spiral. Interestingly, it is seen that the power law exponent of the normalized length obtained here is almost same to that of the bubble fronts in [20], although it's quite different from the others.

   In conclusion, we studied the evolution of a shock impinging heavy gas cylinder and the growth of the normalized length. A self-similar behavior is shown from the snapshots of the evolution of the gas cylinder. The two dimensional numerical results of the negative circulation, the center of vorticity, and the vortex spacing are in a good agreement with the results obtained from the analyses of Picon and Boris [29]. Moreover, the normalized length obeys a power law $\eta \sim t^{\theta}$ with $\theta = 0.43$. Whether the power law is a universal property for a shock accelerated cylinder is under investigation. More investigations of a shock accelerated cylinder with different radii, density ratios, and Mach numbers will be performed in future.

Acknowledgements

This work was sponsored by the National Science Foundation of China under Grants No. 11202195 and No. 11072228 and the Science Foundation of the China Academy of Engineering Physics under Grants No. 2011B0202005 and No. 2011A0201002.

References

[1]   Richtmyer, R.D. (1960) Taylor Instability in a Shock Acceleration of Compressible Fluids. *Communications on Pure and Applied Mathematics*, **13**, 297-319. http://dx.doi.org/10.1002/cpa.3160130207

[2]   Meshkov, E.E. (1968) Instability of the Interface of Two Gases Accelerated by a Shock Wave. *Soviet Fluid Dynamics*, **4**, 101-104. http://dx.doi.org/10.1007/BF01015969

[3]   Lindl, J.D., McCropy, R.L. and Campbell, E.M. (1992) Progress toward Ignition and Propagating Burn in Inertial Confinement Fusion. *Physics Today*, **45**, 32-40. http://dx.doi.org/10.1063/1.881318

[4]   Yang, J., Kubota, T. and Zukoski, E.E. (1994) A Model for Characterization of a Vortex Pair Formed by Shock Passage over a Light-Gas Inhomogeneity. *Journal of Fluid Mechanics*, **258**, 217-244. http://dx.doi.org/10.1017/S0022112094003307

[5]   Arnett, D. (2000) The Role of Mixing in Astrophysics. *The Astrophysical Journal Supplement Series*, **127**, 213-217. http://dx.doi.org/10.1086/313364

[6]   Kumar, S., Orlicz, G., Tomkins, C., Goodenough, C., Prestridge, K., Vorobieff, P. and Benjamin, R. (2005) Stretching of Material Lines in Shock-Accelerated Gaseous Flows. *Physics of Fluids*, **17**, Article ID: 082107. http://dx.doi.org/10.1063/1.2031347

[7]   Moore, D.W. (1975) The Rolling Up of a Semi-Infinite Vortex Sheet. *Proceedings of the Royal Society of London A*, **345**, 417-430. http://dx.doi.org/10.1098/rspa.1975.0147

[8]   Pullin, D.I. (1978) The Large-Scale Structure of Unsteady Self-Similar Rolled-Up Vortex Sheets. *Journal of Fluid Mechanics*, **88**, 401-430. http://dx.doi.org/10.1017/S0022112078002189

[9]   Krasny, R. (1986) A Study of Singularity Formation in a Vortex Sheet by the Point-Vortex Approximation. *Journal of Fluid Mechanics*, **167**, 65-93. http://dx.doi.org/10.1017/S0022112086002732

[10] Klein, R.I., McKee, C.F. and Colella, P. (1994) On the Hydrodynamic Interaction of Shock Waves with Interstellar Clouds. 1: Nonradiative Shocks in Small Clouds. *The Astrophysical Journal*, **420**, 213-236. http://dx.doi.org/10.1086/173554

[11] Hwang, U., Flanagan, K.A. and Petre, R. (2005) Chandra X-Ray Observation of a Mature Cloud-Shock Interaction in the Bright Eastern Knot Region of Puppis A. *Astrophysical Journal*, **635**, 355-364. http://dx.doi.org/10.1086/497298

[12] Haas, J.F. and Sturtevant, B. (1987) Interaction of Weak Shock Waves with Cylindrical and Spherical Gas Inhomogeneities. *Journal of Fluid Mechanics*, **181**, 41-76. http://dx.doi.org/10.1017/S0022112087002003

[13] Jacobs, J.W. (1993) The Dynamics of Shock Accelerated Light and Heavy Gas Cylinders. *Physics of Fluids A*, **5**, 2239-2247. http://dx.doi.org/10.1063/1.858562

[14] Tomkins, C.D., Kumar, S., Orlicz, G. and Prestridge, K.P. (2008) An Experimental Investigation of Mixing Mechanisms in Shock-Accelerated Flow. *Journal of Fluid Mechanics*, **611**, 131-150. http://dx.doi.org/10.1017/S0022112008002723

[15] Picone, J.M. and Boris, J.P. (1983) Vorticity Generation by Asymmetric Energy Deposition in a Gaseous Medium. *Physics of Fluids*, **26**, 365-382. http://dx.doi.org/10.1063/1.864173

[16] Quirk, J.J. and Karni, S. (1994) On the Dynamics of a Shock-Bubble Interaction. NASA CR 194978, ICASE Report No. 94-75.

[17] Weirs, V.G., Dupont, T. and Plewa, T. (2008) Three-Dimensional Effects in Shock-Cylinder Interactions. *Physics of Fluids*, **20**, Article ID: 044102. http://dx.doi.org/10.1063/1.2884787

[18] Shankar, S.K., Kawai, S. and Lele, S.K. (2011) Two-Dimensional Viscous Flow Simulation of a Shock Accelerated Heavy Gas Cylinder. *Physics of Fluids*, **23**, Article ID: 024102. http://dx.doi.org/10.1063/1.3553282

[19] Dimonte, G. and Schneider, M. (2000) Density Ratio Dependence of Rayleigh-Taylor Mixing for Sustained and Impulsive Acceleration Histories. *Physics of Fluids*, **12**, 304-321. http://dx.doi.org/10.1063/1.870309

[20] Alon, U., Hecht, J., Ofer, D. and Shvarts, D. (1995) Power Laws and Similarity of Rayleigh-Taylor and Richtmyer-Meshkov Mixing Fronts at All Density Ratios. *Physical Review Letters*, **74**, 534-537. http://dx.doi.org/10.1103/PhysRevLett.74.534

[21] Sohn, S.-I. (2008) Quantitative Modeling of Bubble Competition in Richtmyer-Meshkov Instability. *Physical Review E*, **78**, Article ID: 017302. http://dx.doi.org/10.1103/PhysRevE.78.017302

[22] Bai, J.S., Wang, T., Li, P., Zou, L.Y, and Liu, C.L. (2009) Numerical Simulation of the Hydrodynamic Instability Experiments and Flow Mixing. *Science in China Series G*, **52**, 2017-2040. http://dx.doi.org/10.1007/s11433-009-0277-9

[23] Colella, P. and Woodward, P.R., (1984) The Piecewise Parabolic Method (PPM) for Gas-Dynamical Simulations. *Journal of Computational Physics*, **54**, 174-201. http://dx.doi.org/10.1016/0021-9991(84)90143-8

[24] Vreman, W. (2004) An Eddy-Viscosity Subgrid-Scale Model for Turbulent Shear Flow: Algebraic Theory and Applications. *Physics of Fluids*, **16**, 3670-3681. http://dx.doi.org/10.1063/1.1785131

[25] Miles, J.W. (1958) On the Disturbed Motion of a Plane Vortex Sheet. *Journal of Fluid Mechanics*, **4**, 538-552. http://dx.doi.org/10.1017/S0022112058000653

[26] Samtaney, R. and Pullin, D.I. (1996) On Initial-Value and Self-Similar Solutions of the Compressible Euler Equations. *Physics of Fluids*, **8**, 2650-2655. http://dx.doi.org/10.1063/1.869050

[27] Jones, B.D. and Jacobs, J.W. (1997) A Membraneless Experimental for the Study of Richtmyer-Meshkov Instability of a Shock-Accelerated Gas Interface. *Physics of Fluids*, **9**, 3078-3085. http://dx.doi.org/10.1063/1.869416

[28] Zhang, S., Zabusky, N.J., Peng, G. and Gupta, S. (2004) Shock Gaseous Cylinder Interactions: Dynamically Validated Initial Conditions Provide Excellent Agreement between Experiments and Numerical Simulations to Late-Intermediate Time. *Physics of Fluids*, **16**, 1203-1216. http://dx.doi.org/10.1063/1.1651483

[29] Picone, J.M. and Boris, J.P. (1988) Vorticity Generation by Shock Propagation through Bubbles in a Gas. *Journal of Fluid Mechanics*, **189**, 23-51. http://dx.doi.org/10.1017/S0022112088000904

[30] Tong, B.G., Yin, X.Y. and Zhu, K.Q. (2009) Theory of Vortex Motion. 2nd Edition, University of Science and Technology of China Press, Hefei.

21

Possible Duplicities of Five Asteroids

Isao Satō[1]*, Hiromi Hamanowa[2], Hiroyuki Tomioka[3], Sadaharu Uehara[4]

[1]Nihon University, Sakata, Yamagata, Japan
[2]Hamanowa Astronomical Observatory, Motomiya, Fukushima, Japan
[3]Hitachi, Ibaraki, Japan
[4]KEK, Tsukuba, Ibaraki, Japan
Email: *satoisao@nifty.com, hamaten@poplar.ocn.ne.jp, tomi-01@sea.plala.or.jp, uehara@post.kek.jp

Abstract

Some evidences of possible duplicities of five asteroids are presented. A satellite of (279) Thule was convincingly detected by a stellar occultation on 2008 April 3 by Thule and also from follow-up light curve observations. The orbital period of the satellite is 3.007 or 6.014 days, and the minimum diameter is estimated to be 52 km. A satellite of (324) Bamberga was detected by secondary drops of the light curve in 2007. The rotation period of the primary is 1.22625 days, and the revolution period of the secondary is 3.00 or 6.00 days. Presumed contact duplicity of the main body of the L4 Trojan (624) Hektor was detected by a stellar occultation by Hektor on 2008 January 24. A possible satellite of (657) Gunlöd was suggested from a secondary occultation by Gunlöd on 2008 November 29. The minimum of the diameter is 7 km. A possible satellite of (739) Mandeville was suggested by stellar occultation observations on 1980 December 10.

Keywords

Asteroid, Occultation, Light Curve, Satellite, Binary

1. Introduction

Since the first discovery of Dactyl, a satellite of (243) Ida, by the Galileo spacecraft in 1993, several hundreds of satellites of asteroids have been detected by various means, such as asteroidal occultations, light curve observations, radar observations, and spacecraft encounters. Discovery of a satellite of an asteroid provides a total mass of the binary system from the third law of Kepler if its evolution period and semi-major axis of the orbit are obtained. We present some evidence of the duplicities of five asteroids detected in Japan through occultation observations.

---

*Corresponding author.

2. Satellite of (279) Thule

2.1. Stellar Occultation on 2008 April 3

The asteroid (279) Thule was discovered by J. Palisa in 1888. The semi-major axis of its orbit is 4.27 au, the eccentricity of the orbit is 0.01, and the revolution period is 8.83 years which is within the mean motion resonance of 4:3 with Jupiter. Asteroids with this resonance are part of a group called the Thule group. The other members of the Thule group known thus far are (185290) 2006 UB219 and (186024) 2001 QG207.Thule is a D-type asteroid [1].

The occultation of the star TYC 497200415 by (279) Thule, the first observed stellar occultation by Thule, was observed on 2008 April 3 from four sites in Japan. The event was predicted by Satō (**Figure 1**). The position of the occulted star is α_{2000} = 13 h 31 m 32.3082 ± 0.020 s, δ_{2000} = −06°59'53". 701 ± 0.022, and the brightness of the star is m_v = 10.82. The assumed diameter of Thule is 127 km and the predicted extinction would be 3.7 mag for 8.6 seconds. The occultation track would go through Kanto and Chubu districts of Japan. The observation results are listed in **Table 1**. Three chords among four are timed by video with GPS clock, and one is visual.

Revealed occultation cross section is (126 ± 1 km) × (123 ± 7 km), the position angle of the ellipse is 80° ± 129° (**Figure 2**). Three chords by Yanagida Astronomical Observatory, Hideo Takashima, and Hideyoshi Karasaki are consistent with an ellipse of the primary, but the chord by Hiroyuki Tomioka is not coincident with it. This specific observation was made with GPS time-inserted video, hence the observation must be considered reliable. The extinction suggests a possible previously unknown satellite with diameter of 52 km. The following-up light curve observations of Thule was obtained after the occultation.

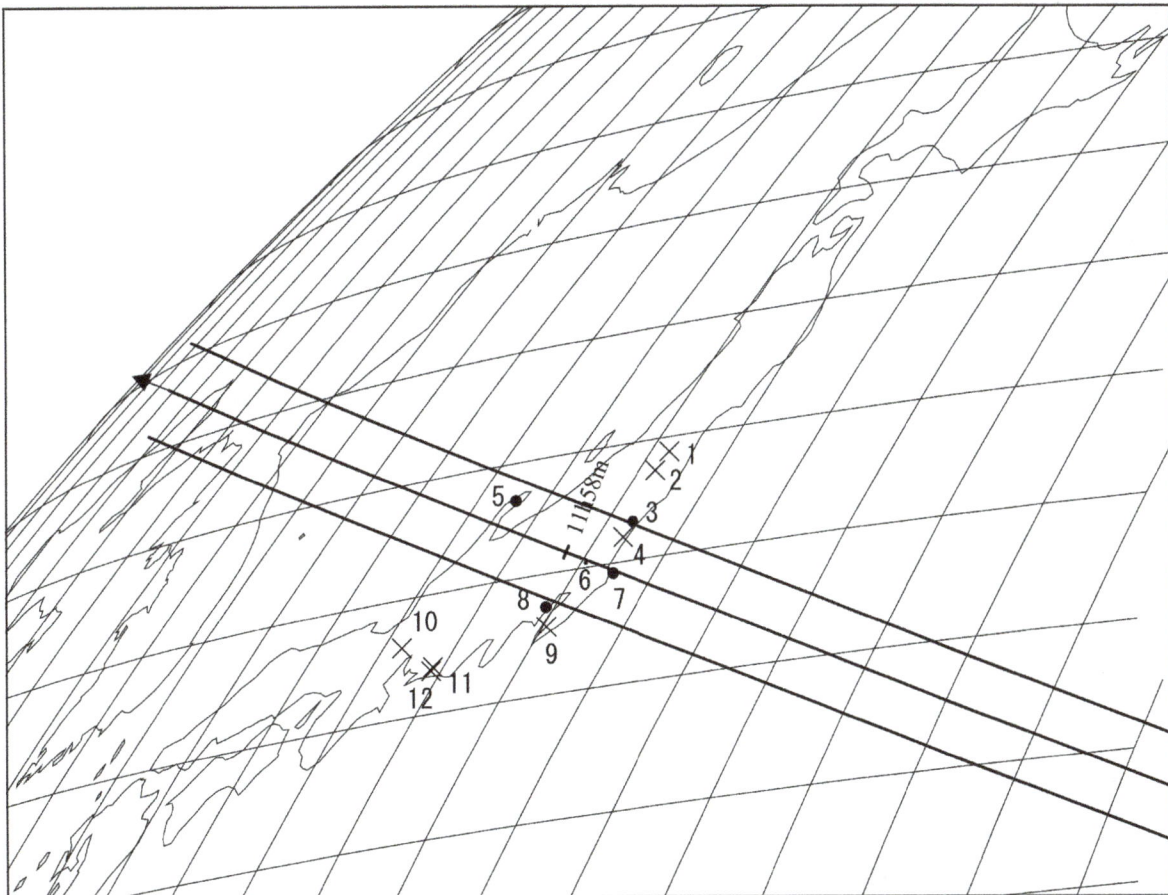

Figure 1. Occultation track and locations of observers of the occultation of TYC 497200415 by (279) Thule on 2008 April 3. The black circles (#3, 5, 7, 8) are successfully times sites and the crosses (#1, 2, 4, 9, 10, 11, 12) are places where no occultation was observed.

Table 1. Observationsdata of the occultation of TYC 497200415 by (279) Thule on 2008 April 3. Geodetic datum is WGS84.

No.	observer	location	longitude	latitude	height	telescope	method	phen	UTC
1	Hikaru Satō	Fukushima, Fukushima, Japan	E140°29'25.3"	N37°44'35.9"	90 m	16 cm N, F = 3.3,	GHS, TiVi video		no occultation
2	Hamanowa Observatory	Koriyama, Fukushima, Japan	E140°25'48.4"	N37°27'37.7"	260 m	40 cm N, F = 4.5	GHS ,TiVi, video		no occultation
3	Kazuyuki Shinozaki	Takahagi, Ibaraki, Japan	E140°40'29.4"	N36°42'19.4"	40 m	35cm SC, F=11	GPS, visual		no occultation
4	Hiroyuki Tomioka	Hitachi, Ibaraki, Japan	E140°41'09"	N36°38'33"	33 m	30 cm C	GHS, TiVi, video	2D	11 h 57 m 57.90 s ± 0.03 s
								2R	11 h 58 m 01.37 s ± 0.03 s
5	Akira Tsuchikawa	Yanagida, Ishikawa, Japan	E137°08'13.4"	N37°20'04.1"	200 m	60 cm C	TEL, video	1D	11 h 58 m 02.50 s ± 0.03 s
								1R	11 h 58 m 10.09 s ± 0.03 s
6	Sadaharu Uehara	Tsukuba, Ibaraki, Japan	E140°07'04.2"	N36°05'11.2"	30 m	20 cm N, F = 4			cloudy
7	Hideo Takashima	Kashiwa, Chiba, Japan	E140°58'07"	N35°49'53"	28 m	20 cm S, F = 2	GHS, TiVi, video	1D	11 h 57 m 56.13 s ± 0.04 s
								1R	11 h 58 m 02.24 s ± 0.04 s
8	Hideyoshi Karasaki	Nerima, Tokyo, Japan	E139°47'00.3"	N35°44'25.4"	46 m	20.3 cm SC	TEL, TiVi, video	1D	11 h 57 m 58.6 s ± 0.6 s
								1R	11 h 57 m 59.9 s ± 0.6 s
9	Mamoru Urabe	Kamogawa, Chiba, Japan	E140°06'05"	N35°06'35"	8 m	30 cm N, F = 4	TEL, video		no occultation
10	Minoru, Owada	Hamamatsu, Shizuoka, Japan	E137°44'08.0"	N34°42'02.5"	8 m	10 cm R, F = 4	GHS, TiVi, video		no occultation
11	Hisashi Suzuki	Hamamatsu, Shizuoka, Japan	E137°42'48.4"	N34°45'53.8"	50 m	25 cm R, F = 5	GHS, TiVi, video		no occultation
12	Akira Arai	Inabe, Mie, Japan	E136°31'24.3"	N35°10'14.6"	187 m	35.5 cm SC, F = 11	GHS, TiVi, video		no occultation

2.2. Light Curve Observations in 2008

A light curve observation was obtained at Hamanowa Observatory during 13 nights from 2008 April 8 to May 6. **Figure 3** shows a total combined light curve. Four instances of secondary drops are seen in the light curve. The Light curve shows at rotation period of 7 h 26 m 24 ± 1 s, peak-to-peak amplitude of 0.081 mag with secondary drops of 0.1 mag. The result is coincident with the previous result of 7.44 ± 0.05 hours [2]. **Figure 4** shows the light curves of four nights with secondary drops, 2008 April 6, 15, 30, and May 3. The secondary drops suggest a satellite with revolution period of 3.007 or 6.014 days. The assumed phase of the secondary drop at the time of the stellar occultation is coincident with the hypothesis that Tomioka's extinction occurred from a satellite with a diameter of 52 km, namely, it is assumed that the possible satellite was during the partial phase of a mutual phenomenon.

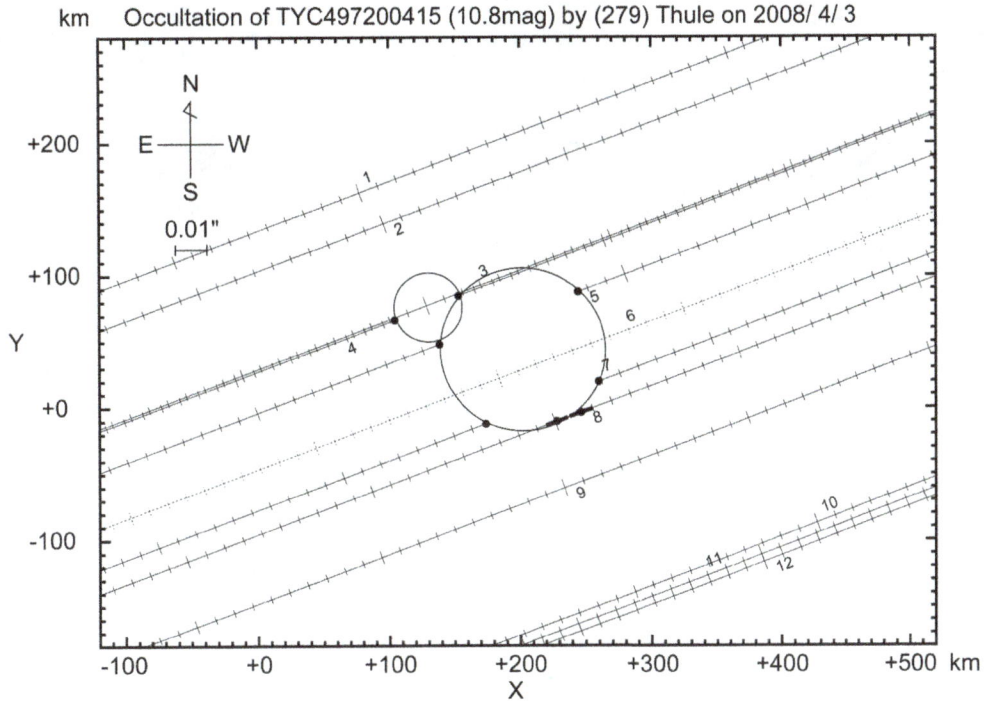

Figure 2. Result of the occultation of TYC 497200415 by (279) Thule on 2008 April 3. An ellipse of (126 ± 1 km) × (123 ± 7 km) is fitted for the primary, and a circle of a diameter of 52 km is fitted for the satellite.

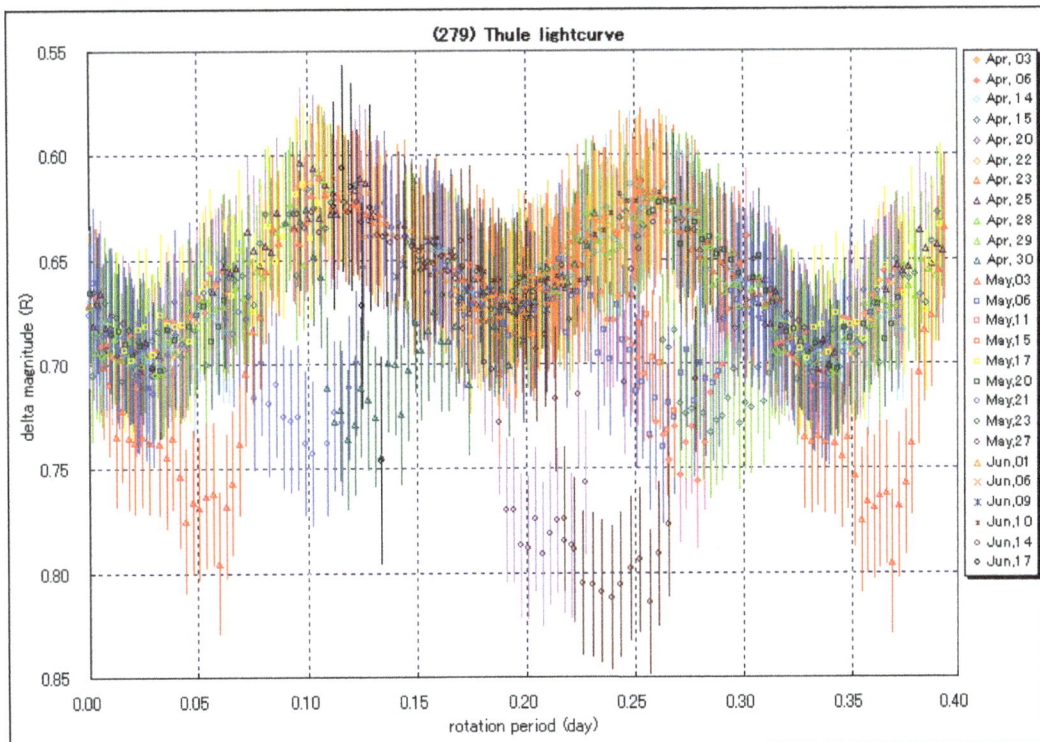

Figure 3. Composed light curve of (279) Thule from 2008 April 3 to May 6 at Hamanowa Astronomical Observatory. The main light curve is consistent with a period of 7.44 ± 0.05 hoursand a peak-to-peak amplitude of 0.081 mag. In addition to the main light curve, secondary drops are shown on four nights (April 6, 15, 30, and May 3).

Observed Long Period.

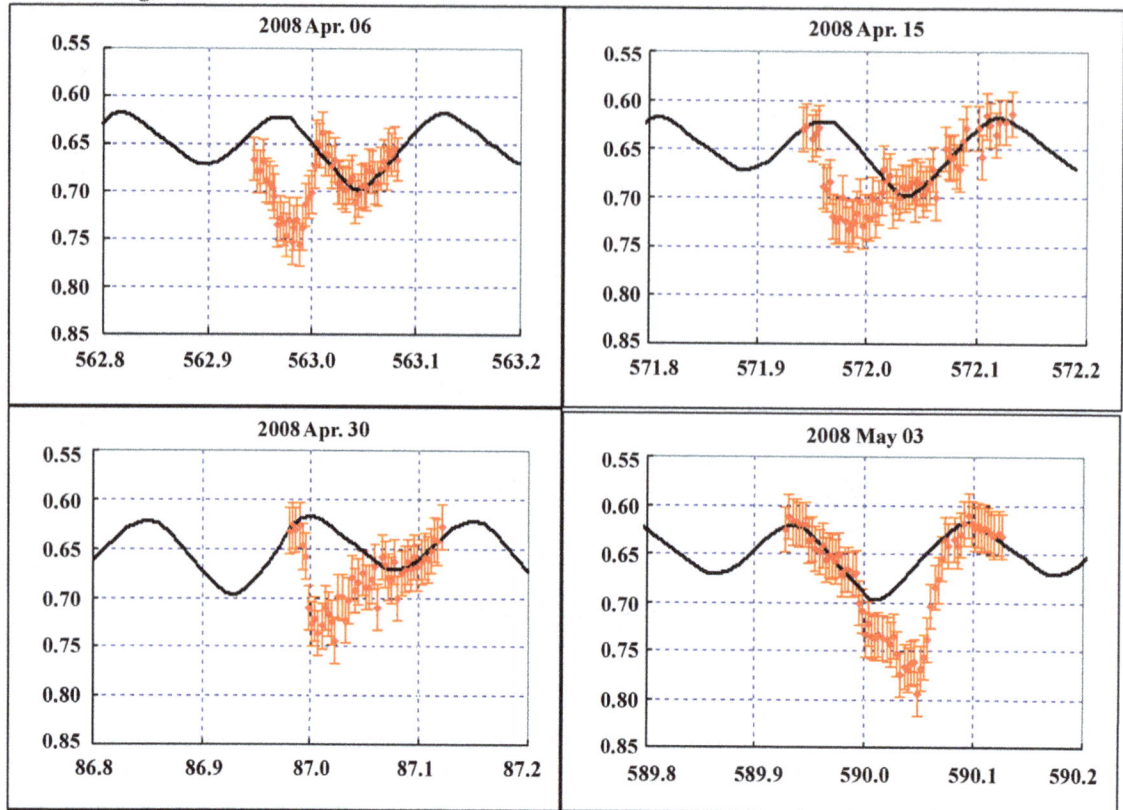

Figure 4. Detailed features of the light curves of (279) Thule on 2008 April 6, 15, 30, and May 3. Each figure shows a secondary drop on the main light curve. The period of the secondary drops is 3.007 or 6.014 days.

3. Satellite of (324) Bamberga

(324) Bamberga was discovered by J. Palisa in 1892. The planet is a main belt asteroid with a revolution period of 4.4 years. The first stellar occultation by (324) Bamberga was observed from the USA, Japan and China on 1987 December 8. The occultation cross section showed a diameter of 227.6 ± 1.8 km [3].

The second event, the occultation of HIP059807 (7.3 mag) by Bamberga, was observed from five sites in Australia on 2007 April 20 (**Table 2**, **Figure 5**). The occultation cross section shows an ellipse of (231 ± 4 km) × (211 ± 12 km), P = 106° ± 8°. Hamanowa Observatory performed light curve observations just after the second event from 2007 April 23 to May 22. As a result, the amplitude is 0.084 ± 0.0030 mag during the rotation period of 29 h 27 m 27 ± 3 s (**Figure 6**). Light curve observations were also obtained by [4]-[6]. The period is coincident with these results.

The light curve (**Figure 6**) also shows secondary drops on the three nights (2007 April 28, May 4 and 7). Expanded light curve from 2007 April 23 to May 8 (**Figure 7**) shows that the observations on the six nights (2007 April 23, 26, 27, 30, May 3, and 8) fit the shorter period (1.2274 days) sine curve but the observations on the three nights (2007 April 28, May 4 and 7) fit the long period (2.96 days) sine curve. This fact suggests an unknown satellite whose revolution period is 2.96 days.

From these two occultation cross sections and light curves, a 3-D model of Bamberga is reconstructed by the statistical inversion method [7]. The results are shown in **Figure 8** and **Figure 9**. North and south pole is (λ, β) = (149° ± 11°, +68° ± 11°) or (329° ± 11°, −68° ± 11°), lengths of the three principal axes are (2a, 2b, 2c) = (246.3 ± 168.6 km) × (228.3 ± 9.5 km) × (220.4 ± 8.0 km). Constructed cross sections are shown in **Figure 10** and **Figure 11**.

4. Duplicity of (624) Hektor

The L4 Trojan (624) Hektor was discovered by A. Kopff in 1907. Hektor has a satellite S/2006 (624) 1 of diameter

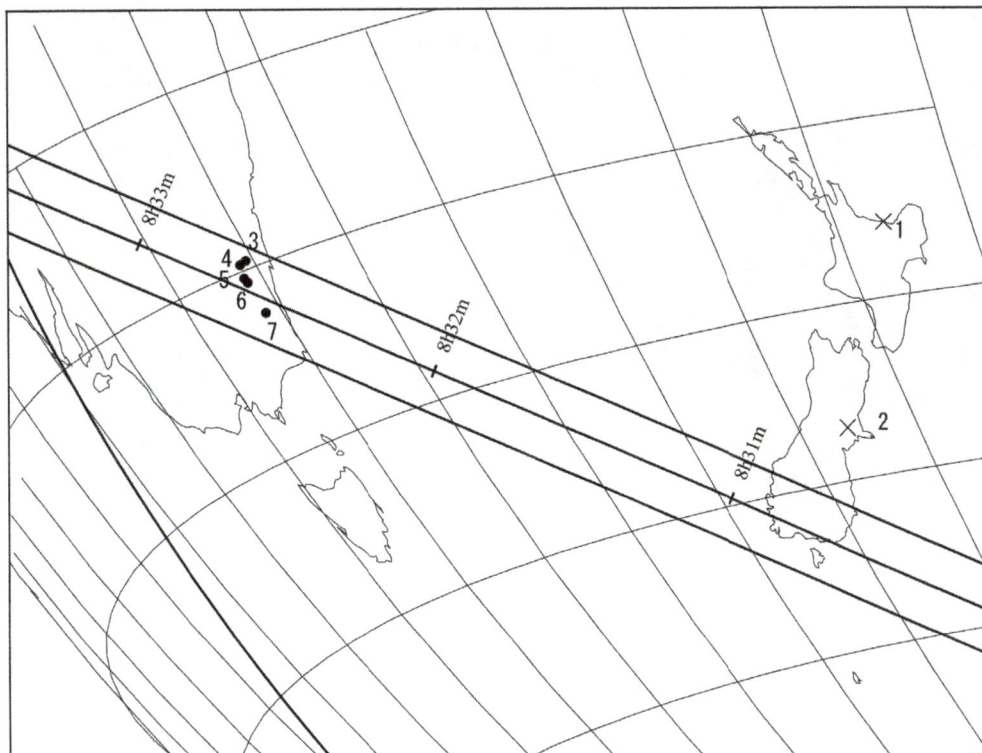

Figure 5. Occultation track and locations of the black circles observers of the occultation of HIP 059807 by (324) Bamberga on 2007 April 20. The black circles (#3 - 7) are sites where the occultation was successfully timed and the crosses (#1, 2) are sites where no occultation occurred.

Table 2. Observation data of the occultation of HIP 59807 by (324) Bamberga on 2007 April 20. Geodetic datum is WGS84 [20].

No.	Observer	location	longitude	latitude	height	telescope	method	phen	UTC
1	D. Waton	Whakatane, NZ	E176°51'50.7"	S37°55'18.5"	3 m	20 cm			no occultation
2	B. Loader	Darfiled, NZ	E172°06'24.4"	S43°28'52.9"	210 m	25 cm			no occultation
3	D. Gault	Goulburn, NSW, Aus	E149°38'29.4"	S34°47'12.9"	702 m	20 cm		D	08 h 32 m 36.46 s
								R	08 h 32 m 47.66 s
4	A. Brakel	Gunning, NSW, Aus	E149°14'31.8"	S34°49'58.5"	620 m	20 cm		D	08 h 32 m 35.9 s ± 0.6 s
								R	08 h 32 m 49.1 s ± 0.4 s
5	M. Nelmes	Spence, ACT, Aus	E149°03'30.6"	S35°11'39.8"	702 m	10 cm		D	08 h 32 m 33.9 s ± 0.3 s
								R	08 h 32 m 48.2 s ± 0.3 s
6	P. Purcell	Red Hill, ACT, Aus	E149°06'51.8"	S35°19'30.6"	733 m	40 cm		D	08 h 32 m 33.1 s± 0.6 s
								R	08 h 32 m 47.73 s ± 0.4 s
7	David Herald	Cooma, NSW, Aus	E149°09'44.8"	S36°14'18.6"	835 m	20 cm		D	08 h 32 m 27.95 s
								R	08 h 32 m 41.48 s

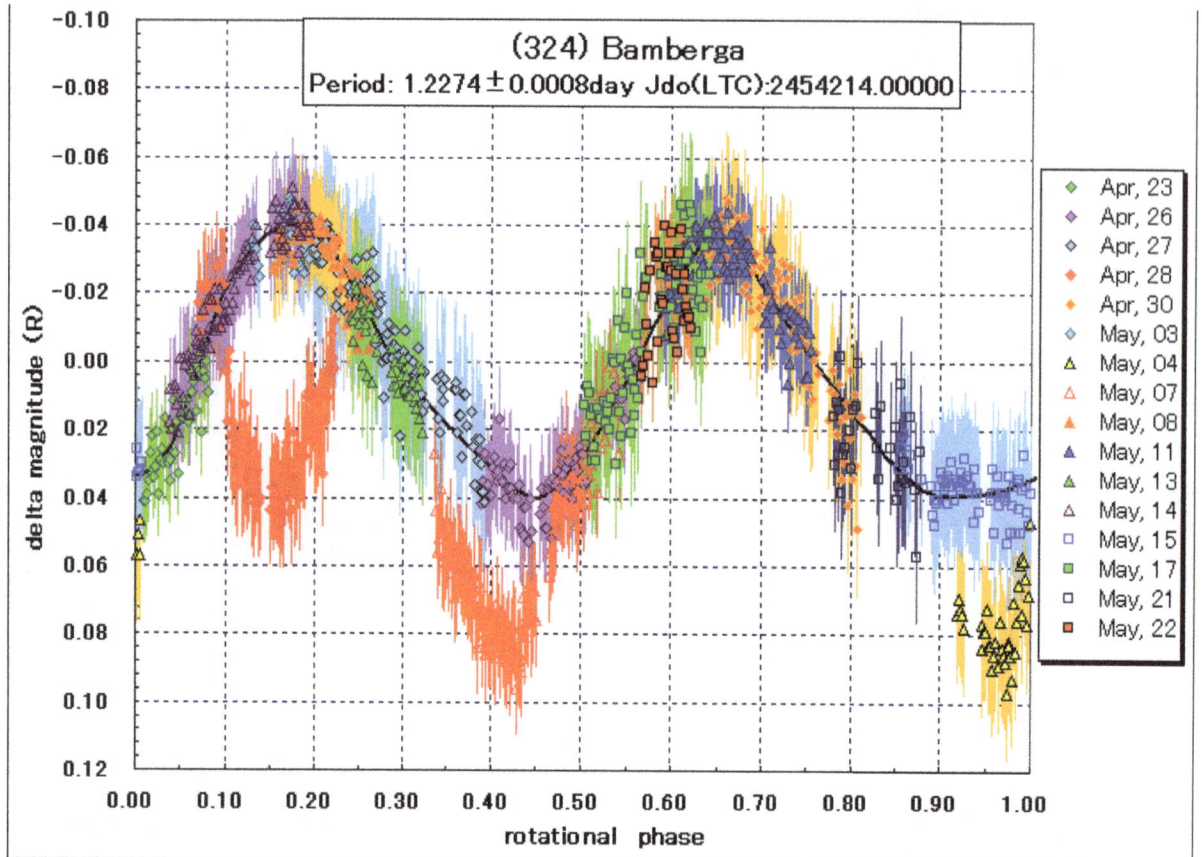

Figure 6. Composed light curve of (324) Bamberga from 2007 April 23 to May 22. The primary period is 1.2274 ± 0.0008 days. In addition to the primary period, secondary drops are shown on 2007 April 28, May 4 and 7.

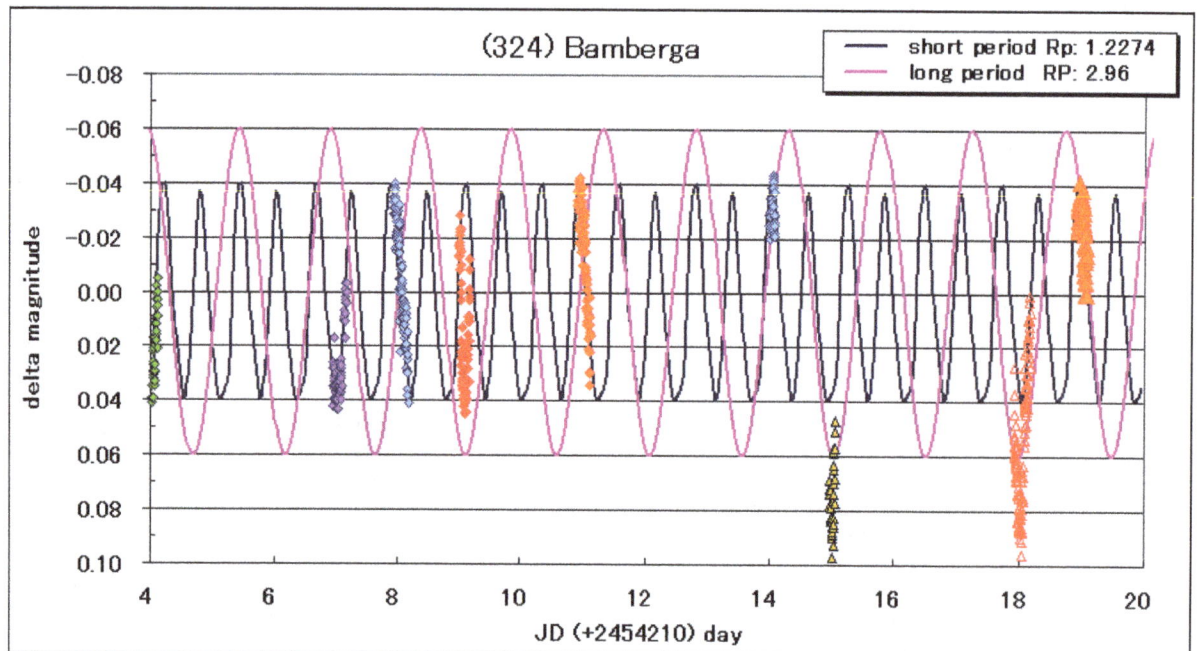

Figure 7. Expanded light curve of (324) Bamberga from 2007 April 23 to May 9. The short period sine curve (blue line) is 1.2274 days, and the long period sine curve (red line) is 2.96 days. The secondary drops on 2007 April 23, May 4 and 7 do not fall on the short period sine curve but are coinsident with the bottom of the long period sine curve.

Probability distribution of orientation of spin vector of (324) Bamberga (ecliptic coordinates)

(occultation : 2007/4/20 , 1987/12/7)

(lightcurve : 2007 , 1978 , 1958)

lambda = 329° + 11°
beta   = -68° + 11°

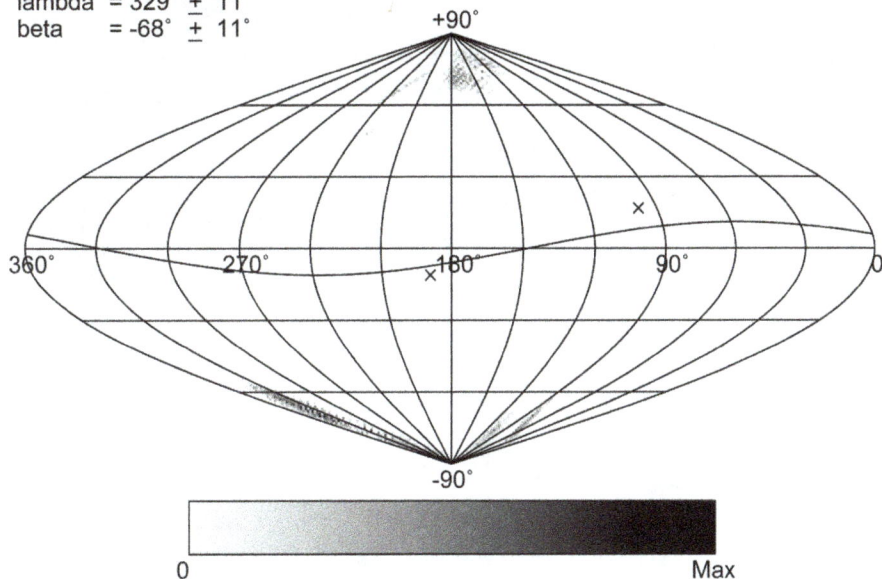

Figure 8. Probability distribution of the spin vector of (324) Bamberga. High probability regions are antipodal $(\lambda, \beta) = (149° \pm 11°, +68° \pm 11°)$ (Ursa Major) or $(\lambda, \beta) = (329° \pm 11°, -68° \pm 11°)$ (Eridanus). Two crosses indicate the positions of the occulted stars in 1987 and 2007 events, respectively.

Probability distribution of lengths of principal axes of (324) Bamberga

(occultation : 2007/4/20 , 1987/12/7)

(lightcurve : 2007 , 1978 , 1958)

2a= 246.3 + 168.6km
2b= 228.3 + 9.5km
2c= 220.9 + 8.0km
D = 231.6 + 37.0km

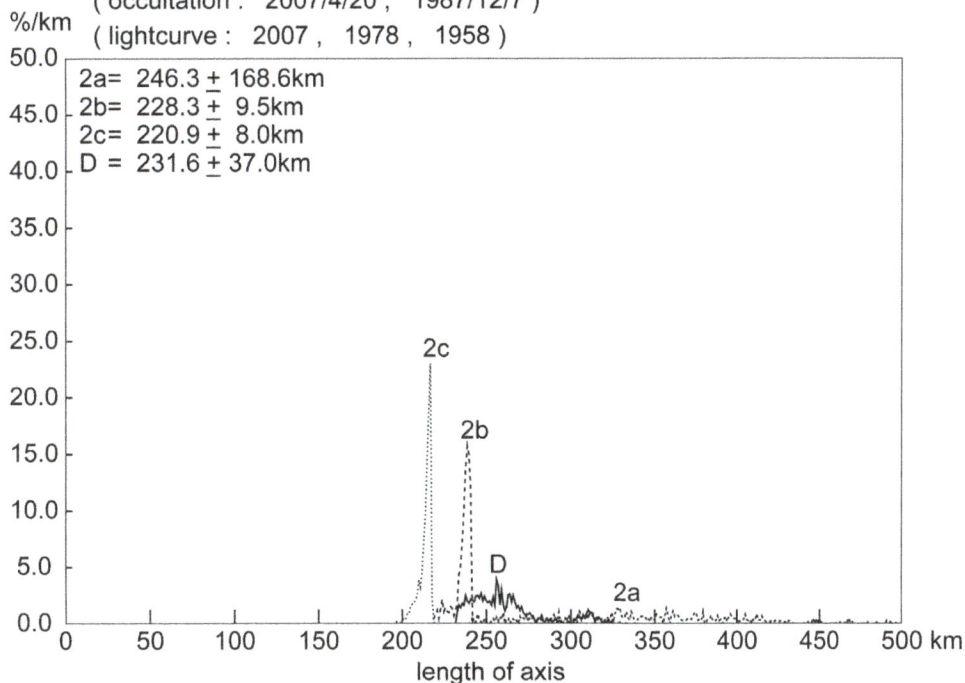

Figure 9. Probability distribution of the lengths of the three principal axes of (324) Bamberga. Since Bamberga is nearly a spheroid, the length of the longest axis a cannot be determined with sufficient accuracy. Assumed lengths are $(2a, 2b, 2c) = (246.3 \pm 168.6 \text{ km}) \times (228.3 \pm 9.5 \text{ km}) \times (220.4 \pm 8.0 \text{ km})$.

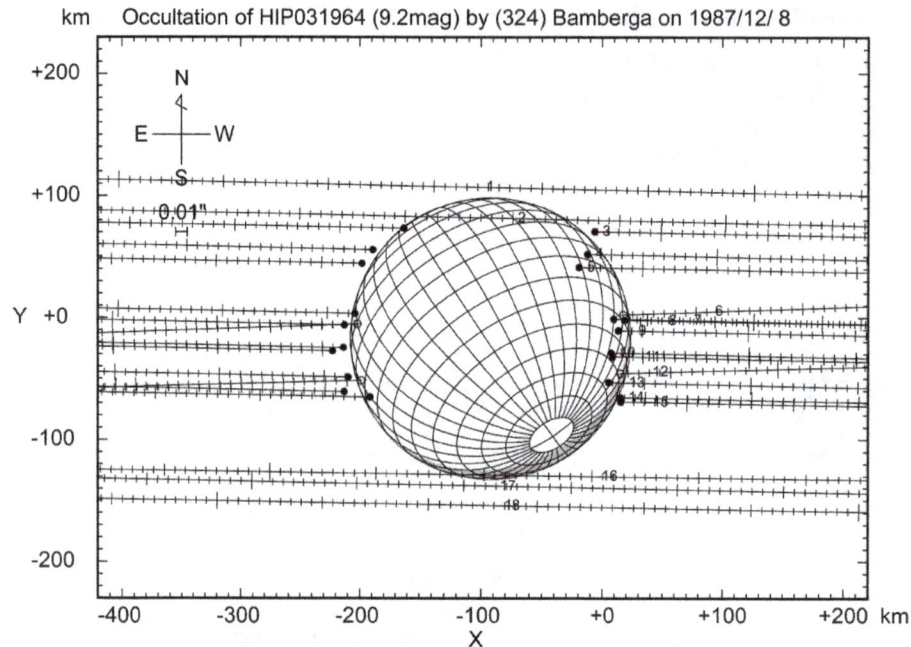

Figure 10. Reconstructed 3-D shape of (324) Bamberga during the occultation of HIP 031964 on 1987 December 8.

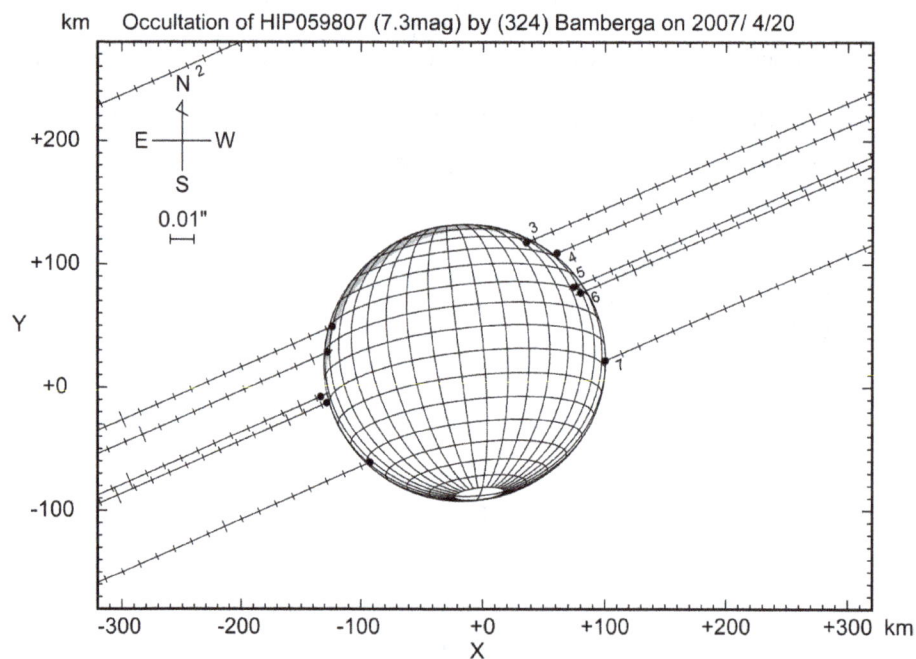

Figure 11. Reconstrunted 3-D shape of (324) Bamberga during the occultation of HIP 059807 on 2007 April 20.

15 km [8]. The stellar occultation by (624) Hektor of TYC057700887 (10.1 mag) was observed from two sites in Japan on 2008 January 24 (**Figure 12**). This is the first stellar occultation observation to be observed by Hektor. Tomioka at Hitachi, Ibaraki, detected an extinction for 4.4 seconds by video with GPS clock. Uehara at Tsukuba, Ibaraki, observed two times extinction, 3.6 and 5.0 seconds, respectively, by visual (**Table 3**, **Figure 13**). His observation suggests the primary body is a close or contact binary. A possible unusual shape of the primary has been suggested by many authors [9]-[19].

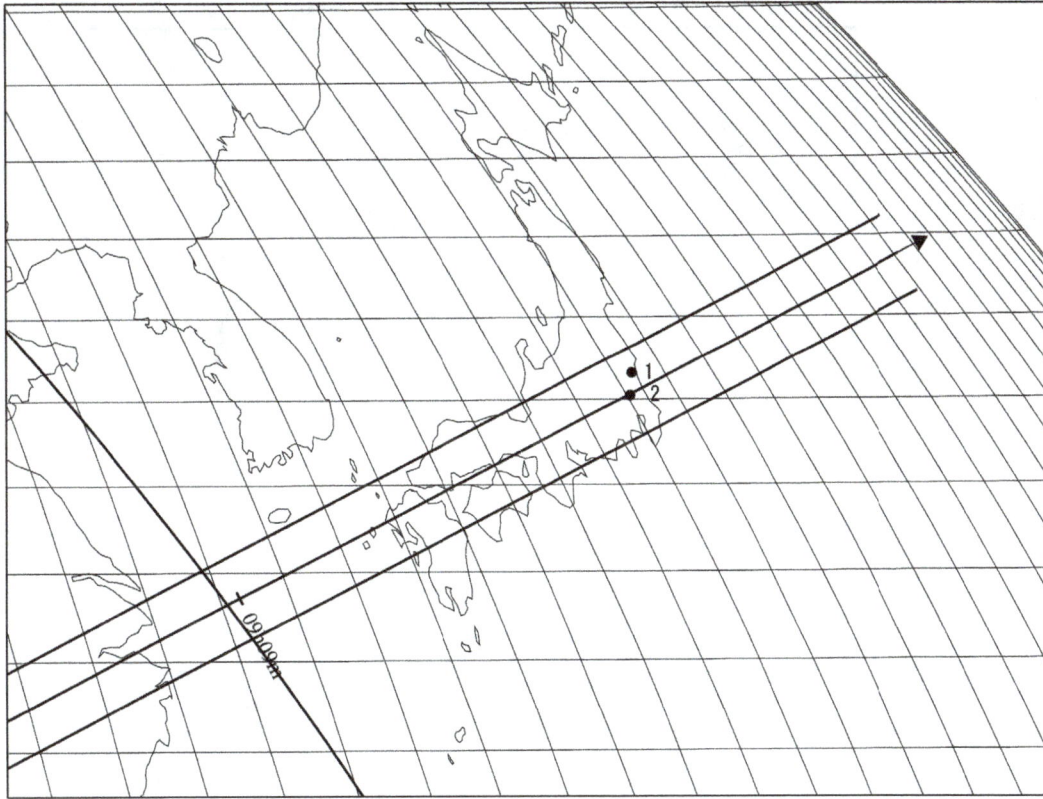

Figure 12. Occultation track and locations of observers of the occultation of TYC 057700817 by (624) Hektor on 2008 January 24. The black circles indicate sites where occultations were successfully timed.

Table 3. Observation data of the occultation of TYC 057700887 by (624) Hektor on 2008 January 24. Geodetic datum is WGS84.

No.	observer	location	longitude	latitude	thgieh	telescope	method	phen	UTC
1	Hiroyuki Tomioka	Hitachi, Ibaraki, Japan	E140°41'09"	N36°38'33"	33 m	30 cm C	GHS, TiVi, video	D	09 h 09 m 36.63 s ± 0.05 s
								R	09 h 09 m 41.03 s ± 0.05 s
2	Sadaharu Uehara	Tsukuba, Ibaraki, Japan	E140°07'04.2"	N36°05'11.2"	30 m	20 cm N, F = 4	visual	1D	09 h 09 m 28.8 s ± 0.5 s
								1R	09 h 09 m 32.4 s ± 0.3 s
								2D	09 h 09 m 35.2 s ± 0.6 s
								2R	09 h 09 m 40.2 s ± 0.4 s

5. Possible Satellite of (657) Gunlöd

(657) Gunlöd was discovered by A. Kopff in 1908. The planet is a main belt asteroid with a revolution period of 4.22 years. The stellar occultation of UCAC2 43078953 (12.0 mag) by (657) Gunlöd was observed from three sites in Japan (**Table 4**, **Figure 14**) on 2008 November 29. Hiroyuki Tomioka at Hitachi, Ibaraki, detected a 3.5 second extinction by video with GPS clock. Akira Yaeza at Hitachi, Ibaraki, also observed a 4.6 seconds visual extinction.

Two chords are coincident with a circle with a cross section of a diameter 45 ± 5 km compared with the assumed diameter of 42 km. These extinctions were caused by the primary. On the other hand, a short extinction of for 0.7 second was detected at Hamanowa Observatory by video with GPS clock located about 100 km north of the occultation track of the primary (**Figure 15**). Compared with the predicted extinction of 2.9 mag for 4.4

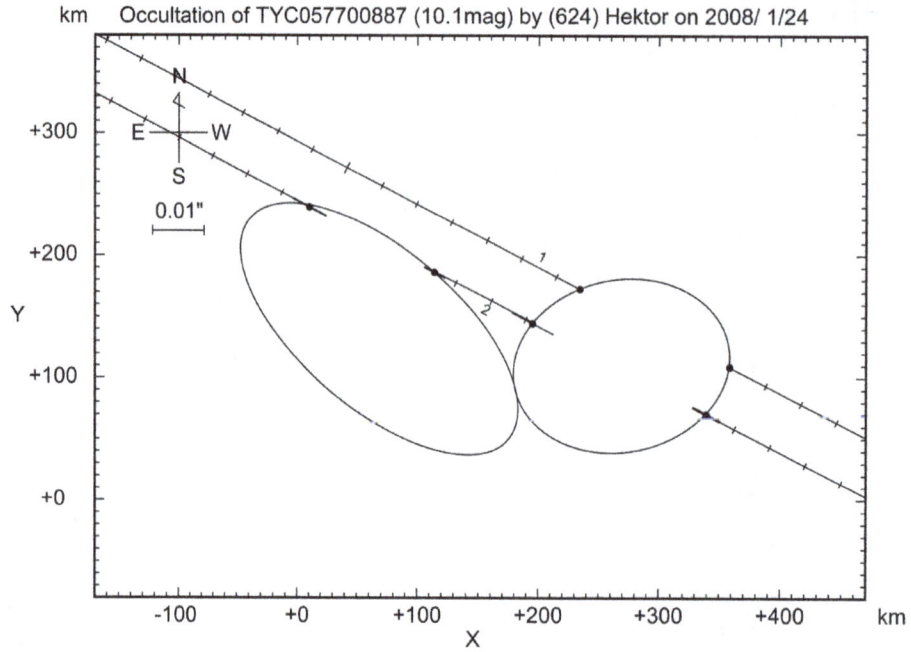

km   Occultation of TYC057700887 (10.1mag) by (624) Hektor on 2008/ 1/24

Figure 13. Result of the occultation of TYC 057700887 by (624) Hektor on 2008 January 24. Two extinctions were visually observed at #2 site. Since the main body is assumed to be a contact binary, two contact ellipses of (280 × 130 km) and (181 × 141 km) are fitted.

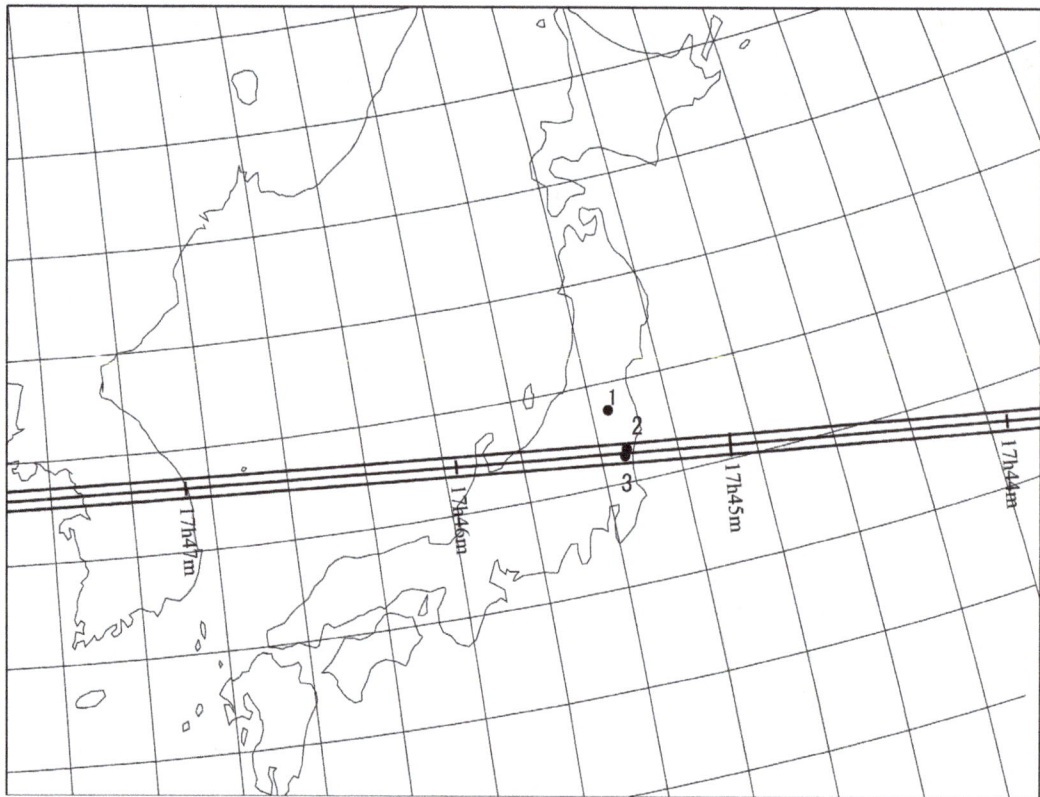

Figure 14. Occultation track and location of observers of the occultation of UCAC2 43078953 by (657) Gunl ŏd on 2008 November 29. The black circles indicate location where occultations were successfully timed. Sites #2 and 3 are where the star was occulted by the primary, and site #1 is where the occultation was by a possible satellite.

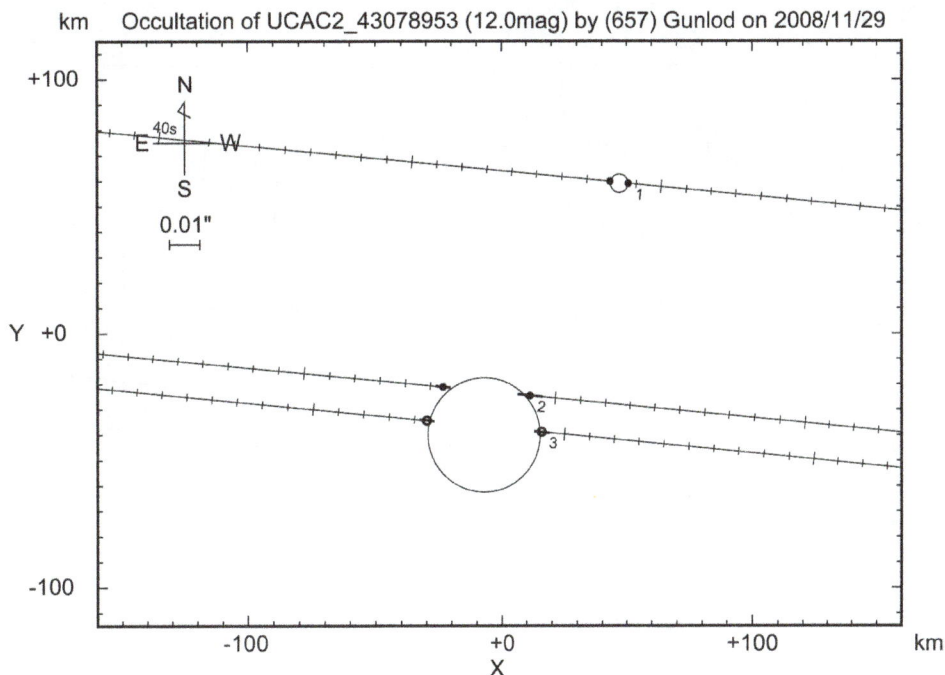

km   Occultation of UCAC2_43078953 (12.0mag) by (657) Gunlōd on 2008/11/29

Figure 15. Result of the occultation of UCAC2 43078953 by (657) Gunlōd on 2008 November 29. Chord #1 and #2 are from timed video and #3 is from a visual timing.

Table 4. Observation data of the occultation of UCAC2 43078953 by (657) Gunlöd on 2008 November 29. Geodetic datum is WGS84.

No.	Observer	location	longitude	latitude	height	telescope	method	phen	UTC
1	Hamanowa Observatory	Koriyama, Fukushima, Japan	E140°25'48.4"	N37°27'37.7"	260 m	40 cm N, F = 4.5	GHS, TiVi, video	D	17 h 45 m 21.34 s ± 0.06 s
									17 h 45 m 22.07 s ± 0.06 s
2	Hiroyuki Tomioka	Hitachi, Ibaraki, Japan	E140°41'09"	N36°38'33"	33 m	30 cm C	GHS, TiVi, video	D	17 45 m 21.0 s ± 0.5 s
								R	17 h 45 m 24.5 s ± 0.3 s
3	Akira Yaeza	Hitachi, Ibaraki, Japan	E140°36'11"	N36°31'28"	230 m	20 cm C, 133x	TEL visual	D	17 h 45 m 25.4 s ± 0.3 s
								R	17 h 45 m 30.0 s ± 0.3 s

seconds, this extinction is possibly caused by an unknown satellite with a diameter of 7 km. Another possible interpretation is that the occulted star is double.

6. Possible Satellite of (739) Mandeville

(739) Mandeville was discovered by J. H. Metcalf in 1913. The planet is a main belt asteroid with a revolution period of 4.53 years. The stellar occultation of PPM177093 = HIP052360 (9.4 mag) by (739) Mandeville was observed from two sites in Japan on 1980 December 10 (**Table 5**, **Figure 16**). The event was the first ever successfully observed asteroid occultation from Japan. Miyoshi Ida at Ohmi-hachiman, Shiga, observed an extinction for five or six second visually with 6.5 cm telescope, but he failed to record his voice, as a result the time and duration of the event are uncertain. The extinction corresponds to a chord from 69 to 83 km. On the other hands, Toshio Hirose at Yaita, Tochigi, visually observed an extinction of 2.1 seconds also with 6.5 cm telescope. The extinction corresponds to a chord of 29 km. Both Ida and Hirose are experienced lunar occultation

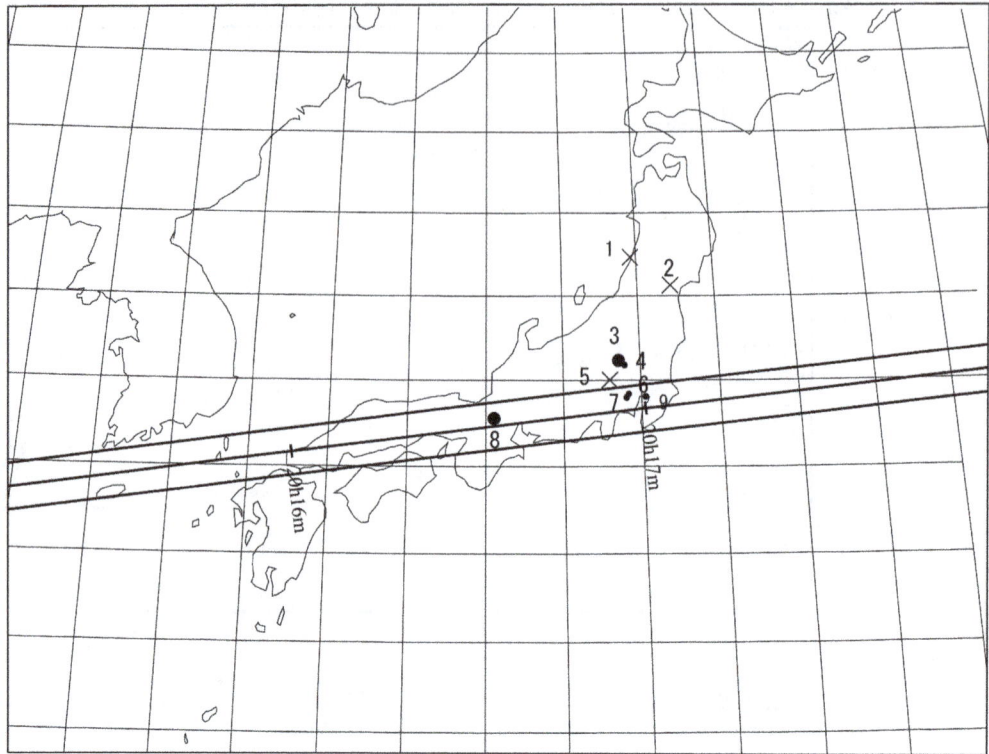

Figure 16. Occultation track and locations of observers of the occultation of HIP 052360 by (739) Mandeville on 1980 December 10. The black circle #3 is visual timing site, and #8 is an occultation from a site where no timing was observed. Crosses (#1, 2, 5) indicate sites where no occultation took place and points (#6, 7, 9) sites where clouds obscured any observation.

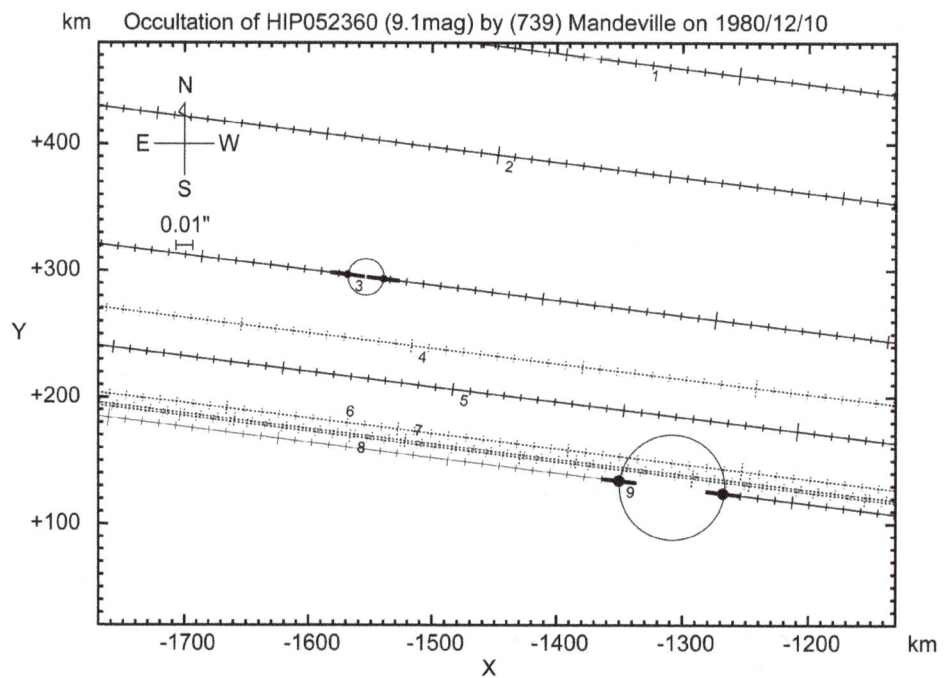

Figure 17. Result of the occultation of HIP 052360 by (739) Mandeville on 1980 December 10. Locations #3 and #8 are where visual occultation were timed but those times are inaccurate. But location #5 between them was no occultation.

Table 5. Observationdata od the occultation of PPM 157093 by (739) Mandeville on 1980 December 10. Geodetic datum is **Tokyo Datum**.

No.	Observer	Location	longitude	latitude	height	telescope	method	phen	UTC
1	Isao Satō	Sakata, Yamagata, Japan	E139°50'44.6"	N38°55'32.8"	4 m	13 cm N, F = 7.7	visual		no occultation
2	Sendai M. O.	Sendai, Miyagi, Japan	E140°51'56"	N38°15'22"	45 m	41 cm C, 15 cm R			no occultation
3	Toshio Hirose	Yaita, Tochigi, Japan	E139°25'48.4"	N36°27'37.7"	260 m	6.5 cm R	visual	D	20 h 16 m 38.5 s ± 1.0 s
								R	20 h 16 m 40.6 s ± 1.0 s
4	Kiyoshi Sakurai	Tanuma, Tochigi, Japan	E139°34'31.5"	N36°21'36.3"	63 m	12 cm N			cloudy
5	Dodaira A.O.	Tokikawa, Saitama, Japan	E139°11'42"	N36°00'22"	879 m	91 cm C	p.e.		no occultation
6	Yoshikazu Ogawa	Nakano, Tokyo, Japan	E139°40'31.1"	N35°41'01.6"	40 m	6.8 cm R			cloudy
7	Yoshihiro Musashi	Kawasaki, Kanagawa, Japan	E139°36'15.0"	N35°35'34.5"	31 m	10 cm N			cloudy
8	Shoji Kawana	Chiba, Chiba, Japan	E140°06'04.1"	N35°41'01.6"	16 m	8 cm R			cloudy
9	Miyoshi Ida	Ohmihachiman, Shiga, Japan	E136°11'53.3"	N35°06'38.5"	128 m	6.5 cm R	visual	D	20 h 16 m 30 s
								R	20 h 16 m 36 s

observers and they convinced the extinctions were real, but the timings were both visual in nature. However, Koichiro Tomita at Dodaira Station of the Tokyo Astronomical Observatory observed no occultation with a 91 cm reflector. The three observations are not consistent as is shown in **Figure 17**. A possible interpretation is duplicity of Mandeville, namely the extinction observed by Ida is by the primary body and the extinction observed by Hirose is by the secondary.

7. Conclusion

Possible satellites of (279) Thule, (324) Bamberga, (657) Gunlōd, (739) Mandeville, and duplicity of the primary body of the possible contact binary Trojan (624) Hektorare suggested from their stellar occultation and light curve observations. Confirmation is needed in order to verify of the nature of the observed extinctions.

References

[1] Neese, C., Ed. (2010) Asteroid Taxonomy V6.0. EAR-A-5-DDR-TAXONOMY-V6.0. NASA Planetary Data System.

[2] Zappala, V., Di Martino, M., Cellino, A., Farinella, P., De Sanctis, G. and Ferreri, W. (1989) Rotation Properties of Outer Belt Asteroids. *Icarus*, **82**, 354-368. http://dx.doi.org/10.1016/0019-1035(89)90043-2

[3] Millis, R.L., Wasserman, L.H., Franz, O.G., Bowell, E., Nye, R.A., Thompson, D.T., White, N.M., Hubbard, W.B., Eplee Jr., R.E., Lebofsky, L.A., Marcialis, R.L., Greenberg, R.J., Hunten, D.M., Reitsema, H.J., Qian, B., Dunham, D.W., Maley, P.D., Klemola, A.R. and Yeomans, D.K. (1989) Observation of the 8 December 1987 Occultation of AG +40° 0783 by 324 Bamberga. *Astronomical Journal*, **98**, 1094-1099

[4] Gehrels, T. and Owings, D. (1962) Photometric Studies of Asteroids IX. Additional Light Curves. *Astrophysical Journal*, **135**, 906-924. http://dx.doi.org/10.1086/147334

[5]   Scaltriti, F., Zappala, V., Stanzel, R., Blanco, C., Catalano, S. and Young, J.W. (1980) Lightcurves and phase relation of asteroid 324 Bamberga. *Icarus*, **43**, 391-398. http://dx.doi.org/10.1016/0019-1035(80)90184-0

[6]   Lagerkvist, C.I., Magnusson, P., Williams, I.P., Buontempo, M.E., Gibbs, P. and Morrison, L.V. (1989) Physical Studies of Asteroids. XIX—Phase Relations and Composite Lightcurves Obtained with the Carlsberg Meridian Circle. *Astronomy & Astrophysics Supplement Series*, **78**, 519-532.

[7]   Satō I., Buie, M., Maley, P.D., Hamanowa, H., Tsuchikawa, A. and Dunham, D.W. (2014) A 3-D Shape of (704) Interamnia from Its Occultations and Lightcurves. *International Journal of Astronomy and Astrophysics*, **4**, 91-118. http://dx.doi.org/10.4236/ijaa.2014.41010

[8]   Marchis, F., Wong, M.H., Berthier, J., Descamps, P., Hestroffer, D., Vachier, F., Le Mignant, D.W.M. and de Pater, I. (2006) S/2006 (624) 1. *IAU Circular* No. 8732.

[9]   Dunlap, J.L. and Gehrels, T. (1969) Minor Planet III. Lightcurves of a Trojan Asteroid. *Astronomical Journal*, **74**, 796-803. http://dx.doi.org/10.1086/110860

[10]   Hartmann, W.K. and Cruikshank, D.P. (1978) The Nature of Trojan Asteroid 624 Hektor. *Icarus*, **36**, 353-366. http://dx.doi.org/10.1016/0019-1035(78)90114-8

[11]   Zappala, V. and Knezevic, Z. (1984) Rotation Axes of Asteroids: Results for 14 Objects. *Icarus*, **59**, 436-455. http://dx.doi.org/10.1016/0019-1035(84)90112-X

[12]   Magnusson, P. (1986) Distribution of Spin Axes and Senses of Rotation for 20 Large Asteroids. *Icarus*, **68**, 1-39. http://dx.doi.org/10.1016/0019-1035(86)90072-2

[13]   Michalowski, T. (1988) Photometric Astronomy Applied to Asteroids: 6, 15, 43, and 624. *Acta Astronomica*, **38**, 455-468.

[14]   Dahlgren, M., Fitzsimmons, A., Lagerkvist, C.-I. and Williams, I.P. (1991) Differential CCD Photometry of Dubiago, Chron and Hektor. *Monthly Notice of the Royal Astronomical Society*, **250**, 115-118. http://dx.doi.org/10.1093/mnras/250.1.115

[15]   De Angelis, G. (1992) New Asteroid Pole Determinations. In: Brahic, A., Gerard, J.-C. and Surdej, J., Eds., *Observations and Physical Properties of Small Solar System Bodies*, Universite de Liege, Institut d'Astrophysique, Liege, 195-201.

[16]   Detal, A., Hainaut, O., Ibrahim-Denis, A., Pospieszalska-Surdej, A., Schils, P., Schober, H.J. and Surdej, J. (1994) Pole, Albedo and Shape of the Minor Planets 624 Hektor and 43 Ariadne: Two Tests for Comparing Four Different Pole Determination Methods. *Astronomy & Astrophysics*, **281**, 269-280.

[17]   De Angelis, G. (1995) Asteroid Spin, Pole and Shape Determinations. *Planetary Space Science*, **43**, 649-682. http://dx.doi.org/10.1016/0032-0633(94)00151-G

[18]   Hainaut-Rouelle, M.-C., Hainaut, O.R. and Detal, A. (1995) Light Curves of Selected Minor Planets. *Astronomy & Astrophysics Supplement Series*, **112**, 125-142.

[19]   Blanco, C. and Riccioli, D. (1998) Pole Coordinates and Shape of 30 Asteroids. *Astronomy and Astrophysics Supplement Series*, **131**, 385-394. http://dx.doi.org/10.1051/aas:1998277

[20]   Dunham, D.W., Herald, D., Frappa, E., Hayamizu, T., Talbot, J. and Timerson, B. (2014) Asteroid Occultations V12.0. EAR-A-3-RDR-OCCULTATIONS-V12.0. NASA Planetary Data System.

22

Collinear Libration Points in the Photogravitational CR3BP with Zonal Harmonics and Potential from a Belt

Jagadish Singh[1], Joel John Taura[2]

[1]Department of Mathematics, Faculty of Science, Ahmadu Bello University, Zaria, Nigeria
[2]Department of Mathematics and Computer Science, Federal University, Kashere, Nigeria
Email: jgds2004@yahoo.com, taurajj@yahoo.com

Abstract

We have studied a reformed type of the classic restricted three-body problem where the bigger primary is radiating and the smaller primary is oblate; and they are encompassed by a homogeneous circular cluster of material points centered at the mass center of the system (belt). In this dynamical model, we have derived the equations that govern the motion of the infinitesimal mass under the effects of oblateness up to the zonal harmonics J_4 of the smaller primary, radiation of the bigger primary and the gravitational potential generated by the belt. Numerically, we have found that, in addition to the three collinear libration points L_i (i = 1, 2, 3) in the classic restricted three-body problem, there appear four more collinear points L_{ni} (i = 1, 2, 3, 4). L_{n1} and L_{n2} result due to the potential from the belt, while L_{n3} and L_{n4} are consequences of the oblateness up to the zonal harmonics J_4 of the smaller primary. Owing to the mutual effect of all the perturbations, L_1 and L_3 come nearer to the primaries while L_{n3} advances away from the primaries; and L_2 and L_{n1} tend towards the smaller primary whereas L_{n2} and L_{n4} draw closer to the bigger primary. The collinear libration points L_i (i = 1, 2, 3) *and* L_{n2} are linearly unstable whereas the L_{n1}, L_{n3} and L_{n4} are linearly stable. A practical application of this model could be the study of motion of a dust particle near a radiating star and an oblate body surrounded by a belt.

Keywords

Circular Restricted Three-Body Problem, Photogravitational, Zonal Harmonic Effect, Potential from the Belt

1. Introduction

In celestial mechanics, one amidst various inspiring subject is the restricted three-body problem (R3BP). The problem entails three bodies: two primary bodies having finite masses moving under their mutual gravitational attraction and the third with a negligible-mass (infinitesimal) body, whose motion is influenced by the primaries. If the primaries move on circular orbits about their common centre of mass, it is termed as the circular R3BP (CR3BP). Then, the objective of this CR3BP is to determine the motion of the infinitesimal mass. [1] and [2] gave a detailed description of the solution of the CR3BP. They showed that if the primary bodies were fixed in a rotating coordinate system, five libration points existed. That is the points where the infinitesimal mass can remain permanent, if placed there with zero velocity. Three of the points L_1, L_2, L_3 are on the line linking the primaries, whereas the other two L_4, L_5 are in equilateral triangular alignment with the primaries. The collinear points L_1, L_2, L_3 are linearly unstable, while the triangular points L_4, L_5 are linearly stable for the mass ratio of the primaries less than 0.03852.

Researches on the sites and stability of the libration points of the CR3BP with perturbations have achieved ample attention in recent times. [3] indicated that small particles were equally influenced by the gravitation and light radiation force as they moved toward luminous celestial bodies. [4] [5] established that the presence of direct solar radiation pressure caused a variation in the sites of the libration points of the CR3BP. He called the CR3BP, photogravitational when one or both of the masses of the primaries were discharges of radiation. Researchers [6]-[10] have examined the existence of libration points and their linear stability in the photogravitational CR3BP.

[11] [12] studied a modified CR3BP by considering the influence from a belt (circular cluster of material points) for planetary systems and found that the likelihood to get libration points around the inner part of the belt was greater than the one nigh the outer part. The impact of the belt makes the configuration of the dynamical system altered such that new libration points emerge under certain condition [13]-[16].

The primaries in CR3BP are generally considered to be spherical in shape, whereas in real situations, numerous celestial bodies are non-spherical (e.g. the Earth, Jupiter, Saturn, Regulus stars are oblate). The oblateness of the planets causes large deviations from a two-body orbit. The most salient instance of disturbance due to oblateness in the solar system is the orbit of the fifth satellite of Jupiter, Amalthea. This planet is extremely oblate and the satellite's orbit is exceptionally small that its line of apsides progresses approximately 900° in one year [17]. This vindicates the incorporation of oblateness of the primaries in the study of CR3BP [18]-[25].

The orbital effects of the oblateness up to the quadrupole, $i.e.\ J_2$, and the octupole, $i.e.\ J_4$, on the orbital motion of a particle in the field of a non-spherical body have been worked out in the general case of an arbitrarily oriented spin axis [26]. [22] certified that the sites of the triangular libration points and their linear stability were influenced by the oblateness up to J_4 of the bigger primary in the CR3BP. [27] examined the effects of photogravitational force and oblateness in the perturbed restricted three-body problem. [15] analyzed analytically and numerically the effects of oblateness up to J_2 of the smaller primary and gravitational potential from the belt on the linear stability of libration points in the photogravitational CR3BP. [16] explored the combined effect of radiation and oblateness up to J_2 of both primaries, together with additional gravitational potential from the circumbinary belt on the motion of an infinitesimal body in the binary stellar systems within the frame work of CR3BP. [9] studied the effects of oblateness up to J_4 of the smaller primary and gravitational potential from a belt, on the linear stability of triangular libration points in the photogravitational CR3BP. [24] looked at the effects of oblateness of both primaries up to zonal harmonic J_4 and gravitational potential from the belt on the linear stability of the triangular libration points in the CR3BP.

Here, our intention is to look into the resultant effect of radiation of the bigger primary, oblateness up to the zonal harmonic J_4 of the smaller primary and gravitational potential from the belt on the sites and stability of collinear libration points in the CR3BP.

The manuscript is structured in five units. Unit 2 deals with the mathematical formulation of the problem, while Unit 3 is dedicated to the determination of the sites of the collinear libration points. The linear stability of collinear points and the conclusion are presented in Units 4 and 5 respectively.

2. Mathematical Formulation of Model

2.1. The Problem

Let m_1 and m_2 be the masses of the primaries with $m_1 > m_2$, and let m be the mass of the infinitesimal body moving in the plane of motion of the primaries. The positions of the primaries are defined with respect to a rotating coordinate frame *oxyz* whose *x*-axis overlaps with the line connecting them and whose origin coincides with the center of mass of m_1 and m_2. The y-axis is perpendicular to the x-axis and the z-axis is normal to the orbital plane of the primaries. Let r_1 be the distance between m and m_1, r_2 the distance between m and m_2; and R the distance between m_1 and m_2. The coordinates of m_1, m_2 and m are $(x_1, 0)$, $(x_2, 0)$ and (x, y) corresponding-ly. Our aim is to find the equations of motion of m under the influence of radiation of m_1, oblateness up to J_4 of the smaller primary, and a circumbinary belt centred at the origin of the coordinate system $oxyz$ (see **Figure 1**).

2.2. The Kinetic Energy

The kinetic energy (K.E) of the infinitesimal body in the barycentric coordinate system $oxyz$ rotating about z-axis with uniform angular velocity n **Figure 1**, is given as

$$K = \frac{1}{2}m\left(\dot{x}^2 + \dot{y}^2\right) + mn\left(x\dot{y} - \dot{x}y\right) + \frac{1}{2}mn^2\left(x^2 + y^2\right), \tag{1}$$

where over dot represents differentiation with respect to time t.

2.3. Force Due to Radiation Pressure

Now, since the radiation pressure force F_p varies with distance by the same law as the gravitational attraction force F_g and works opposite to it, it is likely that this force will lead to a decrease of the effective mass of the bigger primary. Furthermore this decrease relies on the properties of the particle; it is therefore tolerable to talk about a reduced mass. Hence, the consequential force on the particle is [4]

$$F = F_g - F_p = F_g\left(1 - \frac{F_p}{F_g}\right) = qF_g; \tag{2}$$

where $q = \left(1 - \dfrac{F_p}{F_g}\right)$, a constant for a particular particle, is the mass reduction factor. We represent the radiation factor for the bigger primary as $q_1 = 1 - p_1$, $0 < p_1 = \dfrac{F_{p_1}}{F_{g_1}} \ll 1$.

2.4. Potential Due to an Oblate Body

In free space the gravitational potential exterior to an oblate body with its mass distributed symmetrically about its equator, can be expanded in terms of Legendre polynomials in the form

$$V_o\left(r_o, \phi, \theta\right) = -\frac{Gm_o}{r_o}\left[1 - \sum_{n=1}^{\infty} J_{2n} P_{2n}\left(\cos\theta\right)\left(\frac{R_o}{r_o}\right)^{2n}\right] \tag{3}$$

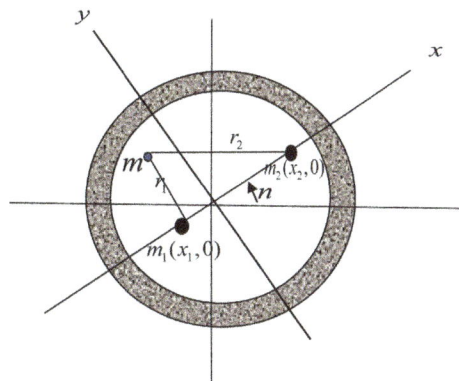

Figure 1. The planar configuration of the problem.

[28]. Equation (3) is expressed in standard spherical coordinates, with ϕ the longitude and θ representing the angle between the body's symmetry axis and the vector to a particle r_o (i.e., the colatitudes). R_o is the mean radius of the oblate body. The terms $P_{2n}(\cos\theta)$ are the Legendre polynomials, given by

$$P_{2n}(x) = \frac{1}{2^{2n}(2n)!}\frac{d^{2n}}{dx^{2n}}(x^2-1)^{2n} \tag{4}$$

J_{2n} are dimensionless coefficients that characterize the size of non spherical components of the potential, called the zonal harmonic coefficients. Since the present study is concerned with planar problem, assuming the equatorial plane of the smaller primary coincides with the plane of motion, then with $\theta = 90°$, Equation (3) becomes

$$V_o(r_o, \phi, \theta) = -Gm_o\left[\frac{1}{r_o} + \frac{J_2 R_o^2}{2r_o^3} - \frac{3J_4 R_o^4}{8r_o^5} + \frac{5J_6 R_o^6}{16r_o^7} + \cdots\right] \tag{5}$$

We denote the oblateness coefficient for the smaller primary as B_i, $0 < B_i = J_{2i}R_0^{2i} \ll 1, i = 1, 2$.

2.5. Potential Due to the Belt

The gravitational potential from belt (circular cluster of material points) centered at the origin of a coordinates system $oxyz$, **Figure 1** as specified by [29] is

$$V_b(r, z) = -\frac{M_b}{\sqrt{r^2 + \left(a + \sqrt{z^2 + b^2}\right)^2}}, \tag{6}$$

where M_b is the total mass of the belt, $r^2 = x^2 + y^2$, a and b are parameters which determine the density profile of the belt. The parameter a controls the flatness of the profile and is known as the *flatness parameter*. The parameter b controls the size of the core of the density profile and is called the *core parameter*. When $a = b = 0$, the potential reduces to the one by a point mass. Restricting ourselves to the xy-plane, Equation (6) becomes

$$V_b(r, 0) = \frac{M_b}{\left(r^2 + T^2\right)^{1/2}}, \quad \text{where } T = a + b. \tag{7}$$

2.6. The Potential Energy of the Infinitesimal Body

The potential energy of the infinitesimal body, under the influence of the oblateness up to J_4 of smaller primary, radiation of the bigger primary and the circumbinary belt, now takes the form

$$V = -Gm\left\{m_1\frac{q_1}{r_1} + m_2\left(\frac{1}{r_2} + \frac{B_1}{2r_2^3} - \frac{3B_2}{8r_2^5}\right) + \frac{M_b}{\left(r^2 + T^2\right)^{1/2}}\right\}, \tag{8}$$

with $r_1^2 = (x - x_1)^2 + y^2$, $r_2^2 = (x - x_2)^2 + y^2$, G is the gravitational constant.

2.7. The Equations of Motion

We start from Lagrangian (L) of the problem which is the kinetic energy minus the potential energy of the infinitesimal body. That is

$$L = \frac{1}{2}m\left(\dot{x}^2 + \dot{y}^2\right) + mn(x\dot{y} - \dot{x}y) + \frac{1}{2}mn^2\left(x^2 + y^2\right) - V.$$

or

$$L = \frac{1}{2}m\left(\dot{x}^2 + \dot{y}^2\right) + mn(x\dot{y} - \dot{x}y) - U, \tag{9}$$

where $U = V - \dfrac{1}{2}mn^2\left(x^2 + y^2\right).$

Subsequently, we obtain the equations of motion of the infinitesimal body as

$$\ddot{x} - 2n\dot{y} = -\frac{1}{m}\frac{\partial U}{\partial x},$$

$$\ddot{y} + 2n\dot{x} = -\frac{1}{m}\frac{\partial U}{\partial y}. \tag{10}$$

To covert the variables to non dimensional, we choose unit for the mass as the sum of the masses of the primaries, the unit of length as the distance between the primaries and unit of time is such that the gravitational constant is unit. Consequently, $m_1 = 1 - \mu$, $m_2 = \mu$ where $0 < \mu = \dfrac{m_2}{m_1 + m_2} \leq \dfrac{1}{2}$ is the mass ratio. Thus, in the dimensionless synodic coordinate system, the equations of motion (10) reduce to

$$\ddot{x} - 2n\dot{y} = \Omega_x, \quad \ddot{y} + 2n\dot{x} = \Omega_y, \tag{11}$$

with

$$\Omega = \frac{n^2\left(x^2 + y^2\right)}{2} + \frac{(1-\mu)q_1}{r_1} + \frac{\mu}{r_2} + \frac{\mu B_1}{2r_2^3} - \frac{3\mu B_2}{8r_2^5} + \frac{M_b}{\left(r^2 + T^2\right)^{1/2}},$$

$$r_1^2 = (x+\mu)^2 + y^2, \quad r_2^2 = (x+\mu-1)^2 + y^2, \tag{12}$$

and n is the mean motion, given by [24] as

$$n^2 = 1 + \frac{3}{2}\left(B_1 - \frac{5}{4}B_2\right) + \frac{2M_b r_c}{\left(r_c^2 + T^2\right)^{3/2}}, \tag{13}$$

r_c is the radial distance of the infinitesimal body in the classical restricted three-body problem.

3. Locations of Collinear Libration Points

We now search for possible collinear libration points of the infinitesimal mass in the rotating reference frame. The libration points are positions of gravitational balance between the primaries. At these points the two finite masses would exert zero net force on the infinitesimal mass, in effect, allowing the infinitesimal mass to have zero velocity in the rotating frame of reference. That is the libration points satisfy $\ddot{x} = \ddot{y} = \dot{x} = \dot{y} = 0$. It thus follows, from Equation (11), that the libration points are the solutions of

$$n^2 x - \frac{(1-\mu)(x+\mu)q_1}{r_1^3} - \frac{\mu(x+\mu-1)}{r_2^3} - \frac{3\mu(x+\mu-1)B_1}{2r_2^5} + \frac{15\mu(x+\mu-1)B_2}{8r_2^7} - \frac{M_b x}{\left(r^2 + T^2\right)^{3/2}} = 0, \tag{14}$$

and

$$n^2 y - \frac{(1-\mu)q_1 y}{r_1^3} - \frac{\mu y}{r_2^3} - \frac{3\mu B_1 y}{2r_2^5} + \frac{15\mu B_2 y}{8r_2^7} - \frac{M_b y}{\left(r^2 + T^2\right)^{3/2}} = 0. \tag{15}$$

Now, an evident solution of Equation (15) is $y = 0$, corresponding to the collinear libration points (the libration points which lie on the x-axis). This deciphers to

$$n^2 x - \frac{(1-\mu)(x+\mu)}{|x+\mu|^3} - \frac{\mu(x+\mu-1)}{|x+\mu-1|^3} - \frac{3\mu(x+\mu-1)B_1}{2|x+\mu-1|^5} + \frac{15\mu(x+\mu-1)B_2}{8|x+\mu-1|^7} - \frac{M_b x}{\left(x^2 + T^2\right)^{3/2}} = 0. \tag{16}$$

Equation (16) reduces to those of [1], in the absence of the perturbations. That is when $q_1 = 1, B_1 = B_2 = M_b = 0$), we have

$$x - \frac{(1-\mu)(x+\mu)}{|x+\mu|^3} - \frac{\mu(x+\mu-1)}{|x+\mu-1|^3} = 0, \tag{17}$$

with three collinear points L_1, L_2 and L_3. Only the collinear point L_2 is located between the primaries (**Figure 2**).

If we consider the effects of the potential from the belt only (*i.e.* $A_1 = A_2 = B_1 = B_2 = 0$), the Equation (17) reduces to

$$n^2 x - \frac{(1-\mu)(x+\mu)}{|x+\mu|^3} - \frac{\mu(x+\mu-1)}{|x+\mu-1|^3} - \frac{M_b x}{(x^2+T^2)^{3/2}} = 0. \tag{18}$$

[16] showed that whenever $T < \sqrt{2}\mu$ and

$$n^2\left(-\frac{T}{\sqrt{2}}\right) - \frac{(1-\mu)\left(-\frac{T}{\sqrt{2}}+\mu\right)}{\left|-\frac{T}{\sqrt{2}}+\mu\right|^3} - \frac{\mu\left(-\frac{T}{\sqrt{2}}+\mu-1\right)}{\left|-\frac{T}{\sqrt{2}}+\mu-1\right|^3} - \frac{M_b\left(-\frac{T}{\sqrt{2}}\right)}{\left(\left(-\frac{T}{\sqrt{2}}\right)^2+T^2\right)^{3/2}} > 0 \text{ in the interval } (-\mu, 0), \text{ Equation}$$

(18) will have five collinear points (**Figure 3**).

Now, using Equation (16) and with the help of the MATLAB (R2007b) software package, we obtain the coordinates of the collinear libration points for different cases as classified in the following order which are portrayed in **Table 1**:

1) Absence of radiation, oblateness and potential from the belt (classical case).
2) Radiation of the bigger primary only.
3) Potential from the belt only.
4) Oblateness of the smaller primary up to J_2 only.
5) Oblateness of the smaller primary up to J_4 only.
6) Radiation of the bigger primary, oblateness of the smaller primary up to J_4 and potential from the belt.

The combined effect of these perturbations on the collinear points is given in **Table 2**.

In the absence of the perturbations (*i.e.* $q_1 = 1, B_1 = B_2 = M_b = 0$) **Table 1** Case 1, it is observed that there are three collinear libration points (L_i, i = 1, 2, 3) which correspond to the classical case of [1]. Owing to the effect of the radiation of the bigger primary only (*i.e.* $q_1 = 0.98, B_1 = B_2 = M_b = 0$) Case 2, L_1 and L_3 stepped closer to the primaries while L_2 moved towards the bigger primary. Nevertheless, on taking into account the effect of the potential from the belt only (*i.e.* $q_1 = 1, B_1 = B_2 = 0, M_b = 0.01$) Case 3, there surface five collinear libration points (L_{n1}, L_{n2} and L_i, i = 1, 2, 3), this confirms those of [14]-[16]. The collinear points L_1 and L_3 shifted nearer to the primaries while L_2 moved away from the bigger primary, due to the potential from the belt. In the presence of the oblateness of the smaller primary up to J_2 only (*i.e.* $q_1 = 1, B_1 = 0.01, B_2 = M_b = 0$) Case 4, the collinear point L_1 sifted away from the primaries while L_2 and L_3 stepped closer to the bigger primary. In Case 5, due oblateness of the smaller primary up to J_4 only (*i.e.* $q_1 = 1, B_1 = 0.01, B_2 = 0.005, M_b = 0$), L_{n1} moved away from the bigger primary while L_{n2} stepped towards it. Similarly, owing to the oblateness of the smaller primary up to J_2 with

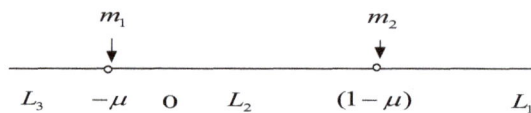

Figure 2. Disposition of the collinear points in the classical case.

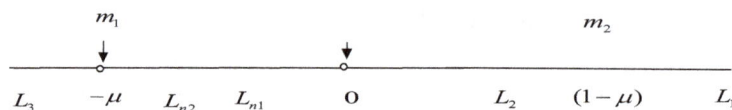

Figure 3. Disposition of the collinear points under the effects of the belt.

Table 1. Positions of the collinear points when $\mu = 0.35$, $q_1 = 0.98$, $B_1 = 0.01$, $B_2 = 0.005$ and $M_b = T = 0.01$, $r_c = 0.8789$.

Case	L_1	L_2	L_3	L_{n1}	L_{n2}	L_{n3}	L_{n4}
1	1.244813	0.213295	−1.142867				
2	1.243714	0.210813	−1.137286				
3	1.239362	0.224700	−1.137090	−0.000451	−0.038855		
4	1.249564	0.205046	−1.138453				
5	1.235582	0.245494	−1.141267			0.961931	0.319350
6	1.228444	0.259431	−1.129916	−0.000441	−0.039247	0.962537	0.314837

Table 2. Combined effects of the perturbations on the collinear points when $\mu = 0.35$, $T = 0.01$, $r_c = 0.8789$.

q_1	B_1	B_2	M_b	L_1	L_2	L_3	L_{n1}	L_{n2}	L_{n3}	L_{n4}
1	0	0	0	1.24481	0.21329	−1.14287	–	–	–	–
0.99	0.001	0.0005	0.01	1.23795	0.22579	−1.13420	−0.000445	−0.03905	0.82421	0.47525
0.98	0.002	0.0006	0.02	1.23240	0.23420	−1.12557	−0.000219	−0.05365	0.83068	0.46863
0.97	0.003	0.0007	0.03	1.22707	0.24145	−1.11719	−0.000144	−0.06391	0.83634	0.46280
0.96	0.004	0.0008	0.04	1.22194	0.24787	−1.10906	−0.000107	−0.07207	0.84138	0.45758

potential from the belt only (*i.e.* $A_1 = A_2 = B_2 = 0, B_1 = M_b = 0.01$) Case 5, collinear points L_1 and L_3 moved nigh to the primaries while L_2 stepped away from the bigger primary; and there emerge additional two new collinear points L_{n3}, L_{n4}. In the presence of all these perturbations (*i.e.* $q_1 = 0.98, B_1 = M_b = 0.01, B_2 = 0.005$) Case 6, there appeared seven collinear points: L_1, L_2, L_3, L_{n1}, L_{n2}, L_{n3}, L_{n4} as shown in **Figure 4**. With increase in these perturbations **Table 2**, the collinear points L_1, L_3 draw closer to the primaries while L_{n3} moves away from the them; L_2, L_{n1} move away from the bigger primary while L_{n2}, L_{n4} tend towards it.

4. Linear Stability of the Collinear Points

To study the stability of a libration point (x_0, y_0), we employ small displacement η, ξ to the coordinates (x_0, y_0). So, the variations η and ξ can take the form: $\eta = x - x_0$ and $\xi = y - y_0$ and the equations of the motion (5) become

$$\ddot{\eta} - 2n\dot{\xi} = \left(\Omega_{xx}^0\right)\eta + \left(\Omega_{xy}^0\right)\xi,$$
$$\ddot{\xi} + 2n\dot{\eta} = \left(\Omega_{yx}^0\right)\eta + \left(\Omega_{yy}^0\right)\xi. \tag{19}$$

The superscript "0" indicates that the partial derivatives have been evaluated at the libration point under consideration (x_0, y_0).

Let solutions of the equations of (19) be $\eta = A\exp(\lambda t)$, $\xi = B\exp(\lambda t)$ where A, B and λ are constants. Then, Equation (19) will have a non –trivial solution for A and B when

$$\begin{vmatrix} \lambda^2 - \Omega_{xx}^0 & -2n\lambda - \Omega_{xy}^0 \\ 2n\lambda - \Omega_{yx}^0 & \lambda^2 - \Omega_{yy}^0 \end{vmatrix} = 0. \tag{20}$$

On expanding the determinant we obtain the characteristic equation equivalent to the variational equations of (19) as

$$\lambda^4 + \left(4n^2 - \Omega_{xx}^0 - \Omega_{yy}^0\right)\lambda^2 + \Omega_{xx}^0\Omega_{yy}^0 - \left(\Omega_{xy}^0\right)^2 = 0 \tag{21}.$$

Now, we obtain the second partial derivatives as:

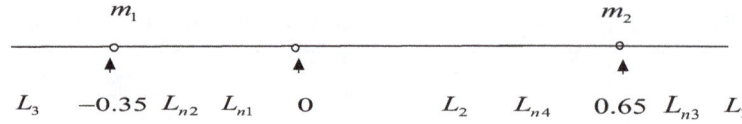

Figure 4. Disposition of the collinear points under the combined effects of the perturbations.

$$\Omega_{xx} = n^2 - \frac{(1-\mu)q_1}{r_1^3} + \frac{3(1-\mu)(x+\mu)^2 q_1}{r_1^5} - \frac{\mu}{r_2^3} + \frac{3\mu(x+\mu-1)^2}{r_2^5} - \frac{3\mu B_1}{2r_2^5} + \frac{15\mu(x+\mu-1)^2 B_1}{2r_2^7}$$

$$+ \frac{15\mu B_2}{8r_2^7} - \frac{105\mu(x+\mu-1)^2 B_2}{8r_2^9} - \frac{M_b}{\left(r^2+T^2\right)^{3/2}} + \frac{3M_b x^2}{\left(r^2+T^2\right)^{5/2}},$$

$$\Omega_{yy} = n^2 - \frac{(1-\mu)q_1}{r_1^3} + \frac{3(1-\mu)q_1 y^2}{r_1^5} - \frac{\mu}{r_2^3} + \frac{3\mu y^2}{r_2^5} - \frac{3\mu B_1}{2r_2^5} + \frac{15\mu y^2 B_1}{2r_2^7}$$

$$+ \frac{15\mu B_2}{8r_2^7} - \frac{105\mu y^2 B_2}{8r_2^9} - \frac{M_b}{\left(r^2+T^2\right)^{3/2}} + \frac{3M_b y^2}{\left(r^2+T^2\right)^{5/2}}, \tag{22}$$

$$\Omega_{xy} = \Omega_{yx} = \frac{3(1-\mu)(x+\mu)q_1 y}{r_1^5} + \frac{3\mu(x+\mu-1)y}{r_2^5} + \frac{15\mu(x+\mu-1)yB_1}{2r_2^7}$$

$$- \frac{105\mu(x+\mu-1)yB_2}{8r_2^9} + \frac{3M_b xy}{\left(r^2+T^2\right)^{5/2}}.$$

The partial derivatives computed at any collinear libration points $(x_0, 0)$, are

$$\Omega_{xx}^0 = n^2 + \frac{2(1-\mu)q_1}{|x_0+\mu|^3} + \frac{2\mu}{|x_0+\mu-1|^3} + \frac{6\mu B_1}{|x_0+\mu-1|^5} - \frac{45\mu B_2}{4|x_0+\mu-1|^7} + \frac{3M_b x_0^2}{\left(x_0^2+T^2\right)^{5/2}} - \frac{M_b}{\left(x_0^2+T^2\right)^{3/2}}, \tag{23}$$

$$\Omega_{yy}^0 = n^2 - \frac{(1-\mu)q_1}{|x_0+\mu|^3} - \frac{\mu}{|x_0+\mu-1|^3} - \frac{3\mu B_1}{2|x_0+\mu-1|^5} + \frac{15\mu B_2}{8|x_0+\mu-1|^7} - \frac{M_b}{\left(x_0^2+T^2\right)^{3/2}}, \tag{24}$$

$$\Omega_{xy}^0 = \Omega_{yx}^0 = 0. \tag{25}$$

Substituting these values in Equation (21), the characteristic equation reduces to

$$\lambda^4 + b\lambda^2 + c = 0 \tag{26}$$

where $b = 4n^2 - \Omega_{xx}^0 - \Omega_{yy}^0$, $c = \Omega_{xx}^0 \Omega_{yy}^0$.

The libration point is stable if all the roots of the characteristic equation (26) are either negative real numbers or distinct pure imaginary numbers or real parts of the complex numbers are negative.

The roots of the characteristic equation (26) for the libration points L_i ($i = 1, 2, 3$), L_{nj} ($j = 1, 2, 3, 4$) of **Table 1** are presented in **Tables 3-9** correspondingly.

Studying **Tables 3-9**, we find that all the collinear libration points L_i ($i = 1, 2, 3$) *and* L_{n2} are unstable (**Table 3, Table 4, Table 5, Table 7**), whereas the additional new collinear points L_{n1}, L_{n3} and L_{n4} are stable (**Table 6, Table 8, Table 9**).

5. Conclusion

The collinear libration points are investigated in a modified CR3BP when the bigger primary is a source of radiation, the smaller primary is an oblate spheroid; and the bodies are surrounded by a belt (circular cluster of material points). We have established the equations that govern the motion of the infinitesimal body under the

Table 3. Stability of L_1.

Case	L_1	Ω_{xx}^0	Ω_{yy}^0	$\lambda_{1,2}$	$\lambda_{3,4}$	Remark
1	1.244813	4.6468	−0.8234	±1.3674	±1.4305i	Unstable
2	1.243714	4.6595	−0.8297	±1.3722	±1.4329i	Unstable
3	1.239362	4.7796	−0.8510	±1.3897	±1.4512i	Unstable
4	1.249564	4.8515	−0.8355	±1.4112	±1.4267i	Unstable
5	1.235582	4.2890	−0.8377	±1.2772	±1.4841i	Unstable
6	1.228444	4.3987	−0.8739	±1.2973	±1.5113i	Unstable

Table 4. Stability of L_2.

Case	L_2	Ω_{xx}^0	Ω_{yy}^0	$\lambda_{1,2}$	$\lambda_{3,4}$	Remark
1	0.213295	16.6783	−6.8391	±3.7405	±2.8552i	Unstable
2	0.210813	16.4862	−6.7431	±3.7147	±2.8383i	Unstable
3	0.224700	18.7266	−7.8271	±3.9966	±3.0293i	Unstable
4	0.205046	17.7676	−7.0603	±3.8738	±2.8912i	Unstable
5	0.245494	8.5669	−5.9936	±2.5451	±2.8155i	Unstable
6	0.259431	7.6595	−6.4396	±2.3914	±2.9368i	Unstable

Table 5. Stability of L_3.

Case	L_3	Ω_{xx}^0	Ω_{yy}^0	$\lambda_{1,2}$	$\lambda_{3,4}$	Remark
1	−1.142867	3.7297	−0.3648	±0.9441	±1.2355i	Unstable
2	−1.137286	3.7334	−0.3667	±0.9463	±1.2364i	Unstable
3	−1.137090	3.8282	−0.3753	±0.9574	±1.2519i	Unstable
4	−1.138453	3.7908	−0.3726	±0.9540	±1.2458i	Unstable
5	−1.141267	3.7523	−0.3675	±0.9476	±1.2392i	Unstable
6	−1.129916	3.8558	−0.3805	±0.9638	±1.2568i	Unstable

Table 6. Stability of L_{n1}.

Case	L_{n1}	Ω_{xx}^0	Ω_{yy}^0	$\lambda_{1,2}$	$\lambda_{3,4}$	Remark
3	−0.000451	−9874.8	−9985.0	±98.6059i	±100.7019i	Stable
6	−0.000441	−9879.7	−9986.0	±98.6019i	±100.7356i	Stable

Table 7. Stability of L_{n2}.

Case	L_{n2}	Ω_{xx}^0	Ω_{yy}^0	$\lambda_{1,2}$	$\lambda_{3,4}$	Remark
3	−0.038855	327.1441	−176.4617	±18.0135	±13.3382i	Unstable
6	−0.039247	319.0101	−171.7775	±17.7859	±13.1617i	Unstable

Table 8. Stability of L_{n3}.

Case	L_{n3}	Ω_{xx}^0	Ω_{yy}^0	$\lambda_{1,2}$	$\lambda_{3,4}$	Remark
5	0.961931	−36.7586	−1.1726	±1.0266i	±6.3953i	Stable
6	0.962537	−36.0006	−1.2218	±1.0453i	±6.3447i	Stable

Table 9. Stability of L_{n4}.

Case	L_{n2}	Ω_{xx}^0	Ω_{yy}^0	$\lambda_{1,2}$	$\lambda_{3,4}$	Remark
5	0.319350	−15.5449	−4.5783	±1.8538i	±4.5507i	Stable
6	0.314837	−11.8748	−5.0872	±1.8490i	±4.2035i	Stable

influence of radiation of the bigger primary, oblateness up to the zonal harmonics J_4 of the smaller primary and gravitational potential from the belt. The equations are affected by the aforementioned perturbations. Numerically, we have determined the positions of the collinear libration points and investigated the resultant effect of the aforesaid perturbations on them. It is found that in count to the three libration points L_1, L_2, L_3 in the classical problem, there emerge four new collinear points which we call L_{n1}, L_{n2}, L_{n3} and L_{n4}. L_{n1} and L_{n2} arise from the effect of the potential from the belt, whereas L_{n3} and L_{n4} stem from the influence of the oblateness up to the zonal harmonics J_4 of the smaller primary. Due to the pooled impact of the aforesaid perturbations, the collinear points L_1 and L_3 advance toward the primaries while L_{n3} moves away from the primaries; and L_2 and L_{n1} tend towards the smaller primary as L_{n2} and L_{n4} come closer to the bigger primary. Despite the influence of radiation of the bigger primary, oblateness up to the zonal harmonics J_4 of the smaller primary and gravitational potential from the belt, the collinear libration points L_i ($i = 1, 2, 3$) as in the classical case, remain unstable. However, all the additional new collinear points are stable except L_{n2}. The existence of stable new collinear points can be utilized as stations for artificial satellites.

References

[1]   Szebehely, V. (1967) Theory of Orbits: The Restricted Problem of Three Bodies. Academic Press, New York.

[2]   Valtonen, M. and Karttunen, H. (2006) The Three-Body Problem. Cambridge University Press, Cambridge. http://dx.doi.org/10.1017/CBO9780511616006

[3]   Poynting, J.H. (1903) Radiation in the Solar System: Its Effect on Temperature and Its Pressure on Small Bodies. *Philosophical Transactions of the Royal Society of London A*, **202**, 525-552. http://dx.doi.org/10.1098/rsta.1904.0012

[4]   Radzievskii, V.V. (1950) The Restricted Problem of Three-Body Taking Account of Light Pressure. *Astronomicheskii-Zhurnal*, **27**, 250-256.

[5]   Radzievskii, V.V. (1953) The Space Photogravitational Restricted Three-Body Problem. *Astronomicheskii-Zhurnal*, **30**, 225.

[6]   Bhatnagar, K.B. and Chawla, J.M. (1979) A Study of the Lagrangian Points in the Photogravitational Restricted Three-Body Problem. *Indian Journal of Pure and Applied Mathematics*, **10**, 1443-1451.

[7]   Simmons, J.F.L., McDonald, J.C. and Brown, J.C. (1985) The Three-Body Problem with Radiation Pressure. *Celestial Mechanics*, **35**, 145-187. http://dx.doi.org/10.1007/BF01227667

[8]   Das, M.K., Narang, P., Mahajan, S. and Yuasa, M. (2008) Effect of Radiation on the Stability of Equilibrium Points in the Binary Stellar Systems: RW-Monocerotis, Krüger 60. *Astrophysics Space Science*, **314**, 261. http://dx.doi.org/10.1007/s10509-008-9765-z

[9]   Singh, J. and Taura, J.J. (2014) Stability of Triangular Libration Points in the Photogravitational Restricted Three-Body Problem with Oblateness and Potential from a Belt. *Journal of Astrophysics and Astronomy*, **35**, 107-119. http://dx.doi.org/10.1007/s12036-014-9299-4

[10]   Singh, J. and Taura, J.J. (2015) Triangular Libration Points in the CR3BP with Radiation, Triaxiality and Potential from a Belt. *Differential Equations and Dynamical System*. http://dx.doi.org/10.1007/s12591-015-0243-0

[11]   Jiang, I.G. and Yeh, L.C. (2004) The Drag-Induced Resonant Capture for Kuiper Belt Objects. *Monthly Notices of the Royal Astronomical Society*, **355**, L29-L32. http://dx.doi.org/10.1111/j.1365-2966.2004.08504.x

[12]   Jiang, I.G. and Yeh, L.C. (2004) On the Chaotic Orbits of Disk-Star-Planet Systems. *The Astronomical Journal*, **128**, 923-932. http://dx.doi.org/10.1086/422018

[13]   Jiang, I.G. and Yeh, L.C. (2003) Bifurcation for Dynamical Systems of Planet-Belt Interaction. *International Journal of Bifurcation and Chaos*, **13**, 617-630. http://dx.doi.org/10.1142/s0218127403006807

[14]   Yeh, L.C. and Jiang, I.G. (2006) On the Chermnykh-Like Problems: II. The Equilibrium Points. *Astrophysics and Space Science*, **306**, 189-200. http://dx.doi.org/10.1007/s10509-006-9170-4

[15]   Kushvah, B.S. (2008) Linear Stability of Equilibrium Points in the Generalized Photogravitational Chermnykh's Problem. *Astrophysics and Space Science*, **318**, 41-50.

[16] Singh, J. and Taura, J.J. (2013) Motion in the Generalized Restricted Three-Body Problem. *Astrophysics and Space Science*, **343**, 95-106. http://dx.doi.org/10.1007/s10509-012-1225-0

[17] Moulton, F.R. (1914) An Introduction to Celestial Mechanics. 2nd Edition, Dover, New York.

[18] Sharma, R.K. (1987) The Linear Stability of Libration Points of the Photogravitational Restricted Three-Body Problem When the Smaller Primary Is an Oblate Spheroid. *Astrophysics and Space Science*, **135**, 271-281. http://dx.doi.org/10.1007/BF00641562

[19] Kalvouridis, T.J. (1997) The Oblate Spheroids Version of the Photo-Gravitational 2+2 Body Problem. *Astrophysics and Space Science*, **246**, 219-227. http://dx.doi.org/10.1007/BF00645642

[20] Singh, J. and Umar, A. (2013) Application of Binary Pulsars to Axisymmetric Bodies in the Elliptic R3BP. *Astrophysics and Space Science*, **348**, 393-402. http://dx.doi.org/10.1007/s10509-013-1585-0

[21] Abdul Raheem, A. and Singh, J. (2006) Combined Effects of Perturbations, Radiation and Oblateness on the Stability of Equilibrium Points in the Restricted Three-Body Problem. *Astronomical Journal*, **131**, 1880-1885. http://dx.doi.org/10.1086/499300

[22] Abouelmagd, E.I. (2012) Existence and Stability of Triangular Points in the Restricted Three-Body Problem with Numerical Applications. *Astrophysics and Space Science*, **342**, 45-53. http://dx.doi.org/10.1007/s10509-012-1162-y

[23] Singh, J. and Taura, J.J. (2014) Effects of Triaxiality, Oblateness and Gravitational Potential from a Belt on the Linear Stability of $L_{4,5}$ in the Restricted Three-Body Problem. *Journal of Astrophysics and Astronomy*, **35**, 729-743. http://dx.doi.org/10.1007/s12036-014-9308-7

[24] Singh, J. and Taura, J.J. (2014) Effects of Zonal Harmonics and a Circular Cluster of Material Points on the Stability of Triangular Equilibrium Points in the R3BP. *Astrophysics and Space Science*, **350**, 127-132. http://dx.doi.org/10.1007/s10509-013-1719-4

[25] Singh, J. and Taura, J.J. (2014) Combined Effect of Oblateness, Radiation and a Circular Cluster of Material Points on the Stability of Triangular Libration Points in the R3BP. *Astrophysics and Space Science*, **351**, 499-506. http://dx.doi.org/10.1007/s10509-014-1860-8

[26] Renzetti, G. (2013) Satellite Orbital Precessions Caused by the Octupolar Mass Moment of a Non-Spherical Body Arbitrarily Oriented in Space. *Journal of Astrophysics and Astronomy*, **34**, 341-348. http://dx.doi.org/10.1007/s12036-013-9186-4

[27] Abouelmagd, E.I. (2013) The Effect of Photogravitational Force and Oblateness in the Perturbed Restricted Three-Body Problem. *Astrophysics and Space Science*, **346**, 51-69. http://dx.doi.org/10.1007/s10509-013-1439-9

[28] Peter, I.D. and Lissauer, J.J. (2007) Planetary Science. Cambridge University Press, New York.

[29] Miyamoto, M. and Nagai, R. (1975) Three-Dimensional Models for the Distribution of Mass in Galaxies. *Publications of the Astronomical Society of Japan*, **27**, 533-543.

23

The Basics of Flat Space Cosmology

Eugene Terry Tatum[1], U. V. S. Seshavatharam[2], S. Lakshminarayana[3]

[1]760 Campbell Ln. Ste. 106 #161, Bowling Green, USA
[2]Honorary Faculty, I-SERVE, Alakapuri, Hyderabad-35, Telangana, India
[3]Department of Nuclear Physics, Andhra University, Visakhapatnam, India
Email: ett@twc.com, seshavatharam.uvs@gmail.com, lnsrirama@gmail.com

Abstract

We present a new model of cosmology which appears to show great promise. Our flat space cosmology model, using only four basic and reasonable assumptions, derives highly accurate Hubble parameter H_0, Hubble radius R_0 and total mass M_0 values for our observable universe. Our model derives a current Hubble parameter of $2.167826 \times 10^{-18} \ \text{sec}^{-1} \cong 66.89 \ \text{km/sec/Mpc}$, in excellent agreement with the newly reported (lower limit) results of the 2015 Planck Survey. Remarkably, all of these derivations can be made with only these basic assumptions and the current CMB radiation temperature $T_0 \cong 2.725 \ \text{K}$. The thermodynamic equations we have generated follow Hawking's black hole temperature formula. We have also derived a variety of other useful cosmological formulae. These include angular velocity and other rotational formulae. A particularly useful hyperbolic equation, $T^2 R = cT^2/\omega \cong 1.0272646 \times 10^{27} \ \text{m} \cdot \text{K}^2$, has been derived, which appears to be an excellent fit for the Planck scale as well as the current observable universe scale. Using the flat space Minkowski relativistic formula for Doppler effect, and a formula for staging our cosmological model according to its average mass-energy density at every Hubble time (universal age) in its expansion, a persuasive argument can be made that the observable phenomena attributed to dark energy are actually manifestations of Doppler and gravitational redshift. Finally, a theory of cosmic inflation becomes completely unnecessary because our flat space cosmology model is always at critical density.

Keywords

Flat Space Cosmology, Cosmic Inflation, Dark Energy, Hubble Parameter, Critical Density, Angular Velocity, Light Speed Expansion, Light Speed Rotation, Redshift, Universe, CMBR

1. Introduction

Modern cosmology has recently struggled with modeling cosmic acceleration [1] and providing a reasonable explanation for the extreme flatness of the current observable universe. The ideas of a force in opposition to attractive gravity (dark energy) and of new physics required in the theory of cosmic inflation have been the source of much debate and consternation among cosmologists and astrophysicists. The authors have recently explored and published reports [2]-[5] of a new model of cosmology which appears to adequately address these problems without requiring new physics. Our model of flat space cosmology according to the Schwarzschild formula [6], Hawking's black hole temperature formula [7] and two other basic assumptions appears to discount the need for dark energy and the theory of cosmic inflation [8] entirely.

2. Basic Assumptions of Flat Space Cosmology

Our basic assumptions of flat space cosmology can be expressed as follows, for any scale from the Planck scale to the scale of our observable universe:

1) Cosmic radius R and total mass M follow the Schwarzschild formula $R \cong \dfrac{2GM}{c^2}$ at all times.

2) The cosmic event horizon translates at speed of light c with respect to its geometric center. Accordingly, the cosmic Hubble parameter H can be expressed as c/R and Hubble time (universal age) can be expressed as R/c for any stage of cosmic expansion.

3) The cosmic linear velocity of rotation is speed of light c at all Hubble times R/c. Thus, angular velocity $\omega \cong c/R \cong H$, the Hubble parameter.

4) Following thermodynamics of Hawking's black hole temperature formula, at any radius R the cosmic temperature T is inversely proportional to the geometric mean of cosmic total mass M and Planck mass.

3. Characteristic Equations of Flat Space Cosmology

The characteristic equations of flat space cosmology resulting from the above assumptions are:
A. Relations between cosmic radius, total mass and angular velocity:

$$\left.\begin{array}{l} M_R \cong \dfrac{Rc^2}{2G} \cong \dfrac{c^2}{2G}\left(\dfrac{c}{\omega_R}\right) \cong \dfrac{c^3}{2G\omega_R} \\[2mm] \left(M_R \middle/ \dfrac{4\pi}{3}R^3\right) \cong \dfrac{3c^2}{8\pi GR^2} \cong \dfrac{3\omega_R^2}{8\pi G} \end{array}\right\} \tag{1}$$

where R, M_R, and ω_R represent the cosmic radius, total mass and angular velocity (Hubble parameter), respectively. Average mass density (critical density) is derived in the second line.
B. Relations between temperature, mass, radius and angular velocity (thermodynamics):

$$k_B T_R \cong \dfrac{\hbar c^3}{8\pi G\sqrt{M_R M_{pl}}} \cong \dfrac{\hbar\sqrt{\omega_R \omega_{pl}}}{4\pi} \cong \dfrac{\hbar c}{4\pi\sqrt{RR_{pl}}} \tag{2}$$

where T_R is the cosmic temperature, $M_{pl} \cong \sqrt{\hbar c/G} \cong 2.176507949\times10^{-8}$ kg is the Planck mass, $R_{pl} \cong 2G\sqrt{\hbar c/G}\big/c^2 \cong 3.23240045\times10^{-35}$ m is the Planck mass-associated cosmic radius and $\omega_{pl} \cong \left(c/R_{pl}\right) \cong \left(c^3\big/2G\sqrt{\hbar c/G}\right) \cong 9.274607607\times10^{42}$ rad·sec^{-1} is the Planck mass-associated angular velocity (also the Planck mass-associated Hubble parameter H_{pl}).

$$\left.\begin{array}{l} \rightarrow RT_R^2 \cong \dfrac{1}{R_{pl}}\left(\dfrac{\hbar c}{4\pi k_B}\right)^2 \cong 1.0272646\times10^{27} \text{ m}\cdot\text{K}^2 \\[3mm] \text{and} \quad \dfrac{T_r^2}{\omega_R} \cong \dfrac{c}{R_{pl}}\left(\dfrac{\hbar}{4\pi k_B}\right)^2 \cong \omega_{pl}\left(\dfrac{\hbar}{4\pi k_B}\right)^2 \\[3mm] \Rightarrow \dfrac{\omega_R}{T_R^2} \cong \dfrac{1}{\omega_{pl}}\left(\dfrac{4\pi k_B}{\hbar}\right)^2 \cong 2.918356766\times10^{-19} \text{ K}^{-2}\cdot\text{sec}^{-1} \end{array}\right\} \tag{3}$$

4. Our Derivations of Current Cosmological Values

Using only our basic assumptions and the equations they generate above, derivations of current values for our observable universe are as follows:

Relations between universal current radius, current temperature, current angular velocity (also current Hubble parameter H_0), current total mass and current average mass density (critical density):

$$R_0 \cong \frac{1}{R_{pl}} \left(\frac{\hbar c}{4\pi k_B} \right)^2 \left(\frac{1}{T_0} \right)^2 \cong \frac{1}{R_{pl}} \left(\frac{\hbar c}{4\pi k_B} \right)^2 \left(\frac{1}{2.72548} \right)^2 \tag{4}$$

$$\cong \underline{1.3829177 \times 10^{26} \text{ m}}$$

$$\omega_0 \cong \frac{1}{\omega_{pl}} \left(\frac{4\pi k_B}{\hbar} \right)^2 T_0^2 \cong \frac{1}{\omega_{pl}} \left(\frac{4\pi k_B}{\hbar} \right)^2 (2.72548)^2 \tag{5}$$

$$\cong \underline{2.167826 \times 10^{-18} \text{ rad/sec} \cong 66.89 \text{ km/sec/Mpc} = H_0}$$

$$M_0 \cong \frac{R_0 c^2}{2G} \cong \frac{c^3}{2G\omega_0} \cong \underline{9.311752 \times 10^{52} \text{ kg.}} \tag{6}$$

$$Rho_c = \frac{3c^2}{8\pi G R_0^2} \cong \frac{3\omega_0^2}{8\pi G} \cong \underline{8.4053137 \times 10^{-27} \text{ kg} \cdot \text{m}^{-3}} \tag{7}$$

The above-derived radius and total mass values correspond to a current observable universe with a radius of 14.6 billion light-years and roughly 2×10^{22} visible stars plus 5x dark matter, or about 10^{53} kg.

All of these derived current cosmological values are consistent with the 2015 Planck Survey data.

As per the 2015 Planck data [9], the current value of the Hubble parameter H_0 is reported to be:

$$\text{Planck TT + low P: } (67.31 \pm 0.96) \text{ km/sec/Mpc}$$
$$\text{Planck TE + low P: } (67.73 \pm 0.92) \text{ km/sec/Mpc}$$
$$\text{Planck TT, TE, EE + low P: } (67.7 \pm 0.66) \text{ km/sec/Mpc}$$

As per the 2015 Planck data, the current value of CMBR temperature is:

$$\text{Planck TT + lowP + BAO: } (2.722 \pm 0.027) \text{ K}$$
$$\text{Planck TT; TE; EE + low P + BAO: } (2.718 \pm 0.021) \text{ K}$$

$$\text{COBE/FIRAS CMBR temperature measurement [10]: } (2.7255 \pm 0.0006) \text{ K}$$

5. Practical Applications of Current Angular Velocity in Our Model

A. Galactic revolving speed:

For our current light speed rotating cosmic model, on the equatorial plane, galactic revolving speed can be expressed as:

$$\left(v_g \right)_{\text{revolving}} \cong r_g H_0 \cong r_g \omega_0 \leq c \tag{8}$$

Here, r_g and $\left(v_g \right)_{\text{revolving}}$ represent the galactic distance from the cosmic center and galactic revolving speed corresponding to the cosmic angular velocity, respectively. The important point is that, even though $\dfrac{\left(v_g \right)_{\text{revolution}}}{c}$ is always less than 1, the proposed velocity refers to galactic "revolution speed" about the cosmic center and the proposed distance refers to galaxy distance from the cosmic center. In contrast, in Hubble's law [11] [12], velocity refers to galactic "receding speed" and distance refers to "distance between galaxy and observer." *Importantly, actual galactic "revolving speeds" have never been confirmed by any direct cosmological observations.* This is for further study.

B. Galactic receding speed:

In our current expanding cosmic model, on the equatorial plane, galactic receding speed can be expressed as:

$$\left(v_g\right)_{\text{receding}} \cong \left(\frac{r_g}{R_0}\right)c \cong r_g\left(\frac{c}{R_0}\right) \cong r_g\omega_0 \cong r_g H_0 \leq c \tag{9}$$

Qualitatively, this relation also resembles the famous Hubble velocity-distance law. The point is that r_g is the distance between galaxy and cosmic center and not the distance between galaxy and observer. *Importantly, actual galactic "receding speeds" have never been confirmed by direct cosmological observations.* This is for further study.

From the above, it is clear that, at the present time, on the equatorial plane, the magnitude of galactic revolving speed equals the magnitude of galactic receding speed. Hubble's law appears to be a physical consequence of flat space cosmology.

C. Galactic centripetal acceleration:

1) For any revolving galaxy, galactic centripetal acceleration a_g can be expressed as:

$$a_g \cong H_0\left(v_g\right)_{\text{revolving}} \cong \omega_0\left(v_g\right)_{\text{revolving}} \cong r_g\omega_0^2 \tag{10}$$

2) For any satellite that is assumed to be revolving at a distance $r_{\text{satellite}}$ from the cosmic center, its centripetal acceleration $a_{\text{satellite}}$ can be expressed as:

$$a_{\text{satellite}} \cong H_0\left(v_g\right)_{\text{revolving}} \cong \omega_0\left(v_g\right)_{\text{revolving}} \cong r_{\text{satellite}}\omega_0^2 \tag{11}$$

Based on the above applications, and by measuring actual galactic "revolving speeds" and galactic "recession speeds," the current cosmic angular velocity can be estimated.

D. Galactic rotational curves:

The current dominant paradigm is that galaxies are embedded in halos of cold dark matter (CDM), made of non-baryonic weakly-interacting massive particles. However, an alternative way to explain the observed rotation curves of galaxies is the postulate that, for gravitational accelerations below a certain value $a_0 \cong (1.2 \pm 0.3) \times 10^{-10}$ m·sec^{-2}, the true gravitational field strength g approaches $\sqrt{g_N g}$, where g_N is the usual Newtonian gravitational field strength (as calculated from the observed distribution of visible matter). This paradigm is known as modified Newtonian dynamics (MOND) [13]-[15]. MOND explains successfully many phenomena in galaxies, among which the following non-exhaustive list: 1) it predicted the shape of rotation curves of low surface-brightness (LSB) galaxies before any of them had ever been measured; 2) tidal dwarf galaxies (TDG),which should be devoid of collision-less dark matter, still exhibit a mass discrepancy in Newtonian dynamics, which is perfectly explained by MOND; 3) the baryonic Tully-Fisher relation, one of the tightest observed relations in astrophysics, is a natural consequence of MOND, both for its slope and its zero-point; 4) the first realistic simulations of galaxy merging in MOND were recently carried out, notably reproducing the morphology of the Antennae galaxies; 5) it naturally explains the universality of "dark" and baryonic surface densities within one core radius in galaxies.

So far, in the MOND model, the origin of acceleration constant $a_0 \cong (1.2 \pm 0.3) \times 10^{-10}$ m·sec^{-2} is purely empirical and is unknown from first principles. By fitting the rotation curves, its magnitude is being determined empirically. The fundamental question to be answered is: Does MOND reflect the influence of cosmology on local particle dynamics at low accelerations? To understand the issue here, the authors assume:

1) The acceleration term a_0 is not a constant but a variable and depends on the galactic revolving speed about the cosmic center. The reasoning behind this guess is that each revolving galaxy will certainly experience a characteristic centripetal acceleration if the universe is rotating. This idea supports MOND concepts to some extent.

2) The magnitude of this acceleration variable can be assumed to be proportional to the current cosmic angular velocity and can be referred to as the "cosmological galactic acceleration".

With reference to the MOND results, empirically, the revolving speed of a star about the galaxy is represented by the following relation:

$$\left(v_{rev}\right)_{star} \cong \sqrt[4]{GM_g a_0} \tag{12}$$

where $a_0 \cong (1.2 \pm 0.3) \times 10^{-10}$ m·sec$^{-2} \approx \left(cH_0/4.6\right)$ to $\left(cH_0/7.7\right)$. M_g is the mass of the galaxy. By considering the galactic revolving speed $\left(v_g\right)_{\text{revolving}}$ about the cosmic center, the magnitude of galactic centripetal acceleration can be assumed to vary as:

$$a_g \cong H_0 \left(v_g \right)_{revolving} \cong \omega_0 \left(v_g \right)_{revolving} \cong r_g \omega_0^2 \tag{13}$$

where r_g is the distance between galaxy and the cosmic center. Now the rotational speed of a star in any galaxy can be represented as follows:

$$\left(v_{star} \right)_{revolution} \propto \sqrt[4]{GM_g \omega_0 \left(v_g \right)_{revolving}} \propto \sqrt[4]{GM_g r_g \omega_0^2} \tag{14}$$

With an assumed universal proportionality ratio of 1, and by knowing the galactic mass and actual revolving speeds of galactic stars, galactic revolving speed and galactic distance from the cosmic center can be approximated in the following way:

$$\left. \begin{array}{l} \left(v_g \right)_{revolving} \cong \dfrac{\left(v_{star} \right)_{revolution}^4}{GM_g \omega_0} \text{ and} \\[3mm] r_g \cong \dfrac{\left(v_g \right)_{revolving}}{\omega_0} \cong \dfrac{\left(v_{star} \right)_{revolution}^4}{GM_g \omega_0^2} \end{array} \right\} \tag{15}$$

By knowing our mother galactic mass and rotational curves, our galactic distance from the cosmic center can be approximated. By considering the different model-dependent proportionality ratios, and correlating all of the data, the correct magnitude of the proportionality ratio can be fitted. This is for further study.

6. Model Equations of Cosmic Redshift in Flat Space Cosmology

Given our stated basic assumptions, our expanding cosmic model shows average mass-energy density to be inversely proportional to R^2. One way to look at this is that, the deeper an observer from Earth looks into space (and time), the greater the average mass-energy density stage of the cosmos one is observing. Thus, since each progressively denser stage of the cosmos is associated with higher average gravitational field strength, there must be associated gravitational time dilation effects. This conclusion is firmly grounded in general relativity. As such, it is conceivable that the progressively higher redshifts we observe with increasing look-back distances may be, in part, a manifestation of gravitational time dilation. In addition, because of this inverse square relationship over very long distances, plots of proximal galactic redshifts per unit of distance observed would be expected to look relatively linear (as seen by the weaker telescopes of the 1920s and 1930s) and deep space galactic redshifts per unit of distance observed would be expected to clearly fall away from linearity, along with decreasing luminosity, as redshifts extend into the infrared range (as seen in 1998 Type 1a supernovae observations) [16].

In this section, in a semi-empirical approach, the authors propose a simple model equation for observed and predicted cosmic redshifts. It is for further research and analysis. The current model equation under study is:

$$Z \cong \left(\frac{\rho_x c^2}{\rho_0 c^2} \right)^{\frac{1}{3}} \left\{ \ln \left[1 + \left(\frac{\rho_x c^2}{\rho_0 c^2} \right)^{\frac{1}{3}} \right] \right\}^{-1} - \frac{1}{\ln(2)} \tag{16}$$

where $\rho_x c^2$ and $\rho_0 c^2$ represent past and current cosmic average mass-energy density, respectively.

The following graph (**Figure 1**), according to the above formula, shows expected observed cosmic redshift as a function of the above-defined average mass-energy density ratio pertaining to a particular astronomical observation. In this manner, increasingly greater redshifts would be expected to correspond with more distant galactic observations. However, notice the apparent near-linearity up to a density ratio of about 10^4, and the increasingly nonlinear appearance with deeper space observations. The authors propose that something like this mathematical relationship could be responsible for the illusion of dark energy and, therefore, useful in modeling the results of progressively deeper space observations.

Of course, one must also factor in redshift as a function of relativistic Doppler effect. Since we are modeling flat space cosmology, the correct model formula is Minkowski's relativistic Doppler formula for flat space:

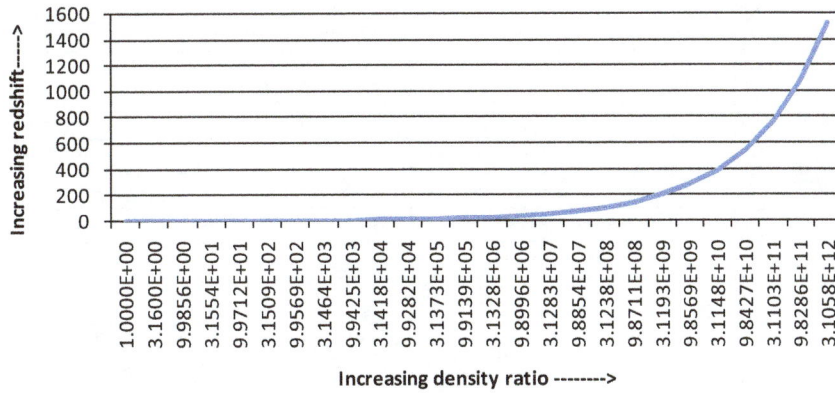

Figure 1. Cosmic redshift vs mass-energy density ratio.

$$Z + 1 \cong \sqrt{\frac{\left[1 + (v/c)\right]}{\left[1 - (v/c)\right]}} \qquad (17)$$

In order to keep scaling similar to **Figure 1**, the velocity term v in the Minkowski formula can be substituted by $\left[1 - (R_x/R_0)\right]c$ where $R_x < R_0$. The simplified relation can be expressed as follows.

$$Z \cong \sqrt{\frac{2 - (R_x/R_0)}{(R_x/R_0)}} - 1 \qquad (18)$$

Figure 2 shows redshift term Z as a function of a log scale of decreasing cosmic radius ratio (R_x/R_0) pertaining to progressively deeper space observations. The reader should note that the CMBR redshift of 1089 corresponds with the place on the horizontal axis corresponding to log value -5.7738. Perhaps more importantly, however, the reader's attention is directed to the place on the horizontal axis corresponding to log value -1.456. A greatly magnified portion of this region of **Figure 2** would show the nonlinearity corresponding to the earliest visible galaxies which are moving away from us at about $0.95c$.

It is now clearly apparent that a combination of gravitational time dilation (**Figure 1**) and flat space relativistic Doppler effect (**Figure 2**) could be the entire explanation for the nonlinearity of deep space Type 1a supernovae observations currently being attributed to "dark energy." Our flat space cosmology model provides a reasonable explanation for current astronomical observations without the need to invoke a new type of force and energy.

7. Summary

Our flat space cosmology model, using only four basic and reasonable assumptions, generates highly accurate Hubble parameter H_0, Hubble radius R_0 and total mass M_0 values for our observable universe. These values are in excellent agreement with the newly reported results of the 2015 Planck Survey and require only our assumptions and the current CMB radiation temperature $T_0 \cong 2.725 \text{ K}$ to generate them. The thermodynamic equations we have generated, following Hawking's black hole temperature formula, in conjunction with our basic assumptions, create a variety of useful cosmological formulae. These include angular velocity and other rotational formulae. Such rotational formulae should be correlated with further galactic observations, perhaps putting further constraints on dark matter.

Our results correlate nicely with a variety of astronomical observations. To take one example, one can roughly estimate the total mass M_0, based upon observational estimates that there are approximately 10^{11} visible galaxies times approximately 2×10^{11} visible stars per galaxy times approximately 10^{30} kg per star, totalling to approximately 2×10^{52} kg of visible (baryonic) mass. Multiplying this number by the roughly 5x expected dark matter gives a total mass observable M_0 of approximately 10^{53} kg. Our model derives a M_0 value of approximately 9.3×10^{52} kg from the Schwarzschild formula, after using a thermodynamic equation to derive $R_0 \cong 1.3829177 \times 10^{26}$ m (14.6 billion light-years). See equations 4 thru 7 for details. This is a remarkable achievement, since the only precise observational data our model requires is current CMB radiation temperature!

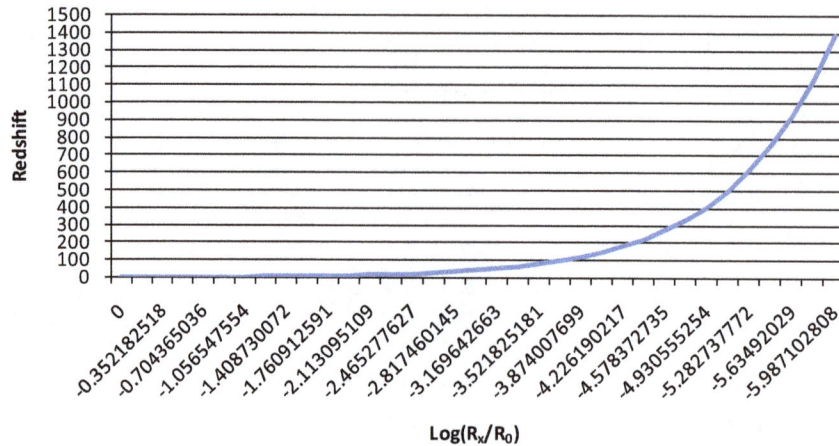

Figure 2. Cosmic redshift vs. decreasing $\log(R_x/R_0)$.

Dr. Stephen Hawking's black hole temperature formula has been extremely useful in this undertaking. Since a picture is worth a thousand words, the following log graph (**Figure 3**) neatly summarizes our model relationships between Hubble time (universal age), Hubble radius, total mass and CMB radiation temperature.

One of the more useful thermodynamic equations generated with the help of Hawking's black hole temperature formula is our hyperbolic equation: $T^2 R \cong cT^2/\omega \cong 1.0272646 \times 10^{27} \ \mathrm{m \cdot K^2}$, which could, theoretically, apply all the way down to the Planck scale. Applicable numbers for current universal observations are shown at the extremes of the four axes.

One of the most interesting features of our model is that, by following the Schwarzschild formula and the assumption that c/R is the appropriate Hubble parameter, the cosmic average mass density is always at critical density. The simple proof of this is that our derived cosmic average mass density formula $\left(3c^2/8\pi GR^2\right)$ is identical to the Friedmann critical density formula $\left(3H^2/8\pi G\right)$ when c/R is taken to be the Hubble parameter. Hence, our cosmic model is always "flat" (as defined by a universe at the Friedmann critical density) at every Hubble time stage of its growth.

The significance of the above revelation cannot be ignored. Ever since physicist Robert Dicke first made the observation [17] in 1969, cosmologists have been deeply puzzled as to how our universe appears to be expanding in a very precise way so as to perfectly balance out the attractive "force" of gravity. This is what is meant by a flat universe. In fact, as it was pointed out at the time, for such an apparent balance to be within observable error in the present, the presumably opposing forces in the very early universe (within a fraction of the first second after the Big Bang) must have been of equal magnitude to within one part in 10^{14}. This has since been referred to as the "cosmological flatness problem." There is an excellent discussion of this problem in Alan Guth's book [18], "The Inflationary Universe". As one of the pioneers and early proponents of the theory of cosmic inflation, Dr. Guth makes it very clear in his book that the flatness problem is the primary reason a theory of cosmic inflation appears to be necessary.

One of the important requirements for a suitable theory of cosmic inflation is that it shows the very early universe to scale at least 25 logs of 10 in a tiny fraction of a second. However, one need only look at our summary log graph to see that such a "hyper-rapid exponential expansion" occurs within 10^{-17} of a second of Hubble time (universal age) in the very early growth of our flat space cosmos. Thus, a flat space cosmology which acts according to our basic assumptions, including light speed expansion and light speed rotation [19] [20], naturally exhibits this cosmic inflation effect without requiring new physics.

Our model also suggests that the observational phenomena attributable to dark energy (especially the 1998 Type 1a supernovae data) may be entirely a manifestation of Minkowski flat space relativistic Doppler effect and gravitational time dilation, as explained in Section 6. Our mathematical model, as graphically represented in **Figure 1** and **Figure 2** of Section 6, clearly shows expected nonlinearity corresponding to our deepest space observations. It should be noted that the appearance of a very nearly perfectly balanced "force" in opposition to attractive gravity, as suggested by the 2015 Planck Survey value for the dark energy equation of state $\left(w = -1.006 \pm 0.045\right)$, could actually be an illusion produced by a constantly flat universe. Observations in support of a flat universe simply imply that no apparent *net* forces are acting on the universal system as a whole.

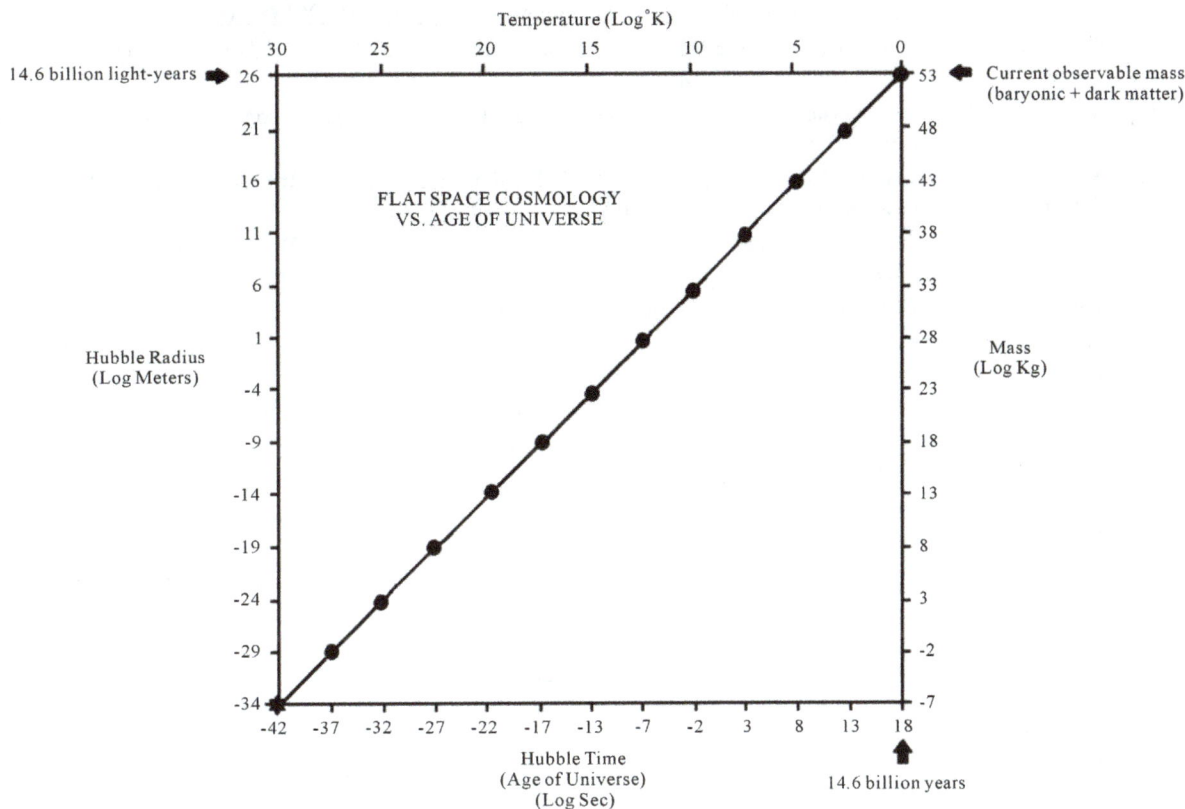

Figure 3. Flat space cosmology vs. age of the universe.

Finally, although currently formulated scientific laws appear to prevent one from ever observing the internal conditions of a black hole, we must at least consider the possibility that our own universe could be a particularly large evolved and evolving black hole (*i.e.*, truly gargantuan in comparison to the known supermassive giant black holes). There appears to be nothing in general relativity which prevents such a possibility, however remote this possibility may seem to the reader at the present time.

8. Conclusion

Flat space cosmology, as introduced here, is one of the most exciting, interesting and productive new theories in cosmology. Given the few basic and reasonable assumptions of our model, it is astounding as to how well the resulting derivations fit with our current observable universe, as detailed in the 2015 Planck Survey results. The authors humbly request that the scientific community explore this fascinating subject in a true scientific spirit. Furthermore, the authors humbly request that the Nobel committee seriously consider honoring Drs. Stephen W. Hawking and Abhas Mitra for their invaluable work on black holes and cosmology in general.

Acknowledgements

The authors express their thanks to Dr. Abhas Mitra for his kind and valuable suggestions in developing this paper. One of the authors, Seshavatharam U.V.S., is indebted to professors K.V. Krishna Murthy, Chairman, Institute of Scientific Research in Vedas (I-SERVE), Hyderabad, India and Shri K.V.R.S. Murthy, former scientist IICT (CSIR), Govt. of India, Director, Research and Development, I-SERVE, for their valuable guidance and great support in developing this subject. Author Dr. E. Terry Tatum would also like to thank Dr. Rudy Schild, Harvard Center for Astrophysics, for his support and encouragement in developing this subject.

References

[1]   Mitra, A. (2013) Energy of Einstein's Static Universe and Its Implications for the ΛCDM Cosmology. *Journal of*

Cosmology and Astroparticle Physics, **03**, 7. http://dx.doi.org/10.1088/1475-7516/2013/03/007

[2]   Tatum, E.T. (2015) Could Our Universe Have Features of a Giant Black Hole? *Journal of Cosmology*, **25**. (Part I, in Press)

[3]   Tatum, E.T. (2015) How a Black Hole Universe Theory Might Resolve Some Cosmological Conundrums. *Journal of Cosmology*, **25**. (Part II, in Press)

[4]   Seshavatharam, U.V.S. and Lakshminarayana, S. (2014) Friedmann Cosmology: Reconsideration and New Results. *International Journal of Astronomy, Astrophysics and Space Science*, **1**, 16-26.

[5]   Seshavatharam, U.V.S. and Lakshminarayana, S. (2015) Primordial Hot Evolving Black Holes and the Evolved Primordial Cold Black Hole Universe. *Frontiers of Astronomy, Astrophysics and Cosmology*, **1**, 16-23.

[6]   Pathria, R.K. (1972) The Universe as a Black Hole. Nature, **240**, 298-299. http://dx.doi.org/10.1038/240298a0

[7]   Hawking, S.W. (1975) Particle Creation by Black Holes. *Communications in Mathematical Physics*, **43**, 199-220. http://dx.doi.org/10.1007/BF02345020

[8]   Steinhardt, P.J. (2011) The Inflation Debate: Is the Theory at Heart of Modern Cosmology Deeply Flawed? *Scientific American*, **304**, 18-25. http://dx.doi.org/10.1038/scientificamerican0411-36

[9]   Planck Collaboration: Planck 2015 Results. XIII. Cosmological Parameters. http://arxiv.org/abs/1502.01589

[10]   Fixsen, D.J. (2009) The Temperature of the Cosmic Microwave Background. *The Astrophysical Journal*, **707**, 916. http://dx.doi.org/10.1088/0004-637X/707/2/916

[11]   Hubble, E.P. (1929) A Relation between Distance and Radial Velocity among Extra-Galactic Nebulae. *Proceedings of the National Academy of Sciences*, **15**, 168-173. http://dx.doi.org/10.1073/pnas.15.3.168

[12]   Hubble, E.P. (1947) The 200-Inch Telescope and Some Problems It May Solve. *Publications of the Astronomical Society of the Pacific*, **59**, 153-167. http://dx.doi.org/10.1086/125931

[13]   Milgrom, M. (1983) A Modification of the Newtonian Dynamics as a Possible Alternative to the Hidden Mass Hypothesis. *Astrophysical Journal*, **270**, 365-370.

[14]   Brownstein, J.R. and Moffat, J.W. (2006) Galaxy Rotation Curves without Non-Baryonic Dark Matter. *The Astrophysical Journal*, **636**, 721-741. http://dx.doi.org/10.1086/498208

[15]   Chadwick, E.A., Hodgkinson, T.F. and McDonald, G.S. (2013) Gravitational Theoretical Development Supporting MOND. *Physical Review D*, **88**, Article ID: 024036.

[16]   Perlmutter, S., Gabi, S., Goldhaber, G., Goobar, A., Groom, D.E., Hook, I.M., *et al.* (1997) Measurements of the Cosmological Parameters Ω and Λ from the First Seven Supernovae at $z \geq 0.35$. *Astrophysical Journal*, **483**, 565-581. http://dx.doi.org/10.1086/304265

[17]   Dicke, R.H. (1970) Gravitation and the Universe. American Philosophical Society, Philadelphia.

[18]   Guth, A.H. (1997) The Inflationary Universe. Basic Books, New York.

[19]   Longo, M.J. (2011) Detection of a Dipole in the Handedness of Spiral Galaxies with Redshifts z~0.04. *Physics Letters B*, **699**, 224-229. http://dx.doi.org/10.1016/j.physletb.2011.04.008

[20]   Sivaram, C. and Arun, K. (2012) Primordial Rotation of the Universe, Hydrodynamics, Vortices and Angular Momenta of Celestial Objects. *The Open Astronomy Journal*, **5**, 7-11. http://dx.doi.org/10.2174/1874381101205010007

24

Time-Frequency Analysis of Asymmetric Triaxial Galaxy Model Including Effect of Spherical Dark Halo Component

Beena R. Gupta[1], Vinay Kumar[2*]

[1]Department of Mathematics, Lakshmibai College, University of Delhi, New Delhi, India
[2]Department of Mathematics, Zakir Husain Delhi College, University of Delhi, New Delhi, India
Email: beenaguptalbc@gmail.com, *krvinayaidso@yahoo.com

Abstract

A method of time-frequency analysis (TFA) based on wavelets is applied to study the phase space structure of three-dimensional asymmetric triaxial galaxy enclosed by spherical dark halo component. The investigation is carried out in the presence and absence of dark halo component. Time-frequency analysis is based on the extraction of instantaneous frequency from the phase of the continuous wavelet transform. This method is comparatively fast and reliable. This method can differentiate periodic from quasi-periodic, chaotic sticky from chaotic non-sticky, ordered from chaotic and also, it can accurately determine the time interval of the resonance trapping and transitions too. Apart from that, the phenomenon of transient chaos can be explained with the help of time-frequency analysis. Comparison with the method of total angular momentum (denoted as L_{tot}) proposed recently is also presented.

Keywords

Time-Frequency Analysis, Triaxial Galactic Potential, Instantaneous Frequency, Total Angular Momentum

1. Introduction

We know that the phase space of nonlinear dynamical systems consists of periodic, quasi-periodic and chaotic trajectories. Chaotic trajectories visit resonance islands, remain there for some time and then escape to the chaotic region during its evolution. To know the time interval of resonance trapping and resonance transition

---

*Corresponding author.

and to visualize the phenomenon of transient chaos are some important questions which compel us to study more about the application of different chaos indicators. Over the last few years, several chaos indicators have been introduced to study those aspects. Moreover, for dynamical system of two degrees of freedom, there are several chaos indicators such as the Poincare Surface of Section (PSS), Largest Lyapunov Characteristic Exponent (LLCE), Smaller Alignment Index (SALI). Fast Liapunov Indicators (FLI), the Generalized Alignment Index (GALI) and the Correlation Dimension (CD) (see [1]-[3]). Visualization of the Poincare surfaces of section (PSS) is very useful for the study of the dynamical system of two degrees of freedom. But in case of three degrees of freedom, it becomes four-dimensional which is difficult to analyze. On the other hand, other indicators mentioned above require the solution of equations of motion and the first order variational equations whose computation is not easy in case of higher dimensional systems. Also, they do not tell about additional qualitative features like whether a trajectory is resonant or non-resonant. Recently in [4], a new indicator L_{tot} (i.e. Total angular momentum) is introduced to study the ordered and chaotic motion of the asymmetric triaxial galaxy, including the effect of spherical dark halo components. L_{tot} (TAM) is used to describe the nature of orbits in this potential and is proved to be a fast and reliable indicator in comparison to a Lyapunov characteristic exponent and P(f)-indicator. The method of total angular momentum can distinguish regular trajectory from chaotic. But there is no clarity about the other aspects such as stickiness, resonance trapping and transition and transient chaos, which must be investigated. In [4], the effect of dark halo component in the chaotic region is shown with the help of the Poincare surface of section by reducing the three-dimensional galactic model to two dimensions. Time-frequency analysis based on the extraction of instantaneous frequency from wavelet transform is computed via two ways. The first method is the computation of instantaneous frequency of the phase of continuous wavelet transform (CWT) and the second method is based on the computation of instantaneous frequency from the amplitude of the continuous wavelet transform. The amplitude-based method is very well described in [5]. Characterization of the phase space of standard map and the Hamiltonian of the hydrogen atom (moving in crossed magnetic and elliptically polarized microwave fields) is presented using this method. On the other hand, phase-based approach is already explained in [6]-[9]. In [10], we can see the application of the phase-based approach to the circular restricted three-body problem (Sun-Jupiter system) to explain the phenomenon of resonance transition and transport condition. The phenomenon of resonance trapping of chaotic trajectory, resonance transition and transient chaos (time dependent system causes an orbit move from regular to chaos and vice-versa (see [11]) can be computed and visualized with the help of TFA. Identification between regular and chaotic, periodic and quasi-periodic, chaotic sticky and chaotic non-sticky is possible with the help of TFA. Also, the computational time for TFA is negligible as compared to other chaos indicators. The aim of this paper is two folded. The first is to establish the method of time-frequency analysis in comparison to total angular momentum. The second is to investigate and characterize the phase space structure of asymmetric triaxial galaxies in the presence and absence of spherical dark halo components.

We have organized the paper as follows:

In Section 2, we have given a brief description of asymmetric triaxial galaxy enclosed by spherical dark halo component (3D). In Section 3, a brief description of TFA based on the phase of CWT and its implementation in Matlab are given. Results and discussion based on the application of TFA to the three-dimensional galactic model are shown in Section 4. The conclusion is given in Section 5.

2. Triaxial Galaxy Enclosed by Spherical Dark Halo Component in 3D

The potential for triaxial galaxy enclosed by spherical dark halo (see [4]) component is given by

$$V_t(x,y,z) = V_g(x,y,z) + V_h(x,y,z), \tag{1}$$

where

$$V_g(x,y,z) = \frac{v_0^2}{2}\ln\left(x^2 - \lambda x^3 + \alpha y^2 + \beta z^2 + c_b^2\right), \tag{2}$$

and

$$V_h(x,y,z) = \frac{-M_h}{\left(x^2 + y^2 + z^2 + c_h^2\right)^{\frac{1}{2}}}. \tag{3}$$

Equation (2) denotes a triaxial galaxy with a bulge and a small asymmetry introduced by the term

$-\lambda x^3, \lambda \ll 1$. The parameters α and β denote the flattening of the galaxy and c_b denotes the scale length of the bulge of the galaxy. The parameter v_0 is used for consistency of galaxy units. Equation (3) presents a spherical dark halo component. Here M_h and c_h are the mass and the scale length of the dark halo component, respectively. Equations (1), (2) and (3) together represents the three-dimensional galactic model. Now we can write the equations of motion as follows,

$$\ddot{x} = -\frac{\partial V_t(x,y,z)}{\partial x}, \quad \ddot{y} = -\frac{\partial V_t(x,y,z)}{\partial y} \quad \text{and} \quad \ddot{z} = -\frac{\partial V_t(x,y,z)}{\partial z}. \tag{4}$$

The Hamiltonian of the potential given by Equation (1) can be expressed as

$$H = \frac{1}{2}\left(p_x^2 + p_y^2 + p_z^2\right) + V_t(x,y,z) = h_3, \tag{5}$$

where p_x, p_y and p_z are the momenta corresponding to coordinates x, y and z respectively. Also, h_3 denotes the numerical value of the Hamiltonian or the energy constant. In Equation (5), if we take $z = 0$, we get the potential for two dimensional triaxial galaxy which can be expressed as

$$H = \frac{1}{2}\left(p_x^2 + p_y^2\right) + V_t(x,y) = h_2, \tag{6}$$

where h_2 stands for numerical value of Hamiltonian in two dimension. In this paper, we use a system of units which is defined as follows:

Unit of length = 1 kpc;
Unit of mass = $2.325 \times 10^7 M_\odot$;
Unit of time = 0.97748×10^8 yr;
Unit of velocity = 10 km·s^{-1};
Unit of energy (per unit mass) = 100 km^2·s^{-2};
$G = 1$ (gravitational constant).

While integrating the equations of motion in (4) for the computation of all the orbits, we use the fixed value of $v_0 = 15, c_b = 2.5, \alpha = 1.5, \beta = 1.8$ and $\lambda = 0.03$ where as M_h and c_h are taken as parameters. We calculate the trajectory by integrating the equations of motion given in Equation (4). It is done using Runge Kutta variable step size Integrator. Accuracy of calculations is maintained up to eight significant figures.

3. Total Angular Momentum and Time-Frequency Analysis Based on Wavelets

3.1. Total Angular Momentum

Total angular momentum for a star of mass $m = 1$ moving in a 3D orbit is defined as

$$L_{tot} = \sqrt{L_x^2 + L_y^2 + L_z^2}, \tag{7}$$

where L_x, L_y and L_z are the components along x, y and z axis respectively, given as

$$L_x = y\dot{z} - \dot{y}z, \quad L_y = z\dot{x} - \dot{z}x \quad \& \quad L_z = x\dot{y} - \dot{x}y. \tag{8}$$

Note: L_{tot} (total angular momentum) in Equation (7) is conserved only for a spherical system (see [4]).

3.2. Time-Frequency Analysis Based on Wavelets

Time-frequency analysis based on phase of continuous wavelet transform is described in this Section. At first we define continuous wavelet transform, instantaneous frequency and the mother wavelet. The continuous wavelet transform is defined in terms of Ψ, called mother wavelet expressed as

$$L_\Psi f(a,b) = \frac{1}{\sqrt{a}} \int_{-\infty}^{\infty} f(t) \bar{\Psi}\left(\frac{t-b}{a}\right) dt, \tag{9}$$

The function $\Psi \in L^2(R)$ must have compact support or decay rapidly to 0 for $|t|$ tending to ∞. Here $\bar{\Psi}$ denotes the usual complex conjugate of Ψ. The wavelet transform depends on two parameters (a,b): a is called the scale and b the time parameter. The wavelet transform produces a complex surface as a function of the variables a and b. The mother wavelet, we use throughout this work is known as the Morlet-Grossman wavelet. It is expressed as

$$\Psi(t) = \frac{1}{\sigma\sqrt{2\Pi}} e^{i2\Pi\eta t} e^{-\frac{t^2}{2\sigma^2}}. \tag{10}$$

Here, σ and σ are the parameters for the mother wavelet (see [8] for detail).

Note: The parameter η and σ can be tuned to improve the resolution. In our case $\eta = 0.8$ and $\sigma = 1$ serves the purpose. Here due to the part $\left(\frac{t-b}{a}\right)$ the length of the window in wavelet transform change according to frequency. Due to this unique feature *i.e.* capability of adaptation of time window according to frequency range gives better localization in frequency and time.

3.3. Instantaneous Frequency and Ridge-Plot

Let us consider an analytic signal $Z_f(x)$ of a real signal $f(x)$ whose real part is $f(x)$ and the complex part is Hilbert transform of $f(x)$, *i.e.*

$$Z_f(x) = f(x) + H(f(x)).$$

Now it's unique polar representation is

$$Z_f(t) = |Z_f(t)| e^{iArg(Z_f(t))}$$

where

$$|Z_f(t)| = \sqrt{RZ_f(t)^2 + ImZ_f(t)^2}$$

and

$$Arg(Z_f(t)) = \frac{ImZ_f(t)}{RZ_f(t)}.$$

where R and Im denote the real and imaginary part of the signal. Also a unique representation of $f(t)$ in canonical form is

$$f(t) = A(t)\cos\phi(t), \quad f(t) = RZ_f(t), \quad A(t) = |Z_f(t)|, \quad \text{and} \quad \phi(t) = Arg(Z_f(t)).$$

Instantaneous frequency is defined as

$$\omega(t) = \frac{d}{dt}\big(Arg(Z_f(t))\big).$$

The Ridge of the wavelet transform of $Z(t)$, $L_\psi f(a,b)$ is the set of points (a,b) in the domain of the transform $Z(t)\overline{\Psi}_{(a,b)}(t)$ is stationary *i.e.* the points which satisfy $t_0(a,b) = b$. Ridge-plot is in fact the time-frequency landscape of the signal. Details can be seen in [8] and [9].

3.4. Implementation of Wavelet-Ridges in Matlab

The algorithm for computing ridges from the phase of continuous wavelet transform is already explained in [6], [7] and [9]. Programs based on the algorithm for finding Ridge-plot can be made with the help of Wavelab routines written in Matlab (see [12]-[14]). Computation of CWT is done by Wavelab routines written in Matlab. Stepwise procedure for finding Ridge-plot is given in **Figure 1**. The value of η, σ, number of voices per octave and number of the octave for the calculation of wavelet transform and its ridges are also given in **Figure 1**.

4. Results and Discussion (Applicaion to Three Dimensional Triaxial Galaxy model)

In this Section, we analyze the data of **Table 1** and **Table 2** and discuss the plots obtained by integrating the

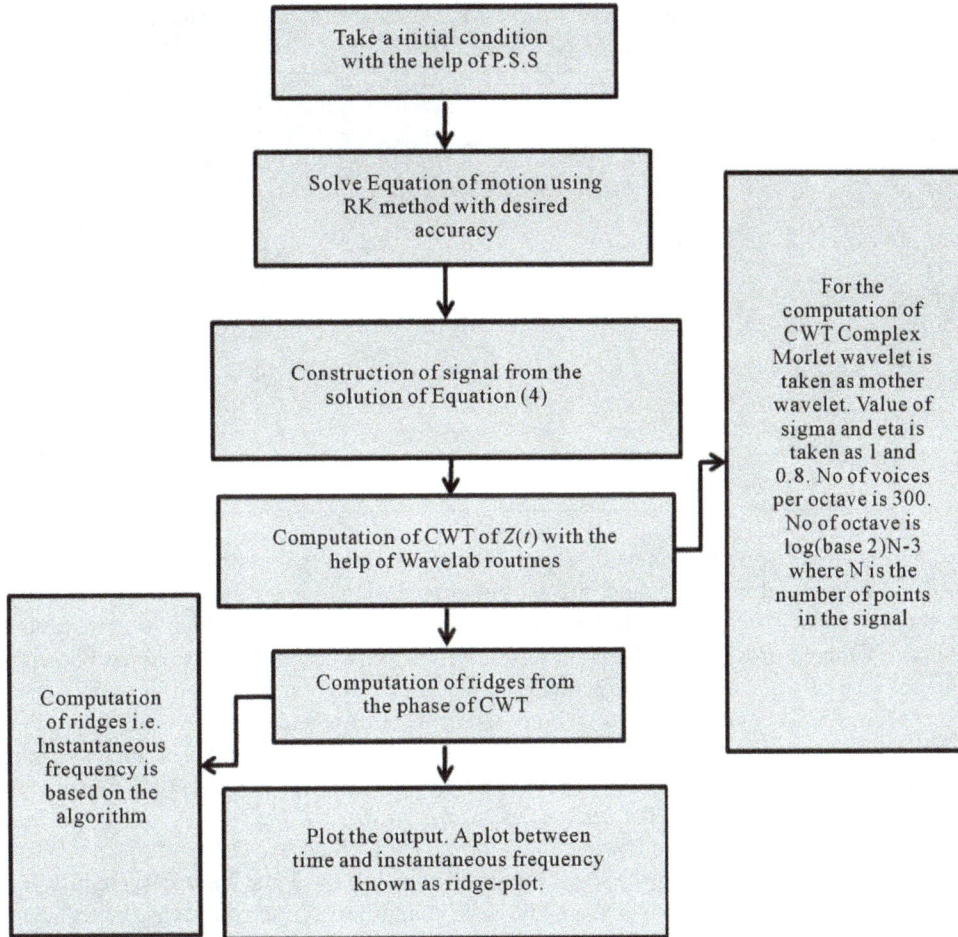

Figure 1. Procedure for finding instantaneous frequency (*i.e.* wavelet ridges) from the solution of equations of motion.

Table 1. Sample of five orbits for the triaxial galaxy model in presence of dark halo component and the corresponding results. The approximate C.P.U time for the execution of the programme for L_{tot} and Ridge-plot on the computer is given in second.

Initial condition $(x_0, y_0, z_0, \dot{x}_0, \dot{y}_0, \dot{z}_0)$	Mass length (Mh)	Scale length (Ch)	Energy const. (h_3)	L_{tot}-plot (in sec)	Ridge-plot (in sec)	Type of orbit
$(-9.55, 0, 0.1, 0, 5.8029, 0)$	10000	18	68	4.32	14.55	Periodic
$(5.5, 0, 0.6, 0, 20.4251, 0)$	10000	18	68	4.19	14.58	Quasi-Periodic
$(-0.5, 0, 0.85, 0, 27.9848, 0)$	10000	18	68	4.30	15.01	Chaotic-sticky
$(-0.9, 0, 0.85, 0, 27.6788, 0)$	10000	18	68	4.44	14.69	Chaotic nonsticky
$(0.7, 0, 0.85, 0, 27.8624, 0)$	10000	18	68	4.31	31.57	Transient chaos and Resonance trapping

equation of motion (4) at the given initial conditions using Runge-Kutta (4/5) variable step-size Integrator. Phase-portrait and L_{tot}-plot is drawn with the help of software Mathematica and Ridge-plot is drawn in Matlab.

Table 2. Sample of orbits of the triaxial galaxy model in the absence of dark halo component and the corresponding results. The C.P.U time for the execution of the program for L_{tot}-plot and Ridge-plot on the computer are given in seconds.

Initial condition $(x_0, y_0, z_0, \dot{x}_0, \dot{y}_0, \dot{z}_0)$	Mass mass (Mh)	Scale length (Ch)	Energy const. (h_3)	L_{tot}-plot (in sec)	Ridge-plot (in sec)	Type of orbit
(−7.5, 0, −1, 0, 7.4008, 0)	0	8	516	4.23	14.57	Periodic
(3, 0, 0.5, 0, 20.5996, 0)	0	8	516	4.33	14.55	Quasi-periodic
(0.1, 0, 0.5, 0, 24.5701, 0)	0	8	516	4.25	14.58	Chaotic sticky
(1.7, 0, −1, 0, 22.2880, 0)	0	8	516	4.31	15.01	Chaotic non-sticky
(0.01, 0, 0.1, 0, 24.8801, 0)	0	8	516	4.37	32.67	Resonant transition and Transient chaos

At first, we discuss the results obtained using the Poincare surface of section in **Figure 2**. "Section condition" taken for the Poincare plot is $y = 0$ and $\dot{y} > 0$. Initial conditions are taken as $\left(x(0), 0, 0, \dot{y}(0)\right)$ on x-axis on the interval (−10, 10) with step-size 0.1 and integrated at each initial condition up to 2500 time units. $\dot{y}(0)$ is calculated with the help of Equation (5). Both results of PSSs are in confirmation of the Poincare plots given in [4]. In the absence of the dark halo component, almost all phase plane is filled up with a hazy collection of points except few small regular regions which consist of a set of islands. On the other hand, in the presence of the dark halo component we notice the substantial reduction in chaotic regions and increase in regular regions. Thus, we can say that the dark halo component serves as chaos controller in triaxial galaxies with small asymmetries. These observations confirm the result of [4].

We have selected a sample of five representative orbits for both cases. Time interval considered for the TFA is (1, 32768) except two figures (**Figure 3** and **Figure 4**). For **Figure 3** and **Figure 4**, the interval of time unit is taken as (1, 65536) to present the results in a better way. It adds an additional expense of eighteen seconds in the C.P.U time for the execution of the program. In **Figure 5** and **Figure 6**, we consider the orbits at initial conditions (−9.55, 0, 0.1, 0, 5.8029, 0) and (−7.5, 0, −1, 0, 7.4008, 0) respectively. In Ridge-plots of both figures, we notice a completely flat Ridge throughout the motion. We know that frequency remains constant for the periodic orbit and hence both orbits are periodic.

Note: In [4], at same initial condition (−9.55, 0, 0.1, 0, 5.8029, 0) the orbit is termed as quasiperiodic on the basis of L_{tot}-plot. But, according to Ridge-plot this trajectory is periodic. Also, the phase portrait and L_{tot}-plot in **Figure 5** is exactly similar to the figure of [4].

Sample of orbits considered in **Figure 7** and **Figure 8** is at initial conditions (5.5, 0, 0.6, 0, 20.4251, 0) and (3, 0, 0.5, 0, 20.5996, 0) respectively. We observe a very little variation in Ridge curve (or instantaneous frequency) of both figures. In the plots of L_{tot}, we notice symmetric peaks indicating regular motion. On the basis of both Ridge-plots, we term both trajectories as quasi-periodic. Also L_{tot} plots of both orbits are confirming the result of L_{tot}-plot in [4].

Now we consider the orbits at initial conditions (−0.5, 0, 0.85, 0, 27.9848, 0) and (0.1, 0, 0.5, 0, 24.5701, 0) are shown in **Figure 9** and **Figure 10** respectively. If we look at both Ridge-plots, we can say that the trajectories are chaotic in nature. Also, highly asymmetric features and large deviation between maxima and between minima of L_{tot}-plot confirms the chaotic behavior of trajectories. But in addition to this, we find very little variation in both Ridge-plots in the time intervals (0, 20000) and (0, 12000) respectively. It means that both trajectories are trapped on resonance islands for long time intervals and then escape to the chaotic region. This phenomenon is known as stickiness. Thus, we term both trajectories as chaotic sticky. But in case of L_{tot} plots, it is not clear.

Now, we consider the sample of trajectories at initial conditions (−0.9, 0, 0.85, 0, 27.6788, 0) and (1.7, 0, −1, 0, 22.2880, 0) presented in **Figure 11** and **Figure 12** respectively. If we look at L_{tot}-plots of both figures, we find similar results as in the previous case. But, if we look at the Ridge-plots of both figures we notice a continuous variation in the instantaneous frequency throughout the motion and hence we call these trajectories

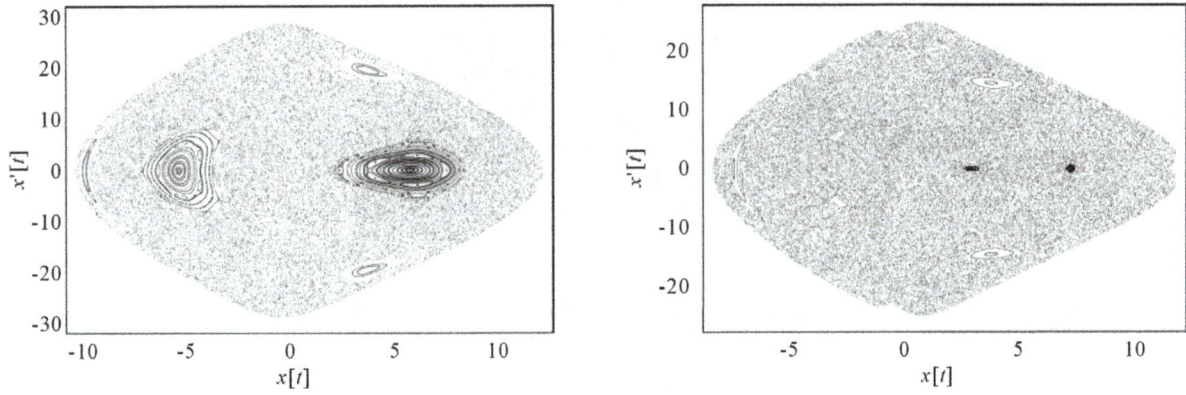

Figure 2. Poincare surface of sections of triaxial galaxy model in presence (left side) and absence (right side) of dark halo component with parameteric values ($Mh = 10000$, $Ch = 18$, $H = 68$) and ($Mh = 0$, $Ch = 8$, $H = 516$) respectively.

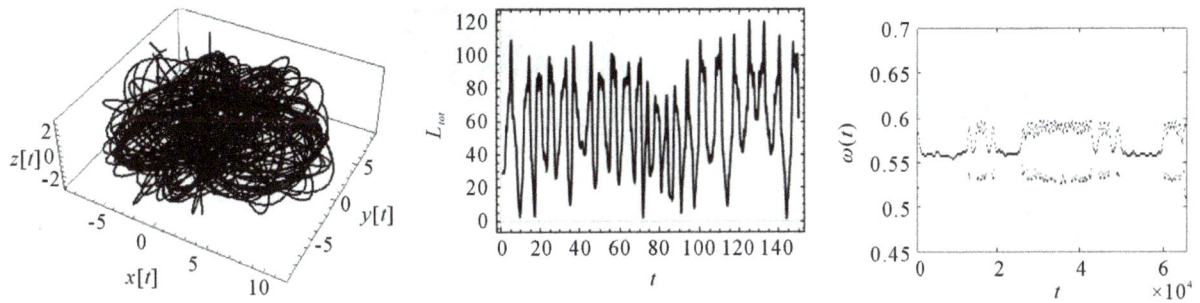

Figure 3. Phase portrait (extreme left), L_{tot}-plot (middle) and Ridge-plot (extreme right) of a transient trajectory in presence of dark halo component at initialcondition (0.7, 0, 0.85, 0, 27.8624, 0).

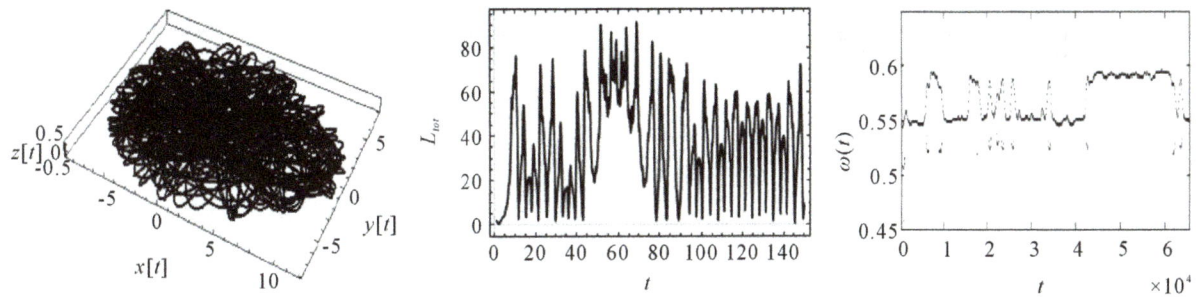

Figure 4. Phase portrait (extreme left), L_{tot}-plot (middle) and Ridge-plot (extreme right) of a trajectory presenting transient chaos and resonance transition in absence of dark halo component at initial condition (0.01, 0, 0.1, 0,24.8801, 0).

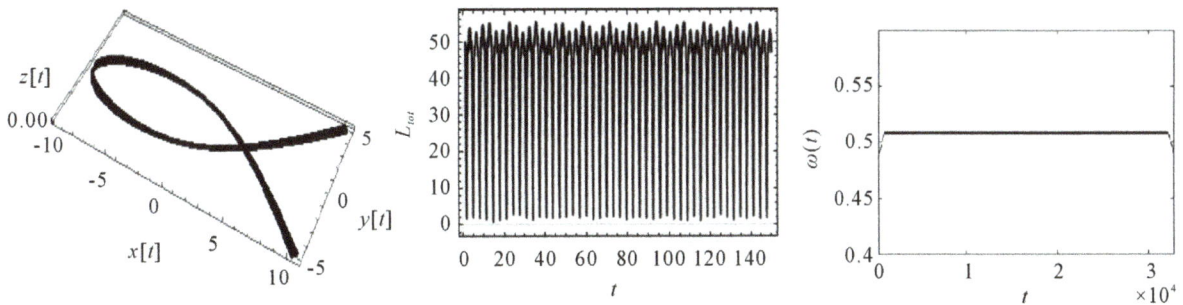

Figure 5. Phase portrait (extreme left), L_{tot}-plot (middle) and Ridge-plot (extreme right) of a periodic trajectory in presence of dark halo component at initial condition (−9.55, 0, 0.1, 0, 5.8029, 0).

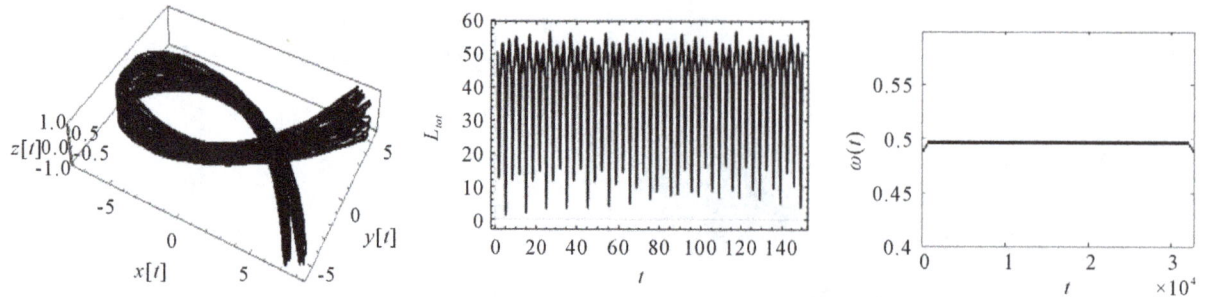

Figure 6. Phase portrait (extreme left), L_{tot}-plot (middle) and Ridge-plot (extreme right) of a periodic trajectory in absence of dark halo component at initial condition $(-7.5, 0, -1, 0, 7.4008, 0)$.

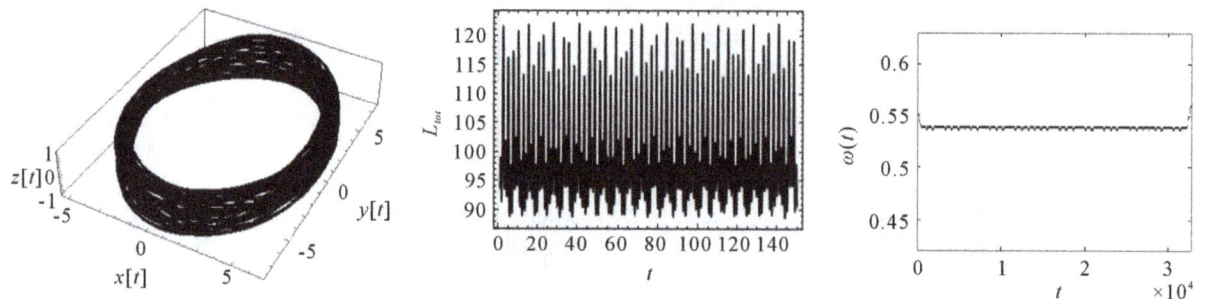

Figure 7. Phase portrait (extreme left), L_{tot}-plot (middle) and Ridge-plot (extreme right) of a quasi-periodic trajectory in presence of dark halo component at initial condition $(5.5, 0, 0.6, 0, 20.4251, 0)$.

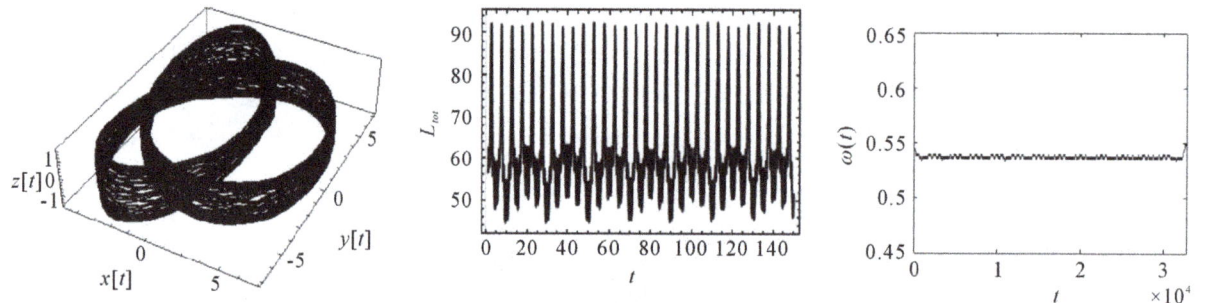

Figure 8. Phase portrait (extreme left), L_{tot}-plot (middle) and Ridge-plot (extreme right) of a quasi-periodic trajectory in absence of dark halo component at initial condition $(3, 0, 0.5, 0, 20.5996, 0)$.

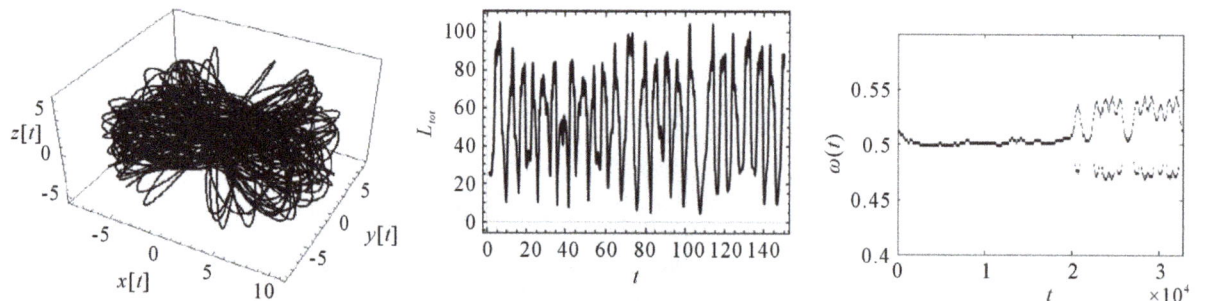

Figure 9. Phase portrait (extreme left), L_{tot}-plot (middle) and Ridge-plot (extreme right) of a chaotic sticky trajectory in presence of dark halo component at initial condition $(-0.5, 0, 0.85, 0, 27.9848, 0)$.

as chaotic non-sticky.

At last, we consider sample of two orbits at initial conditions $(0.7, 0, 0.85, 0, 27.8624, 0)$ and $(0.01, 0, 0.1, 0, 24.8801, 0)$ shown in **Figure 3** and **Figure 4** respectively. Again in the plots of L_{tot}, we observe the same

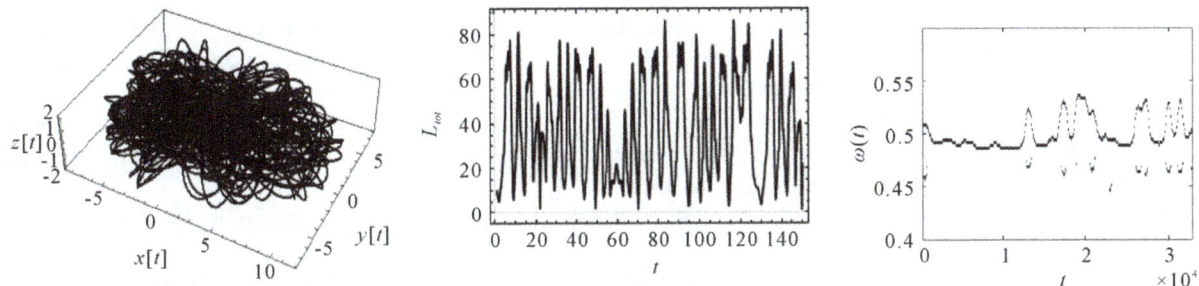

Figure 10. Phase portrait (extreme left), L_{tot}-plot (middle) and Ridge-plot (extreme right) of a chaotic sticky trajectory in absence of dark halo component at initial condition $(0.1, 0, 0.5, 0, 24.5701, 0)$.

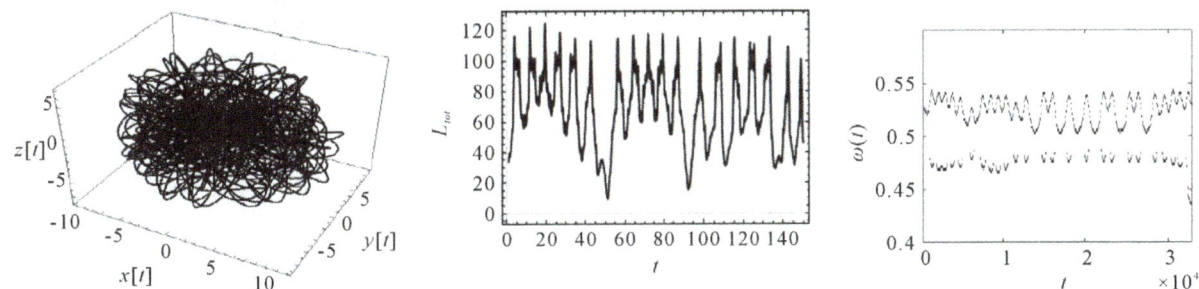

Figure 11. Phase portrait (extreme left), L_{tot}-plot (middle) and Ridge-plot (extreme right) of a chaotic non-sticky trajectory in presence of dark halo component at initial condition $(-0.9, 0, 0.85, 0, 27.6788, 0)$.

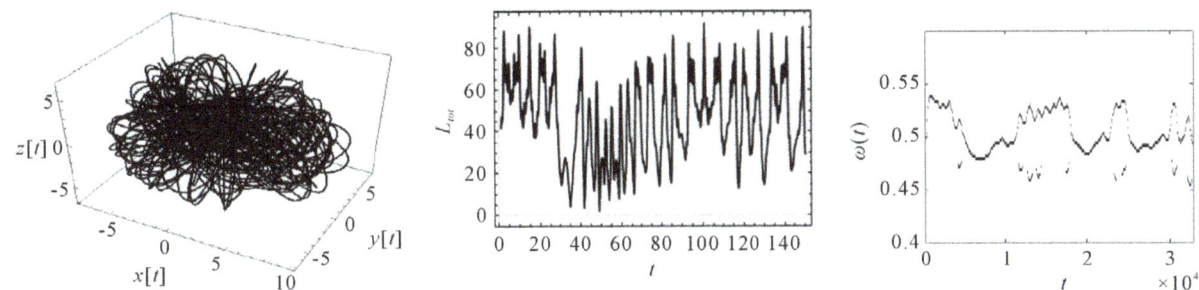

Figure 12. Phase portrait (extreme left), L_{tot}-plot (middle) and Ridge-plot (extreme right) of a chaotic non-sticky trajectory in absence of dark halo component at initial condition $(1.7, 0, -1, 0, 22.2880, 0)$.

phenomenon as in the previous case and hence we can say that these trajectories are chaotic. But in case of Ridge-plots scene is little different. We observe following things in the Ridge-plots:

1) Resonance trapping: In **Figure 3** at time intervals $(0, 13000)$, $(19000, 25000)$ and $(50000, 60000)$ approx we find a little variation in the frequency and besides that there is rapid change in frequency. Similarly, in **Figure 4** at time intervals $(0, 5000)$, $(10000, 15000)$, $(25000, 32000)$ and $(41000, 60000)$ there is very little change in instantaneous frequency and besides that rapid change in frequency takes place. This is the case of resonance trapping which can be visualized and also accurately determined with the help of TFA.

2) Resonance transition: In **Figure 4**, we notice that at time unit 42,000 approx there is a jump from one type of resonance to another type of resonance and remains there for approx 18,000-time units. This phenomenon is known as resonance transition which is clearly observed in the Ridge-plot.

3) Transient Chaos: We know that in a time-dependent system a chaotic trajectory move from regular to chaos and vice versa (see [11]). We call this phenomenon as transient chaos. We can visualize this phenomenon in the Ridge-plots.

5. Conclusions

As we have already discussed, the aim of the present work is to show the advantage of TFA in comparison to

TFA and to explore some additional information of phase space structures of asymmetric triaxial galaxies in the presence and absence of the spherical dark halo component. Based on the discussion of Section 4, we can conclude that TFA has several advantages in comparison to TAM. We conclude following things:

1) TFA based on wavelets is comparatively fast and more reliable in comparison to TAM (C.P.U time taken for the computation of Ridge-plots for 32,768-time units is 15 seconds (maximum) whereas the time taken by TAM for 150-time units is 5 seconds (maximum)).

2) TFA can identify between periodic and quasi-periodic, chaotic sticky and non-sticky, and ordered and chaotic motion.

3) With the help of TFA, we can accurately determine and also visualize the event of trapping of a chaotic trajectory around resonance island (see **Figure 3** and **Figure 4**).

4) The phenomenon of resonance transition and transient chaos can also be explained with the help of Ridge-plot (see **Figure 3** and **Figure 4**).

5) Computational effort needed for programming of TFA based on wavelets is not easy in comparison to TAM. This is an important drawback of TFA based on wavelets. But once it is done, we can perform other computational works in comparatively negligible time.

6) We always search for an indicator which is applicable to higher-dimensional nonlinear dynamical systems. TFA is independent of the degree of freedom and requires the only solution of equations of motion which can be computed. Our present work is also an important example of the application to higher dimensional systems.

Thus, we can say that Time-frequency analysis based on wavelets can be given preference for the study of nonlinear dynamical systems for two or more degrees of freedom.

References

[1]   Racoveanu, O. (2014) Comparison of Chaos Detection Methods in the Circular Restricted Three-Body Problem. *Astronomische Nachrichten*, **335**, 877-885. http://dx.doi.org/10.1002/asna.201212110

[2]   Smith, R.H. (1991) The Onset of Chaotic Motion in the Restricted Problem of Three Bodies. Ph.D. Thesis, University of Texas at Austin, Austin.

[3]   Manos, T., Skokos, C.H. and Antonopoulos, C.H. (2008) Probing the Local Dynamics of Periodic Orbits by the Generalized Alignment Index (GALI) Method. *International Journal of Bifurcation and Chaos*, **22**, 1-17. http://arxiv.org/pdf/1103.0700v3

[4]   Caranicolas, N.D. and Zotos, E.E. (2011) Dark Halos Acting as Chaos Controllers in Asymmetric Triaxial Galaxy Models. *Research in Astronomy and Astrophysics*, **11**, 811-823. http://dx.doi.org/10.1088/1674-4527/11/7/006

[5]   Chandre, C., Wiggins, S. and Uzer, T. (2003) Time-Frequency Analysis of Chaotic Systems. *Physica D*, **181**, 171-196.

[6]   Deplart, N., Escudie, B., Guillemain, P., Kronland Martinet, R., Tchamichian, P. and Torresani, B. (1992) Asymptotic Wavelet and Gabor Analysis, Extraction of Instantaneous Frequency. *IEEE Transactions on Information Theory*, **38**, 644-664.

[7]   Vela-Arevalo, L.V. (2002) Time-Frequency Analysis Based on Wavelets for Hamiltonian Systems. Ph.D. Dissertation, California Institute of Technology, Pasadena. http://resolver.caltech.edu/CaltechETD:etd-03302004-115559

[8]   Vela-Arevalo, L.V. and Wiggins, S. (2001) Time-Frequency Analysis of Classical Trajectories of Polyatomic Molecules. *International Journal of Bifurcation and Chaos*, **11**, 1359-1380. http://dx.doi.org/10.1142/S0218127401002766

[9]   Todorovska, M. (2001) Estimation of Instantaneous Frequency of Signals Using the Continuous Wavelet Transform. Department of Civil Engineering, University of Southern California, Report CE 01-07, 2001.

[10] Vela-Arevalo, L.V. (2004) Time-Frequency Analysis of the Restricted Three-Body Problem: Transport and Resonance Transitions. *Classical and Quantum Gravity*, **21**, S351-S375.

[11] Kandrup, H.H., Vass, I.M. and Sideris, I.V. (2003) Transient Chaos and Resonant Phase Mixing in Violent Relaxation. *Monthly Notice of the Royal Astronomical Society*, **341**, 927-936. http://dx.doi.org/10.1046/j.1365-8711.2003.06466.x

[12] http://www-stat.stanford.edu/~wavelab/

[13] The MathWorks, Inc. (2010) Matlab and Statistics Toolbox Release. The MathWorks, Inc., Natick.

[14] Mallat, S. (1999) A Wavelet Tour of Signal Processing. Academic Press, San Diego.

25

Bulk Viscous Bianchi Type V Space-Time with Generalized Chaplygin Gas and with Dynamical G and Λ

Shubha S. Kotambkar[1], Gyan Prakash Singh[2], Rupali R. Kelkar[3]

[1]Department of Applied Mathematics, Laxminarayan Institute of Technology, Rashtrasant Tukadoji Maharaj Nagpur University, Nagpur, India
[2]Department of Mathematics, Visvesvaraya National Institute of Technology, Nagpur, India
[3]Department of Applied Mathematics, S. B. Jain Institute of Technology, Management and Research, Nagpur, India
Email: shubha.kotambkar@rediffmail.com, gpsingh@mth.vnit.ac.in, rupali.kelkar@yahoo.com

Abstract

In this paper, bulk viscous Bianchi type V cosmological model with generalized Chaplygin gas, dynamical gravitational and cosmological constants has been investigated. We are assuming the condition on metric potential $\frac{\dot{R_1}}{R_1} = \frac{\dot{R_2}}{R_2} = \frac{m}{t^n}$. To obtain deterministic model, we have considered physically plausible relations like $P = p + \Pi$, $\eta = \eta_0 \rho^r$ and the generalized Chaplygin gas is described by equation of state $p = \frac{-B}{\rho^\alpha}$. A new set of exact solutions of Einstein's field equations has been obtained in Eckart theory, truncated theory and full causal theory. Physical behavior of the models has been discussed.

Keywords

Bianchi Type V, Gravitational Constant, Cosmological Constant, Bulk Viscosity, Chaplygin Gas

1. Introduction

Recent cosmology is on Fridman-Lemaitra-Robertson-Walkar (FLRW) which is completely homogeneous and isotropic. But it is widely believed that FLRW model does not give a correct matter description in the early stage of universe. The theoretical argument [1] and the recent experimental data support the existence of an anisotropic phase, which turns into an isotropic one during the evolution of the universe. Anisotropic model plays significant role in description of evolution of the early phase of the universe and also helps in finding more general cosmological models than the isotropic FRW models. This motivates researcher for obtaining exact anisotropic solution for Einstein's field equations as a cosmologically accepted physical models for the universe (in the early stages). The study of Bianchi type V cosmological model being anisotropic generalization of open FRW models is important to study old universe. A number of authors have investigated Bianchi type V cosmological model in general relativity in different context [2]-[15]. Rajbali and Seema Tinkar have discussed Bianchi type V bulk viscous Barotropic fluid cosmological model with variable G and Λ. Recently Yadav and Sharma [16] and Yadav [17] have discussed about transit universe in Bianchi type V space-time with variable G and Λ.

It has been widely discussed in the literature that during the evolution of the universe, bulk viscosity can arise in many circumstances and can lead to an effective mechanism of galaxy formation [18]. It is known that real fluids behave irreversibly and therefore it is important to consider dissipative processes both in cosmology and in astrophysics. To consider more realistic models, one must take in to account the viscosity mechanism. Bulk viscosity leading to an accelerated phase of the universe today has been studied by Fabris *et al.* [19]. Very recently Kotambkar *et al.* [20] have investigated anisotropic cosmological models with quintessence considering the effect of bulk viscosity.

A wide range of observations strongly suggest that the universe possesses non zero cosmological term [21]. The astronomical observations [22] [23] support that the expansion of the universe is accelerated. It suggests that there exists a new component in universe named as dark energy with negative pressure. A natural explanation for the accelerated expansion is due to a positive small cosmological constant. An attention has been paid to cosmological models with non zero cosmological term Λ [21] [24], whose existence is favored by supernovae SNe Ia observations (refer to [22] [23]) which are consistent with the recent anisotropy measurements of the cosmic microwave background (CMB) made by the WAMAP experiment [25]. Sahni and Starobinski [26] have presented detailed discussion on current observational situation focusing on cosmological tests on Λ.

Time varying G has many interesting consequences in astrophysics. Cunuto and Narlikar [27] have shown that G-varying cosmology is consistent with what so ever cosmological observations available at present. A new approach is appealing; it assumes the conservation of the energy momentum tensor which consequently gives G and Λ as coupled fields similar to the case of G in original Brans-Dicke theory. The cosmological model with variable G and Λ has been investigated by several researchers [28]-[32]. A number of researchers have discussed various anisotropic cosmological models with variable G and Λ [33]-[37].

According to recent observational evidence, the expansion of the universe is accelerated, which is dominated by a smooth component with negative pressure, the so called dark energy. To avoid problems associated with Λ and quintessence models, recently, it has been shown that Chaplygin gas may be useful. The unification of the dark matter and dark energy component creates a considerable theoretical interest, because on the one hand, model building becomes reasonably simpler, and on the other hand such unification implies existence of an era during which the energy densities of dark matter and dark energy are strikingly similar. For representation of such a unification, the generalized Chaplygin gas (GCG) with exotic condition of state $p = \dfrac{-B}{\rho^{\alpha}}$ is considered, where constant B and α satisfy $B > 0$ and $0 < \alpha \leq 1$ respectively. Due to observational evidence, cosmological models based on CG-EOS are very encouraging. Chaplygin gas and generalized Chaplygin gas cosmological models are first time proposed by Kamenshchik *et al.* [38]. WMAP constraints on the generalized Chaplygin gas model have been investigated by Bento *et al.* [39].

Motivated by above work we thought that it was worthwhile to study bulk viscous Bianchi type V space-time with generalized Chaplygin gas and dynamical G and Λ.

2. Metric and Field Equations

The spatially homogeneous and anisotropic space-time metric is given by

$$ds^2 = -dt^2 + R_1^2 dx^2 + R_2^2 e^{2kx} dy^2 + R_3^2 e^{2kx} dz^2 \tag{1}$$

where R_1, R_2, R_3 are functions of t alone.

Einstein field equation with time dependent Λ and G may be written as

$$R_{ij} - \frac{1}{2} R g_{ij} = -8\pi G T_{ij} + \Lambda g_{ij} \tag{2}$$

where G and Λ are time dependent gravitational and cosmological constants. T_{ij} is energy momentum tensor of cosmic fluid in the presence of bulk viscosity defined as

$$T_{ij} = (\rho + P) u_i u_j - (P) g_{ij} \tag{3}$$

$$P = p + \Pi \tag{4}$$

where p is equilibrium pressure, Π is bulk viscous stress together with $u_i u^j = 1$.

Einstein's field Equation (2) for the metric (1) takes form

$$\frac{\ddot{R}_1}{R_1} + \frac{\ddot{R}_3}{R_3} + \frac{\dot{R}_1}{R_1}\frac{\dot{R}_3}{R_3} - \frac{1}{R_1^2} = -8\pi G(p + \Pi) + \Lambda, \tag{5}$$

$$\frac{\ddot{R}_2}{R_2} + \frac{\ddot{R}_3}{R_3} + \frac{\dot{R}_2}{R_2}\frac{\dot{R}_3}{R_3} - \frac{1}{R_1^2} = -8\pi G(p + \Pi) + \Lambda \tag{6}$$

$$\frac{\ddot{R}_1}{R_1} + \frac{\ddot{R}_2}{R_2} + \frac{\dot{R}_1}{R_1}\frac{\dot{R}_2}{R_2} - \frac{1}{R_1^2} = -8\pi G(p + \Pi) + \Lambda \tag{7}$$

$$\frac{\dot{R}_1}{R_1}\frac{\dot{R}_2}{R_2} + \frac{\dot{R}_2}{R_2}\frac{\dot{R}_3}{R_3} + \frac{\dot{R}_3}{R_3}\frac{\dot{R}_1}{R_1} - \frac{3}{R_1^2} = 8\pi G\rho + \Lambda, \tag{8}$$

$$2\frac{\dot{R}_1}{R_1} - \frac{\dot{R}_2}{R_2} - \frac{\dot{R}_3}{R_3} = 0. \tag{9}$$

By the divergence of Einstein's tensor *i.e.* $\left(R_{ij} - \frac{1}{2} R g_{ij} \right)_{;j} = 0$ which lead to

$\left(8\pi G T_{ij} - \Lambda g_{ij} \right)_{;j} = 0$, then yields

$$8\pi \dot{G}\rho + \dot{\Lambda} + 8\pi G \left[\dot{\rho} + (\rho + p + \Pi)\left(\frac{\dot{R}_1}{R_1} + \frac{\dot{R}_2}{R_2} + \frac{\dot{R}_3}{R_3} \right) \right] = 0 \tag{10}$$

The energy momentum conservation equation $\left(T_{;j}^{ij} = 0 \right)$ splits Equation (10) into two equations.

$$\dot{\rho} + (\rho + p)\left(\frac{\dot{R}_1}{R_1} + \frac{\dot{R}_2}{R_2} + \frac{\dot{R}_3}{R_3} \right) = 0, \tag{11}$$

$$8\pi \dot{G}\rho + \dot{\Lambda} = -8\pi G\Pi \left(\frac{\dot{R}_1}{R_1} + \frac{\dot{R}_2}{R_2} + \frac{\dot{R}_3}{R_3} \right). \tag{12}$$

For the full causal non-equilibrium thermodynamics the causal evolution equation for bulk viscosity is given by [40]

$$\Pi + \tau\dot{\Pi} = -\eta \left(\frac{\dot{R}_1}{R_1} + \frac{\dot{R}_2}{R_2} + \frac{\dot{R}_3}{R_3} \right) - \frac{\varepsilon\tau\Pi}{2}\left(\frac{\dot{R}_1}{R_1} + \frac{\dot{R}_2}{R_2} + \frac{\dot{R}_3}{R_3} + \frac{\dot{\tau}}{\tau} - \frac{\dot{\eta}}{\eta} - \frac{\dot{T}}{T} \right). \tag{13}$$

$T \geq 0$ absolute temperature, η is bulk viscosity coefficient which cannot become negative, τ denote the relaxation time for transient bulk viscous effects. Causality requires $\tau > 0$. When $\varepsilon = 0$, Equation (13) reduces to evolution equation for truncated theory. For $\varepsilon = 1$ Equation (13) reduces to evolution equation for full causal theory and for $\tau = 0$ Equation (13) reduces to evolution equation for non-causal theory (Eckart's theory).

3. Cosmological Solutions

It can be easily seen that we have five Equations (5)-(9) with eight unknowns $R_1, R_2, R_3, \rho, p, G, \Lambda$ and η. Hence to solve the system of equations completely we need three additional physically plausible relations among these variables.

3.1. Case I: Non-Causal Cosmological Solution

For non causal solution $\tau = 0$, therefore the evolution Equation (13) takes the form of

$$\Pi = -\eta \left(\frac{\dot{R}_1}{R_1} + \frac{\dot{R}_2}{R_2} + \frac{\dot{R}_3}{R_3} \right) = -3\eta H \tag{14}$$

To find the complete solution of the system of equations, following relations are taken into consideration. The power law relation for bulk viscosity is taken as

$$\eta = \eta_0 \rho^r, \tag{15}$$

where $\eta_0 \geq 0$ and r is a constant.

We consider an exotic background fluid, the Chaplygin gas, described by the equation of state

$$p = \frac{-B}{\rho^\alpha}, \tag{16}$$

where B is constant and $0 < \alpha \leq 1$

To obtain the deterministic scenario of the universe, we assume the condition

$$\frac{\dot{R}_1}{R_1} = \frac{\dot{R}_2}{R_2} = \frac{m}{t^n} \tag{17}$$

From Equation (9) and (17), one can get

$$\frac{\dot{R}_3}{R_3} = \frac{m}{t^n}, \tag{18}$$

From Equations (17)-(18), one can easily calculate

$$R_1 = K_1 e^{\frac{mt^{1-n}}{1-n}}, \quad R_2 = K_2 e^{\frac{mt^{1-n}}{1-n}}, \quad R_3 = K_3 e^{\frac{mt^{1-n}}{1-n}}. \tag{19}$$

Using Equations (17) and (18), Equation (11) yields

$$\dot{\rho} + \left(\rho - \frac{B}{\rho^\alpha} \right) \left(\frac{3m}{t^n} \right) = 0. \tag{20}$$

By solving Equation (20), we get

$$\rho = \left[B + C e^{-Dt^{1-n}} \right]^{\frac{1}{1+\alpha}}. \tag{21}$$

where $D = \frac{3m(1+\alpha)}{1-n}$, and where C is constant of integration.

From **Figure 1** one can easily see that energy density is decreasing with evolution of the universe. On differentiating Equation (21), we get

$$\dot{\rho} = -\frac{3mC}{t^n} e^{-Dt^{1-n}} \left[B + C e^{-Dt^{1-n}} \right]^{\frac{1}{1+\alpha}-1}. \tag{22}$$

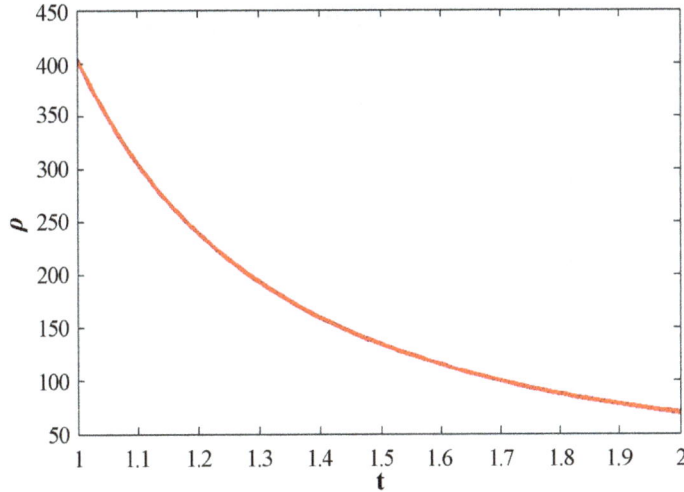

Figure 1. This figure shows variation of energy density ρ with respect to cosmic time t. Here we consider $B = 1$, $C = 1$, $n = 1.5$, $m = 2$ and $\alpha = 1$.

Now with the help of Equations (17)-(19) and (21), Equation (8) becomes

$$8\pi G\rho + \Lambda = \frac{3m^2}{t^{2n}} - \frac{3}{K_1^2}\exp\left\{\frac{-2mt^{1-n}}{1-n}\right\}. \tag{23}$$

Which on differentiation yields

$$8\pi\dot{G}\rho + 8\pi G\dot{\rho} + \dot{\Lambda} = \frac{-6m^2 n}{t^{2n+1}} + \frac{6m}{K_1^2 t^n}\exp\left\{\frac{-2mt^{1-n}}{1-n}\right\}, \tag{24}$$

With the help of Equations (12), (14), (17)-(18) and (21), Equation (24) becomes

$$8\pi G\left[\dot{\rho} + \frac{9m^2\eta}{t^{2n}}\right] = \frac{6m}{t^n}\left[\frac{1}{K_1^2}\exp\left\{-\frac{2mt^{1-n}}{1-n}\right\} - \frac{mn}{t^{n+1}}\right], \tag{25}$$

By use of Equations (15), (21) and (22) in Equation (25), we get

$$G = \frac{1}{4\pi}\left[\frac{1}{K_1^2}\exp\left\{\frac{-2mt^{1-n}}{1-n}\right\} - \frac{mn}{t^{n+1}}\right]\left[-Ce^{-Dt^{1-n}}\left(B + Ce^{-Dt^{1-n}}\right)^{\frac{1}{1+\alpha}-1} + \frac{3m\eta_0}{t^n}\left(B + Ce^{-Dt^{1-n}}\right)^{\frac{r}{\alpha+1}}\right]^{-1}, \tag{26}$$

From **Figure 2** it can be seen that G is increasing with evolution of the universe.
Now using Equations (21) and (26) in Equation (23) gives

$$\Lambda = \frac{3m^2}{t^{2n}} - \frac{3}{K_1^2}\exp\left\{\frac{-2mt^{1-n}}{1-n}\right\} - 2\left[\frac{1}{K_1^2}\exp\left\{\frac{-2mt^{1-n}}{1-n}\right\} - \frac{mn}{t^{n+1}}\right]$$

$$\cdot\left[-Ce^{-Dt^{1-n}}\left(B + Ce^{-Dt^{1-n}}\right)^{-1} + \frac{3m\eta_0}{t^n}\left(B + Ce^{-Dt^{1-n}}\right)^{\frac{r-1}{\alpha+1}}\right]^{-1} \tag{27}$$

Figure 3 shows that cosmological constant is decreasing with the evolution of the universe.
On solving Equations (21) and (15) we can obtain the expression for bulk viscosity coefficient as

$$\eta = \eta_0\left[B + Ce^{-Dt^{1-n}}\right]^{\frac{r}{\alpha+1}}. \tag{28}$$

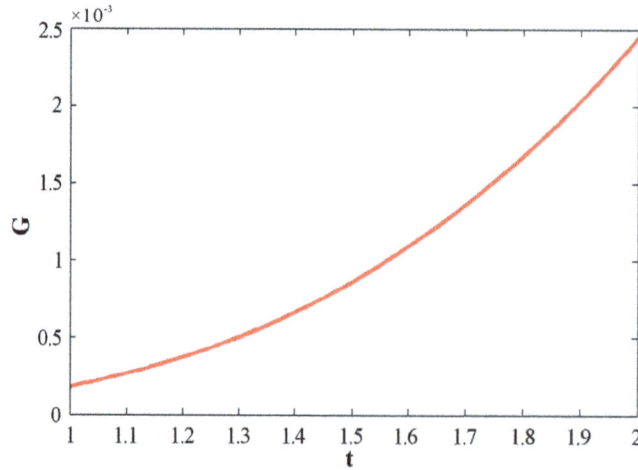

Figure 2. This figure shows variation of gravitational constant with respect to cosmic time t. Here we consider $B = 1$, $C = 1$, $n = 1.5$, $m = 2$, $\alpha = 1$, $r = 1.5$, $a = 1$ and $\eta_0 = 1$.

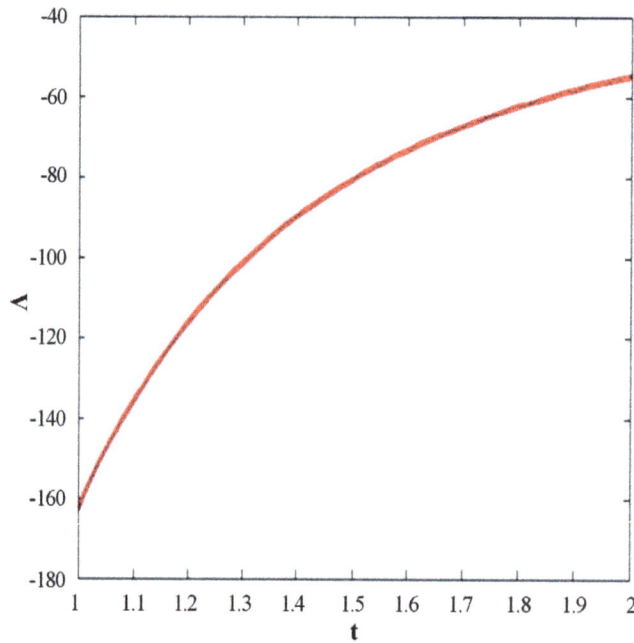

Figure 3. This figure shows variation of cosmological constant with respect to cosmic time t. Here we consider $B = 1$, $C = 1$, $n = 1.5$, $m = 2$, $\alpha = 1$, $r = 1.5$, $a = 1$ and $\eta_0 = 1$.

Figure 4 shows that bulk viscosity coefficient is decreasing with evolution of the universe. Thus the metric (1) reduces into the form

$$ds^2 = -dt^2 + K^2 \exp\left\{\frac{2mt^{1-n}}{1-n}\right\}\left(dx^2 + e^{2k}dy^2 + e^{2k}dz^2\right). \tag{29}$$

The deceleration parameter is given by

$q = -1 - \dfrac{\dot{H}}{H^2}$, for this model deceleration parameter is

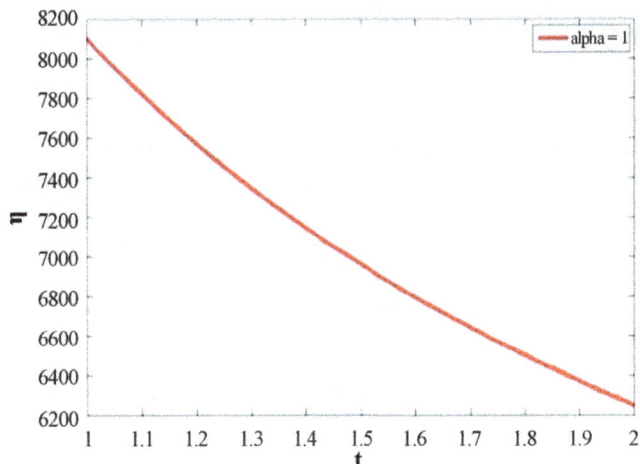

Figure 4. This figure shows variation of bulk viscosity coefficient with respect to cosmic time t. Here we consider $B = 1$, $C = 1$, $n = 1.5$, $m = 2$, $\alpha = 1$, $r = 1.5$ and $\eta_0 = 1$.

$$q = -1 + \frac{n}{mt^{1-n}} \tag{30}$$

Expansion scalar, Shear coefficient, relative anisotropy for this model is given by

$$\Theta = \frac{3\dot{H}}{H} = \frac{3m}{t^n} \tag{31}$$

$$\sigma^2 = \frac{1}{2}\left[\left(\frac{\dot{R}_1}{R_1}\right)^2 + \left(\frac{\dot{R}_2}{R_2}\right)^2 + \left(\frac{\dot{R}_3}{R_3}\right)^2\right] - \frac{\Theta^2}{6}$$

$$\sigma^2 = 0 \tag{32}$$

$$\text{Relative anisotropy} = \frac{\sigma^2}{\rho} = 0 \tag{33}$$

The critical energy density and the critical vacuum energy density are respectively given by

$$\rho_c = \frac{3H^2}{8\pi G}, \quad \rho_v = \frac{\Lambda}{8\pi G}$$

for the anisotropic Bianchi type V model can be expressed respectively as

$$\rho_c = \frac{3m^2 t^{-2n}}{2\left[\frac{1}{K_1^2}\exp\left\{\frac{-2mt^{1-n}}{1-n}\right\} - \frac{mn}{t^{n+1}}\right]\left[-Ce^{-Dt^{1-n}}\left(B + Ce^{-Dt^{1-n}}\right)^{\frac{1}{1+\alpha}-1} + \frac{3m\eta_0}{t^n}\left(B + Ce^{-Dt^{1-n}}\right)^{\frac{r}{\alpha+1}}\right]^{-1}}, \tag{34}$$

$\rho_v =$

$$\frac{\frac{3m^2}{t^{2n}} - \frac{3}{K_1^2}\exp\left\{\frac{-2mt^{1-n}}{1-n}\right\} - 2\left[\frac{1}{K_1^2}\exp\left\{\frac{-2mt^{1-n}}{1-n}\right\} - \frac{mn}{t^{n+1}}\right]\left[-Ce^{-Dt^{1-n}}\left(B + Ce^{-Dt^{1-n}}\right)^{-1} + \frac{3m\eta_0}{t^n}\left(B + Ce^{-Dt^{1-n}}\right)^{\frac{r-1}{\alpha+1}}\right]^{-1}}{2\left[\frac{1}{K_1^2}\exp\left\{\frac{-2mt^{1-n}}{1-n}\right\} - \frac{mn}{t^{n+1}}\right]\left[-Ce^{-Dt^{1-n}}\left(B + Ce^{-Dt^{1-n}}\right)^{\frac{1}{1+\alpha}-1} + \frac{3m\eta_0}{t^n}\left(B + Ce^{-Dt^{1-n}}\right)^{\frac{r}{\alpha+1}}\right]^{-1}}, \tag{35}$$

Mass density parameter and the density parameter of the vacuum are given by

$$\Omega_M = \frac{\rho}{\rho_c}, \quad \Omega_\Lambda = \frac{\rho_v}{\rho_c}$$

for the anisotropic Bianchi type V model can be expressed respectively as

$$\Omega_M = \frac{2t^{2n}\left[B+Ce^{-Dt^{1-n}}\right]^{\frac{1}{1+\alpha}}\left[\frac{1}{K_1^2}\exp\left\{\frac{-2mt^{1-n}}{1-n}\right\}-\frac{mn}{t^{n+1}}\right]}{3m^2\left[-Ce^{-Dt^{1-n}}\left(B+Ce^{-Dt^{1-n}}\right)^{\frac{1}{1+\alpha}-1}+\frac{3mn_0}{t^n}\left(B+Ce^{-Dt^{1-n}}\right)^{\frac{r}{\alpha+1}}\right]} \tag{36}$$

$$\Omega_\Lambda = \frac{t^{2n}}{3m^2}\left(\frac{3m^2}{t^{2n}}-\frac{3}{K_1^2}\exp\left\{\frac{-2mt^{1-n}}{1-n}\right\}-2\left[\frac{1}{K_1^2}\exp\left\{\frac{-2mt^{1-n}}{1-n}\right\}-\frac{mn}{t^{n+1}}\right]\right.$$

$$\left.\cdot\left[-Ce^{-Dt^{1-n}}\left(B+Ce^{-Dt^{1-n}}\right)^{-1}+\frac{3mn_0}{t^n}\left(B+Ce^{-Dt^{1-n}}\right)^{\frac{r-1}{\alpha+1}}\right]^{-1}\right) \tag{37}$$

The State finder parameters $r=\dfrac{\ddot{R}}{RH^3}$ and $s=\dfrac{r-1}{3\left(q-\dfrac{1}{2}\right)}$.

For this model

$$r = 1 - \frac{3n}{mt^{1-n}} + \frac{n(n+1)}{m^2t^{2n-2}} \tag{38}$$

$$s = \frac{2n(n+1)t^{n-1}-6mn}{-9m^2t^{1-n}-6mn} \tag{39}$$

3.2. Case II: Causal Cosmological Solution

In addition to physically plausible relations (15)-(17), in this case we assume

$$\Lambda = \beta H^2. \tag{40}$$

where H is Hubble parameter, given by

$$H = \frac{\dot{R}}{R} \quad \text{and} \quad R = \left(R_1 R_2 R_3\right)^{1/3}. \tag{41}$$

From Equation (17)-(19) and (41), the Hubble parameter is given by

$$H = \frac{m}{t^n} \tag{42}$$

Using equations (17)-(19), (40) and (42) in equation (8), we get

$$8\pi G\rho = \frac{(3-\beta)m^2}{t^{2n}} - \frac{3}{K_1^2}\exp\left\{\frac{-2mt^{1-n}}{1-n}\right\}, \tag{43}$$

From Equations (21) and (43),

$$G = \frac{1}{8\pi}\left[B+C\exp\left(-Dt^{1-n}\right)\right]^{\frac{-1}{\alpha+1}}\left[\frac{(3-\beta)}{t^{2n}}-\frac{3}{K_1^2}\exp\left(D_1t^{1-n}\right)\right] \tag{44}$$

where $D_1 = \dfrac{-2m}{1-n}$.

From **Figure 5** one can easily see that gravitational constant is increasing with cosmic time. Substitute the values from Equations (17)-(19), (40) and (44) in Equation (5), we get

$$\Pi = \left[\frac{(3-\beta)m^2}{t^{2n}} - \frac{2mn}{t^{n+1}} - \frac{1}{K_1^2} \exp\left(D_1 t^{1-n}\right) \right] \cdot \frac{-1}{8\pi G} + \frac{B}{\rho^\alpha}, \tag{45}$$

By use of Equation (21), Equation (44) gives

$$\Pi = \frac{B}{\left[B + C\exp\left(-Dt^{1-n}\right)\right]^{\frac{\alpha}{\alpha+1}}} - \frac{U_1(t)}{U_2(t)}\left[B + C\exp\left(-Dt^{1-n}\right)\right]^{\frac{1}{\alpha+1}} \tag{46}$$

where $U_1(t) = \left[\frac{(3-\beta)m^2}{t^{2n}} - \frac{2mn}{t^{n+1}} - \frac{1}{K_1^2}\exp\left(D_1 t^{1-n}\right)\right]$, $U_2(t) = \left[\frac{(3-\beta)}{t^{2n}} - \frac{3}{K_1^2}\exp\left(D_1 t^{1-n}\right)\right]$

Figure 6 shows that bulk viscous stress is decreasing with the evolution of the universe.

3.2.1. Sub Cease (i): Evaluation of Bulk Viscosity in Truncated Causal Theory

Now we study variation of bulk viscosity coefficient η and relaxation time τ with respect to the cosmic time. It has already been mentioned that for truncated theory $\varepsilon = 0$ and hence Equation (13) reduces to

$$\Pi + \tau\dot{\Pi} = -3\eta H. \tag{47}$$

In order to have exact solution of the system of equations one more physically plausible relation is required. Thus, we consider the well known relation

$$\tau = \frac{\eta}{\rho}. \tag{48}$$

Using Equations (17)-(19), (46) and (48) in Equation (47) one can obtain coefficient of bulk viscosity as

$$\eta = \cfrac{-B\left[B + C\exp\left(-Dt^{1-n}\right)\right]^{\frac{-\alpha}{\alpha+1}} + U_1(t)\left[U_2(t)\right]^{-1}\left[B + C\exp\left(-Dt^{1-n}\right)\right]^{\frac{1}{\alpha+1}}}{-\cfrac{U_1'(t)}{U_2(t)} + \cfrac{U_1(t)U_2'(t)}{\left[U_2(t)\right]^2} - \cfrac{U_1(t)}{U_2(t)}\frac{3mC}{t^n}\exp\left(-Dt^{1-n}\right)\left[B + C\exp\left(-Dt^{1-n}\right)\right]^{-1}} \tag{49}$$

$$-\frac{3\alpha BCm}{t^n}\exp\left(-Dt^{1-n}\right)\left[B + C\exp\left(-Dt^{1-n}\right)\right]^{-2} + \frac{3m}{t^n}$$

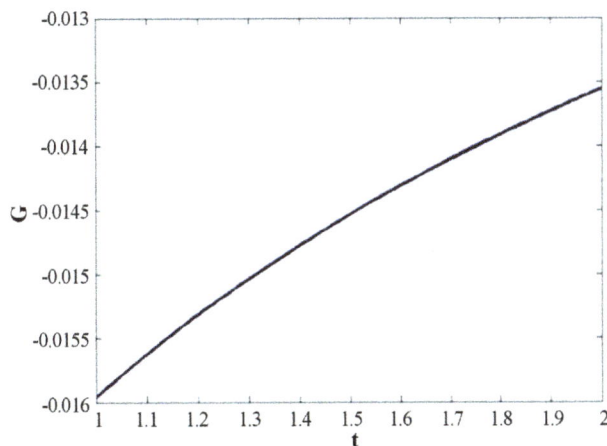

Figure 5. This figure shows variation of gravitational constant with respect to cosmic time t. Here we consider $B = 1, C = 1, n = 1.5, m = 2, \alpha = 1, \beta = 1$.

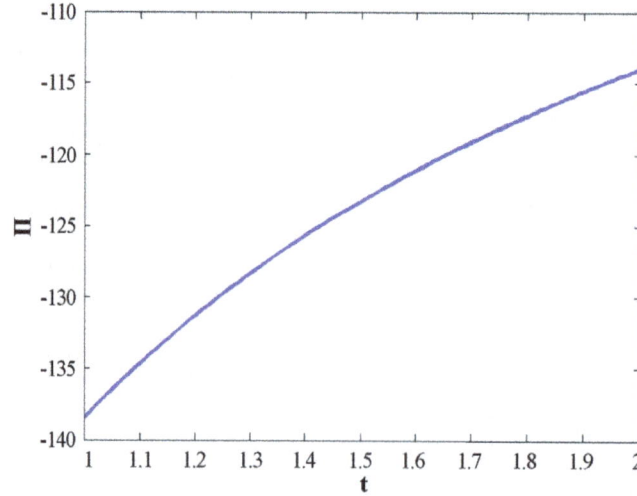

Figure 6. This figure shows variation of bulk viscous stress with respect to cosmic time t. Here we consider $B = 1$, $C = 1$, $n = 1.5$, $m = 2$, $\alpha = 1$, $\beta = 1$, and $K_1 = 1$.

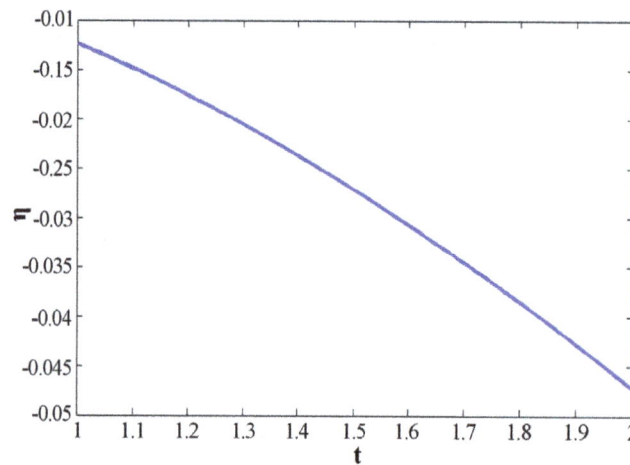

Figure 7. This figure shows variation of bulk viscosity coefficient with respect to cosmic time t. Here we consider $B = 1$, $C = 1$, $n = 1.5$, $m = 2$, $\alpha = 1$.

From **Figure 7** one can see that bulk viscosity coefficient is decreasing with time.

3.2.2. Sub Caese (ii). Evaluation of Bulk Viscosity in Full Causal Theory

It has already been mentioned that for full causal theory $\varepsilon = 1$ and hence Equation (13) reduces to

$$\Pi + \tau\dot{\Pi} = -3\eta H - \frac{\tau\Pi}{2}\left(3H - \frac{\dot{\tau}}{\tau} - \frac{\dot{\eta}}{\eta} - \frac{\dot{T}}{T}\right).$$

(50)

On the basis of Gibb's inerrability condition, Maartens [40] has suggested the equation of state for temperature as

$$T \propto \exp\int\frac{dp}{\rho + p},$$

(51)

which with the help of Equation (21) gives

$$T = T_0 \left[1 - B\rho^{-(\alpha+1)} \right]^{\frac{\alpha}{\alpha+1}}. \tag{52}$$

Figure 8 shows that temperature is decreasing with evolution of the universe. using Equations (21), (42), (48) and (52) in Equation (50) one can obtain

$$\Pi + \frac{\eta}{\rho}\dot{\Pi} = -\eta\frac{2m_1 + m_2}{t^n} - \frac{\eta\Pi}{2\rho}\left[\frac{2m_1 + m_2}{t^n} - \frac{\dot{\rho}}{\rho} - \frac{\dot{T}}{T} \right],$$

which on simplification yields the expression for bulk viscosity as

$$\eta = \frac{-B\left[B + C\exp\left(-Dt^{1-n}\right)\right]^{\frac{-\alpha}{\alpha+1}} + U_1(t)\left[U_2(t)\right]^{-1}\left[B + C\exp\left(-Dt^{1-n}\right)\right]^{\frac{1}{\alpha+1}}}{\dfrac{3m}{t^n} + \dfrac{\dot{\Pi}}{\rho} + \left[\dfrac{B}{\left(B + C\exp\left(-Dt^{1-n}\right)\right)} - \dfrac{U_1(t)}{U_2(t)}\right]\left[\dfrac{3m(1+\alpha)}{t^n} + \dfrac{3mC(1+\alpha)\exp\left(-Dt^{1-n}\right)}{t^n\left(B + C\exp\left(-Dt^{1-n}\right)\right)}\right]} \tag{53}$$

where

$$\frac{\dot{\Pi}}{\rho} = -\frac{U_1'(t)}{U_2(t)} + \frac{U_1(t)U_2'(t)}{\left[U_2(t)\right]^2} - \frac{U_1(t)}{U_2(t)}\frac{3mC}{t^n}\exp\left(-Dt^{1-n}\right)\left[B + C\exp\left(-Dt^{1-n}\right)\right]^{-1}$$

$$-\frac{3\alpha BCm}{t^n}\exp\left(-Dt^{1-n}\right)\left[B + C\exp\left(-Dt^{1-n}\right)\right]^{-2}$$

Figure 9 shows that bulk viscosity coefficient decreasing with evolution of universe.

4. Conclusion

In this paper, we have studied bulk viscous Bianchi type V space-time geometry with generalized Chaplygin gas and varying gravitational and cosmological constants. We have obtained a new set of exact solutions of Einstein's equations by considering $\dfrac{\dot{R_1}}{R_1} = \dfrac{\dot{R_2}}{R_2} = \dfrac{m}{t^n}$. For $n > 1$, the deceleration parameter $q < 0$ for $t > (n/m)^{\frac{1}{1-n}}$.

When $n \to 1$ considering present day limit for deceleration parameter $q = -0.53^{+0.17}_{-0.13}$ [40] suggests $1.56 \leq m \leq 2.94$. It is observed that in case I energy density, bulk viscosity and cosmological constant decrease where as gravitational constant G(t) is increasing with time. In case II, bulk viscosity η, bulk viscous stress Π and temperature T decrease with evolution of the universe which agrees with cosmic observations. In order to

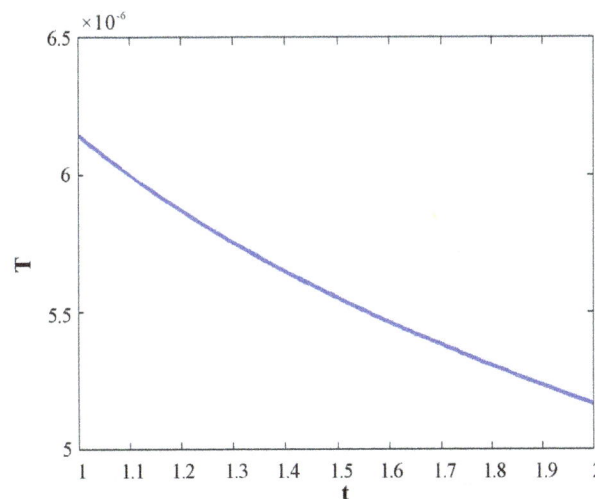

Figure 8. This figure shows variation of temparature with respect to cosmic time t. Here we consider $B = 1$, $C = 1$, $n = 1.5$, $m = 2$ and $\alpha = 1$.

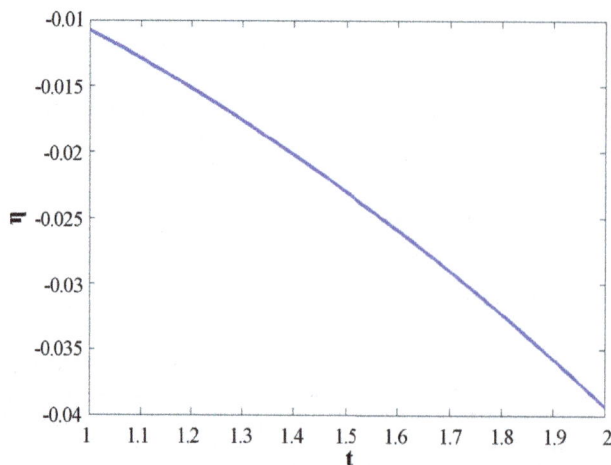

Figure 9. This figure shows variation of bulk viscosity coefficient with respect to cosmic time t. Here we consider B = 1, C = 1, n = 1.5, m = 2, $\alpha = 1$.

have clear idea of variation in behavior of cosmological parameters, relevant graphs have been plotted; all graphs are in fair agreement with cosmological observations.

References

[1]   Misner, C.W. (1968) The Isotropy of the Universe. *Astrophysical Journal*, **151**, 431-457. http://dx.doi.org/10.1086/149448

[2]   Farnsworth, D.L. (1967) Some New General Relativistic Dust Metrics Possessing Isometries. *Journal of Mathematical Physics*, **8**, 2315. http://dx.doi.org/10.1063/1.1705157

[3]   Collins, C.B. (1974) Tilting at Cosmological Singularities. *Communications in Mathematical Physics*, **39**, 131-151. http://dx.doi.org/10.1007/BF01608392

[4]   Maartens, R. and Nel, S.D. (1978) Decomposable Differential Operators in a Cosmological Context. *Communications in Mathematical Physics*, **59**, 273-283. http://dx.doi.org/10.1007/BF01611507

[5]   Wainwright, J., Ince, W.C. and Marshman, W. (1979) Spetially Himigeneous and Inhomogeneous Cosmologies with Equation of State $p = \mu$. *General Relativity and Gravitation*, **10**, 259-271. http://dx.doi.org/10.1007/BF00759860

[6]   Roy, S.R. and Singh, J.P. (1983) LRS Bianchi Type-V Universes Filled with Matter and Radiation. *Astrophysics and Space Science*, **96**, 303-312. http://dx.doi.org/10.1007/BF00651674

[7]   Banerjee, A. and Sanyal, A.K. (1988) Irrotational Bianchi Type V Viscous Fluid Cosmology with Heat Flux. *General Relativity and Gravitation*, **20**, 103-113. http://dx.doi.org/10.1007/BF00759320

[8]   Coley, A.A. (1990) Conformal Killing Vectors and FRW Spacetimes. *General Relativity and Gravitation*, **22**, 241-251. http://dx.doi.org/10.1007/BF00756274

[9]   Roy, S.R. and Prasad, A. (1994) Some L.R.S. Bianchi Type V Cosmological Models of Local Embedding Class One. *General Relativity and Gravitation*, **19**, 939-950. http://dx.doi.org/10.1007/BF02106663

[10] Nayak, B.K. and Sahoo, B.K. (1989) Bianchi Type V Model with the Matter Distribution Admitting Anisotropic Pressure and Heat Flow. *General Relativity and Gravitation*, **21**, 211-225. http://dx.doi.org/10.1007/bf00764095

[11] Bali, R. and Meena, B.L. (2004) Conformally Flat Tilted Bianchi Type-V Cosmological Models in General Relativity. *Pramana: Journal of Physics*, **62**, 1007-1014.

[12] Bali, R. and Yadav, M.K. (2005) Bianchi Type-IX Viscous Fluid Cosmological Model in General Relativity. *Pramana: Journal of Physics*, **64**, 187-196.

[13] Bali, R. and Singh, D.K. (2005) Bianchi Type V Bulk Viscous Fluid String Dust Cosmological Model in General Relativity. *Astrophysics and Space Science*, **300**, 387-394.

[14] Bali, R. and Tinker, S. (2009) Bianchi Type III Bulk Viscous Barotropic Fluid Cosmological Models with Variable G and Lambda. *Chinese Physics Letters*, **26**, Article ID: 029802. http://dx.doi.org/10.1088/0256-307X/26/2/029802

[15] Rajbali, S.T. (2008) Bianchi Type V Bulk Viscous Barotropic Fluid Cosmological Models with Variable G and Lambda. *Chinese Physics Letters*, **25**, 3090-3093. http://dx.doi.org/10.1088/0256-307X/25/8/095

[16] Yadav, A.K. (2013) Anisotropic Massive Strings in the Scalar Tensor Theory of Gravitation. *Research in Astronomy and Astrophysics*, **13**, 772-782. http://dx.doi.org/10.1088/1674-4527/13/7/002

[17] Yadav, A.K., Pradhan, A. and Singh, A.K. (2012) Bulk Viscous LRS Bianchi Type I Universe with Variable G and Lambda. *Astrophysics and Space Science*, **337**, 379-385. http://dx.doi.org/10.1007/s10509-011-0814-7

[18] Ellis, G.F.R. (1979) In: Schas, R., Ed., *General Relativity and Cosmology*, Enrico Fermi Course, Academic Press, New York, 47. Misner, C.W. (1968) The Isotropy of the Universe. *Astrophysical Journal*, **151**, 431.

[19] Fabris, J.C., Concalves, S.V.B. and de Sa'Ribeiro, R. (2006) Bulk Viscosity Driving the Acceleration of the Universe. *General Relativity and Gravitation*, **38**, 495-506. http://dx.doi.org/10.1007/s10714-006-0236-y

[20] Kotambkar, S., Singh, G.P. and Kelkar, R.K. (2014) Anisotropic Cosmological Models with Quintessence. *International Journal of Theoretical Physics*, **53**, 449-460. http://dx.doi.org/10.1007/s10773-013-1829-3

[21] Krauss, L.M. and Turner, M.S. (1995) The Cosmological Constant Is Back. *General Relativity and Gravitation*, **27**, 1137-1144. http://dx.doi.org/10.1007/bf02108229

[22] Riess, A.G., Filippenko, A.V., Challis, P., Clocchiatti, A., Diercks, A., Garnavich, P.M., *et al.* (1998) Observational Evidence from Supernovae for an Accelerating Universe and a Cosmological Constant. *The Astronomical Journal*, **116**, 1009-1038. http://dx.doi.org/10.1086/300499

[23] Perlmutter, S., Aldering, G., Goldhaber, G., Knop, R.A., Nugent, P., Castro, P.G., *et al.* (1999) Measurements of Omega and Lambda from 42 High Redshift Supernovae. *The Astronomical Journal*, **517**, 565-586. http://dx.doi.org/10.1086/307221

[24] Lima, J.A.S. and Trodden, M. (1996) Decaying Vacuum Energy and Deflationary Cosmology in Open and Closed Universe. *Physical Review D*, **53**, 4280-4286. http://dx.doi.org/10.1103/physrevd.53.4280

[25] Bennett, C.L., Halpern, M., Hinshaw, G., Jarosik, N., Kogut, A., Limon, M., *et al.* (2003) First Year Wilkinson Microwave Anisotropy Prob (WMAP) Observations—Preliminary Maps and Basic Results. *Astrophysical Journal Supplement*, **148**, 1-27.

[26] Sahni, V. and Starobinski, A. (2000) The Case for a Positive Cosmological Lambda Term. *International Journal of Modern Physics D*, **9**, 373-444.

[27] Canuto, V.M. and Narlikar, J.V. (1980) Cosmological Tests of the Hoyle-Narlikar Conformal Gravity. *The Astrophysical Journal*, **236**, 6-23. http://dx.doi.org/10.1086/157714

[28] Singh, G.P. and Kotambkar, S. (2001) Higher Dimensional Cosmological Model with Gravitational and Cosmological "Constants". *General Relativity and Gravitation*, **33**, 621-630. http://dx.doi.org/10.1023/A:1010278213135

[29] Singh, G.P. and Kotambkar, S. (2003) Higher Dimensional Dissipative Cosmology with Varying G and Lambda. *Gravitation and Cosmology*, **9**, 206-210.

[30] Singh, G.P., Kotambkar, S. and Pradhan, A. (2008) A New Class of Higher Dimensional Cosmological Models of Universe with Variable G and Lambda Term. *Romanian Journal of Physics*, **53**, 607-618.

[31] Pradhan, A. and Pandey, P. (2006) Some Bianchi Type I Viscous Fluid Cosmological Models with a Variable Cosmological Constant. *Astrophysics and Space Science*, **301**, 127-134.
Pradhan, A., Singh, A.K. and Otarod, S. (2007) FRW Universe with Variable G and Lambda Term. *Romanian Journal of Physics*, **52**, 445-458.

[32] Singh, C.P. and Kumar, S. (2006) Bianchi Type II Cosmological Models with Constant Deceleration Parameter. *International Journal of Modern Physics D*, **15**, 419-438. http://dx.doi.org/10.1142/s0218271806007754

[33] Singh, C.P., Kumar, S. and Pradhan, A. (2007) Early Viscous Universe with Variable Gravitational and Cosmological Constant. *Classical and Quantum Gravity*, **24**, 455-474. http://dx.doi.org/10.1088/0264-9381/24/2/011

[34] Singh, J.P. and Tiwari, S.K. (2008) Perfect Fluid Bianchi Type I Cosmological Constant Models with Time Varying G and Lambda. *Pramana*, **70**, 565-574.

[35] Singh, G.P. and Kale, A.Y. (2009) Anisotropic Bulk Viscous Cosmological Models with Variable G and Lambda. *International Journal of Theoretical Physics*, **48**, 1177-1185. http://dx.doi.org/10.1007/s10773-008-9891-y

[36] Bali, R. and Tinkar, S. (2009) Bianchi Type III Bulk Viscous Barotropic Fluid Cosmological Models with Variable G and Lambda. *Chinese Physics Letters*, **26**, Article ID: 029802. http://dx.doi.org/10.1088/0256-307X/26/2/029802

[37] Verma, M.K. and Ram, S. (2011) Spatially Homogeneous Bulk Viscous Fluid Models with Time-Dependent Gravitational Constant and Cosmological Term. *Advanced Studies in Theoretical Physics*, **5**, 387-398.

[38] Kamenshchik, A., Moschella, U. and Pasquier, V. (2001) An Alternative to Quintessence. *Physics Letters B*, **511**, 265-268. http://dx.doi.org/10.1016/S0370-2693(01)00571-8

[39] Bento, M.C., Bertolami, O. and Sen, A.A. (2003) WMAP Constraints on the Generalized Chaplygin Gas Model. *Physics Letters B*, **575**, 172-180. http://dx.doi.org/10.1016/j.physletb.2003.08.017

[40] Maartens, R. (1995) Dissipative Cosmology. *Classical and Quantum Gravity*, **12**, 1455-1465. http://dx.doi.org/10.1088/0264-9381/12/6/011

26

Mach's Principle of Inertia Is Supported by Recent Astronomical Evidence

Morley B. Bell

National Research Council of Canada, Ottawa, Canada
Email: morley.bell@nrc-cnrc.gc.ca

Abstract

Inertial mass is detected on Earth only when matter is accelerated or decelerated. Recently evidence has been reported for a low-level velocity oscillation with a period of 39 ± 1 Mpc (127 ± 3 Myr) superimposed on the Hubble flow. Like the Hubble flow, this oscillation is assumed to be an expansion and contraction of space itself. If space is oscillating as it expands and the Hubble flow contains a superimposed velocity ripple, matter on Earth will experience alternating accelerations and decelerations relative to the rest of the matter in the Universe. The acceleration curve can be obtained from the velocity oscillation curve simply by taking the magnitude of the derivative of the velocity curve and the acceleration curve is found here to have a period of 63.5 ± 1.5 Myr. Evidence has also been claimed recently for a ubiquitous ~62 ± 3 Myr periodic fluctuation superimposed on general trends in the fossil biodiversity on Earth. The periods of the acceleration curve oscillation and fossil biodiversity fluctuations are thus identical within the errors. A second, weaker fluctuation is also detected in both the Hubble flow and fossil biodiversity trends. They too have identical periods of ~140 Myr. From this excellent agreement, it is argued here that it is the oscillation in the Hubble flow, through an inertia-like phenomenon involving all the matter in the universe that has produced the fluctuations in the fossil biodiversity on Earth. This may represent the first instance where observational evidence supporting Mach's Principle of Inertia has been found.

Keywords

Galaxies, Active-Galaxies, Distances and Redshifts

1. Introduction

Ernst Mach argued that inertia could only be explained if all the masses in the universe were somehow connected, and the means by which such action-at-a-distance can occur has had a long history in physics [1] [2].

Phipps [3] claimed that Philipp Frank, a Hitler-refugee philosopher of science and physicist who succeeded Einstein in his academic appointment at Prague, attributed to Mach himself the following wording to describe his principle: "When the subway train jerks, it's the fixed stars that throw you down". Phipps further points out that if the fixed stars are doing what Mach claims, they are doing it right now via action-at-a-distance. They are not waiting millennia for signals of any kind to propagate. However, from studying electromagnetic and gravitational field propagation, we conclude that neither of these proceeds instantaneously. Although the Weak Equivalence Principle finds gravitational mass similar to inertial mass, neither is well understood. However, perhaps the most important thing to note is that Inertial Mass is tied to accelerated motion while the former is not. A possible explanation for instantaneous action-at-a-distance that relates to advanced and retarded potentials has been discussed by [4].

To date, there have been no observations found that would suggest that there might be a direct link between two physical phenomena through Mach's Principle of inertia. This may not be surprising since, in Mach's Principle, at least one of the two variables must be all of the rest of the matter in the universe. Recently, evidence was reported for a low-level oscillation with a period of 39 ± 1 Mpc superimposed on top of the Hubble flow [5] when intrinsic redshift components were identified and removed from the redshifts of the SNeIa sources studied by [6]. This result is the culmination of many years of work on intrinsic redshifts in quasars and galaxies by us [7]-[13], and by Tifft [14] [15], whose work on galaxies finds similar intrinsic components. When the intrinsic components have been identified, they can be removed. If the source distances are accurate, when the Vcmb velocities of the sources are plotted versus distance, any superimposed oscillation then becomes visible if it is larger than the source peculiar velocities. This can be seen here in **Figure 1** and in Fig. 3 and Fig. 4 of [5] and [16] where the oscillation can be seen to be visible over 6 consecutive cycles.

In the Big Bang model, the cosmological component of the redshift of a source is due to the expansion of space and not to the motion of the object through space. Because of this, the cosmological component of the redshift increases by Ho km/s^{-1}/Mpc^{-1}, where Ho is the Hubble constant. This produces the Hubble slope. If the

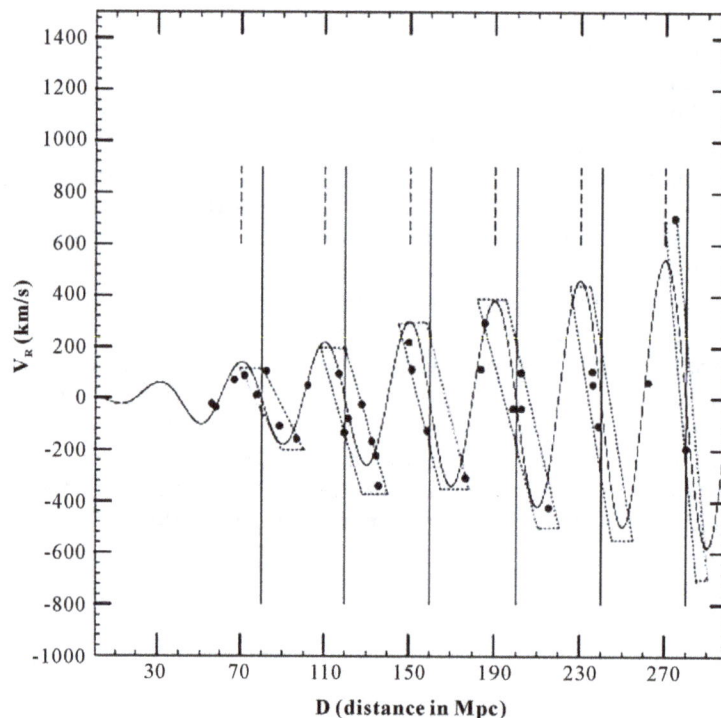

Figure 1. Plot of residual velocity, V_r, versus distance for SNeIa galaxies after removal of intrinsic redshifts and Hubble slope of 57.9 km/s^{-1}/Mpc^{-1} [5] [16]. Regions of high source density are indicated by the dotted rectangles and are centered near 80, 120, 160, 200, 240 and 260 Mpc, as indicated by the solid vertical lines, and are in phase quadrature with the velocity peaks indicated by the dashed vertical lines.

velocity oscillation that we have observed is truly an oscillation in the Hubble flow, as proposed here, for a constant amplitude ripple, its observed amplitude also must increase with distance. In **Figure 1**, which has been reproduced from Fig. 4 of [16], it is clear that the amplitude of the oscillation does increase with distance. As can be seen from **Figure 1**, the Hubble constant of the ripple is Hr = 540/270 = 2 km/s^{-1}/Mpc^{-1}. Although this increase is expected to continue to much larger distances, it may not be possible to identify the intrinsic components at large distances because of increasing uncertainties in the cosmological parameters. If the intrinsic components cannot be identified and removed, the ripple cannot be detected. Now, for the first time, the oscillations in the Hubble flow will appear to be something that represents all the matter in the universe, as is required to check Mach's Principle. However, to check Mach's Principle, we also require some matter on Earth that may be affected by these oscillations.

Recently a ubiquitous ~62 Myr periodic fluctuation superimposed on general trends in fossil biodiversity has been reported [17]-[20]. Since the age of the Earth is generally accepted to be 4.5 Gyr, detecting fluctuations in the fossil records with a 62 Myr period should not be hindered by the lack of an adequate baseline. Possible associations between this 62 Myr periodicity and the variation of the Sun around the galaxy have been claimed, invoking the modulation of cosmic rays, gamma rays, and comet impact frequency. A major problem with these models is obtaining accurate observational data from them that can be used to compare with the fossil biodiversity fluctuations. An excellent review of what has been done in this area is given by [21], who remain unconvinced that the fossil biodiversity fluctuations are likely to be related to these or other similar Galaxy-related explanations.

In case it is possible that all the matter in the Universe is somehow interconnected, as has been suggested in the inertia phenomenon, one can speculate that the fluctuations in the velocity of the Hubble flow with time might then influence life on Earth in real time, and it is therefore of interest to examine these results more closely. This is especially true in the case of the velocity oscillation where matter on Earth will be alternatively accelerated and decelerated relative to all the rest of the matter in the Universe during each velocity cycle. We now examine if the velocity oscillation we have observed to be superimposed on the Hubble flow might be the true source of the ~62 Myr periodic fluctuation superimposed on general trends in fossil biodiversity.

2. Analysis

First we note that the period of the velocity oscillation found to be superimposed on the Hubble flow was 127 Myr, while the fluctuations in the trends in fossil biodiversity on Earth have a period of 62 Myr, which differs by approximately a factor of two. However, since inertia is related to acceleration and deceleration, or the rate of change of velocity, of matter, it is more likely, if inertia is involved, that it is the rate of change of velocity of the Hubble flow oscillation (its derivative), that has affected the survival rate of Earth-based genera. Furthermore, it is unlikely that the direction of the change (acceleration or deceleration) would affect things differently and it is therefore more likely that the effects would be related to the magnitude of this parameter.

One cycle of the fluctuations seen in the velocity of the Hubble flow with a period of 127 Myr (39 Mpc) is represented in **Figure 2** by the solid curve. The dashed curve shows the magnitude of the derivative of this curve, which represents the magnitude of the rate of change of the velocity. The curve in **Figure 3** is the Fourier Transform of the dashed curve in **Figure 2** and shows clearly that the dominant frequency present in the dashed curve is 63.5 Myr (0.0157 cycles per Myr). This is essentially identical to the ubiquitous 62 ± 3 Myr periodic fluctuation found to be superimposed on the general trends in fossil biodiversity and, in turn, confirms, as speculated above, that the two are likely to be related. Thus, for the first time, the Hubble oscillation model provides us with observational data involving all the matter in the Universe and a baseline that can be compared directly with that covered in the fossil biodiversity case. When the magnitude of the derivative of the Hubble oscillation is used, which is necessary if inertia is involved, the periods found for the oscillations in the two cases can be seen to be identical within the uncertainties.

3. Weaker Fluctuations

Rohde and Muller [17] (see their Fig. 1e) reported that a second fluctuation with a period near 140 ± 15 Myrs might also be present in the fossil biodiversity trends. It was also reported previously by us [5] that a weaker fluctuation in the Hubble flow might also be present with a period near 87 Mpc, or 283 Myrs. Again, if it is the magnitude of the derivative of this fluctuation that is affecting the trends in fossil biodiversity we would expect

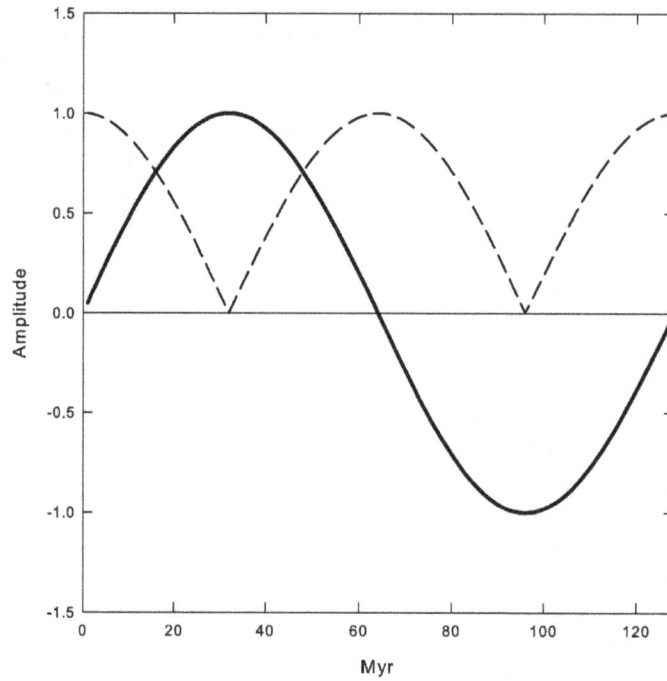

Figure 2. One cycle of the fluctuations seen in the density of the Universe and in the Hubble flow with a period of 127 Myr (39 Mpc) is represented here by the solid curve. The dashed curve shows the magnitude of the derivative of this curve, which then represents the magnitude of the rate of change of the velocity of all the mass in the Universe relative to the matter on Earth.

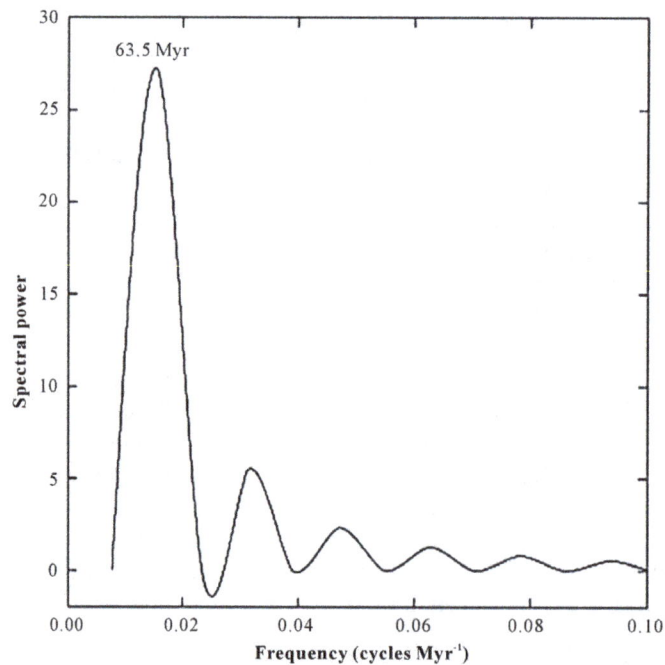

Figure 3. Fourier Transform of the dashed curve in **Figure 2** showing clearly that the strongest frequency present in the dashed curve is 0.0157 cycles per Myr, which corresponds to a period of 63.5 Myr. This is essentially identical to the ubiquitous 62 ± 3 Myr periodic fluctuation found to be superimposed on the general trends in fossil biodiversity.

it to appear there with a period of one-half this, or 141 Myr. These results are shown in **Figure 4** where the curve has been obtained by subtracting best-fit sinusoids from the SNeIa data, recalculating the RMS value, as described previously [5], and converting the fluctuation period from Mpc to Myr using the relation Py (Myr) = [3.26/2] Pp (Mpc). Here the factor 3.26 converts from Mpc to Myr and the factor 2 takes into account the fact that it is the magnitude of the derivative of the fluctuation that is important. In **Figure 4** there is excellent agreement between the two periods found in the biodiversity trends (62 and 140 Myrs shown by the vertical dashed lines) and the periods (63.5 and 141 Myrs) found in the Hubble flow assuming that it is the magnitude of the derivative of the oscillation that is the driving force. In both cases (biodiversity and Hubble flow) the short period fluctuations are stronger than the long period ones. This excellent agreement in both period and relative amplitude suggests strongly that it is the oscillation in the Hubble flow that is the source of the fluctuations detected in the fossil biodiversity. The excellent agreement, in both period and relative amplitude, makes it very unlikely that this has occurred by chance.

4. Instantaneous Communication

Previously, we [5] [16] used the source distance (look-back time) to show from changes in the density and velocity fields that the Hubble flow was oscillating as it was expanding. This assumed only that the radiation used to detect the sources was propagating at the speed of light. From the results obtained here, if the periodicity seen in the fossil biodiversity is related to changes in the density or velocity fields in the Universe through some kind of inertia-like phenomenon, the effects of these changes would have to be communicated instantaneously. If a communications travel time were involved, such as the speed of light, the time-related effects from matter located at different distances would introduce a smearing effect that would prevent any significant periodic effect on matter located on Earth from being detected. This observation then has the same instantaneous communication requirement as Mach's Principle with the changes in the density and velocity being instantaneously felt by all life types (genera) on Earth.

If this action-at-a-distance is related to the quantum entanglement phenomenon, which requires that all the matter in the Universe at one time had to be connected, we might argue that this picture then agrees well with the Big Bang model. For further discussion see [22].

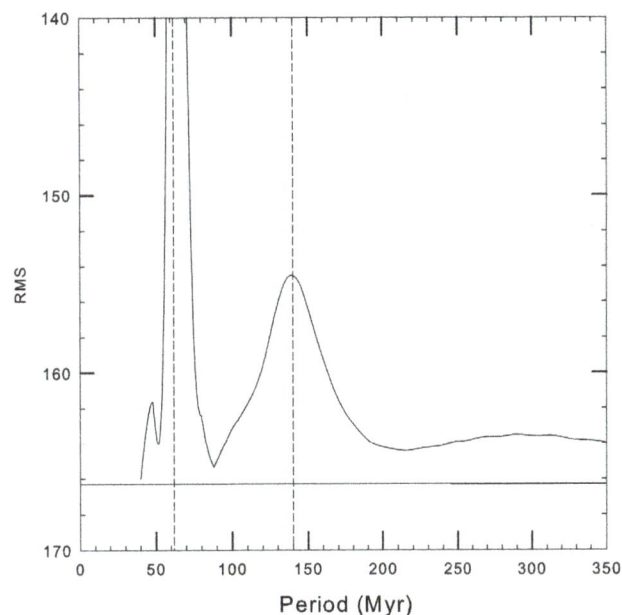

Figure 4. The solid curve represents the ripple peaks found previously for the SNeIa galaxies expressed in millions of years, assuming it is the magnitude of the derivative of the Hubble flow that is the driving force. See text for an explanation of how this curve was obtained. Vertical dashed lines show the periods of 62 and 140 Myr that have already been reported in fossil biodiversity trends.

5. Conclusion

It is shown here that the two dominant periods present in the fluctuations reported in the trends in fossil biodiversity of Earth-based genera (62 ± 3 and 140 ± 15 Myr) are essentially identical to the periods of the derivatives of the two strongest oscillations seen to be superimposed on the velocity of the Hubble flow (63.5 and 141 Myr). We conclude that it is the oscillation in the Hubble flow which has produced the fossil biodiversity fluctuations. For these effects to be somehow related requires that there be an instantaneous, or action-at-a-distance, relation between the matter on Earth and all the rest of the matter in the universe, as is the case for Mach's Principle. Finally, it is concluded that the results obtained here may be the first empirical evidence supporting Mach's Principle of inertia.

Acknowledgements

I thank Simon Comeau for comments on the text and assistance with the figures.

References

[1]   Sciama, D.W. (1953) On the Origin of Inertia. *Monthly Notices of the Royal Astronomical Society*, **113**, 34-42. http://dx.doi.org/10.1093/mnras/113.1.34

[2]   Hoyle, F. and Narlikar, J.V. (1995) Cosmology and Action-at-a-Distance Electrodynamics. *Reviews of Modern Physics*, **67**, 113-155. http://dx.doi.org/10.1103/RevModPhys.67.113

[3]   Phipps, T.E. (1999) Meditations on Action-at-a-Distance. In: Chubykalo, A.E., Pope, V. and Smirnov-Rueda, R., Eds., *Instantaneous Action-at-a-Distance in Modern Physics*: *Pro and Contra*, Nova Science, Commack, 137-156.

[4]   Narlikar, J.V. (1999) Actions at a Distance in Electrodynamics and Inertia. In: Chubykalo, A.E., Pope, V. and Smirnov-Rueda, R., Eds., *Instantaneous Action-at-a-Distance in Modern Physics*: *Pro and Contra*, Nova Science, Commack, 19-34.

[5]   Bell, M.B. (2013) Interesting Evidence for a Low-Level Oscillation Superimposed on the Local Hubble Flow. *Astrophysics and Space Science*, **344**, 471-477. http://dx.doi.org/10.1007/s10509-012-1344-7

[6]   Freedman, W.L., Madore, B.F., Gibson, B.K., Ferrarese, L., Kelson, D.D., Sakai, S., Mould, J.R., Kennicutt, R.C., Ford, H.C., Graham, J.A., Huchra, J.P., Hughes, S.M.G., Illingworth, J.D., Macri, L.M. and Stetson, P.B. (2001) Final Results from the Hubble Space Telescope Key Project to Measure the Hubble Constant. *Astrophysical Journal*, **553**, 47-72. http://dx.doi.org/10.1086/320638

[7]   Bell, M.B. (2002) Evidence for Large Intrinsic Redshifts. *Astrophysical Journal*, **566**, 705-711. http://dx.doi.org/10.1086/338272

[8]   Bell, M.B. (2002) Quasar Distances and Lifetimes in a Local Model. *Astrophysical Journal*, **567**, 801-810. http://dx.doi.org/10.1086/338754

[9]   Bell, M.B. (2002) Evidence that an Intrinsic Component that is a Harmonic of $z = 0.062$ May be Present in Every Quasar Redshift. arXiv:0208320.

[10] Bell, M.B. (2002) Discrete Intrinsic Redshifts from Quasars to Normal Galaxies. arXiv:0211091.

[11] Bell, M.B. and Comeau, S.P. (2003) Intrinsic Redshifts and the Hubble Constant. http://arxiv.org/abs/astro-ph/0305060

[12] Bell, M.B., Comeau, S.P. and Russell, D.G. (2004) Discrete Components in the Radial Velocities of ScI Galaxies. http://arxiv.org/abs/astro-ph/0407591

[13] Bell, M.B. (2007) Further Evidence That the Redshifts of AGN Galaxies May Contain Intrinsic Components. *Astrophysical Journal Letters*, **667**, L129-L132. http://dx.doi.org/10.1086/522337

[14] Tifft, W.G. (1996) Global Redshift Periodicities and Periodicity Structure. *Astrophysical Journal*, **468**, 491-518. http://dx.doi.org/10.1086/177710

[15] Tifft, W.G. (1997) Global Redshift Periodicities and Variability. *Astrophysical Journal*, **485**, 465-483. http://dx.doi.org/10.1086/304443

[16] Bell, M.B. and Comeau, S.P. (2014) More Evidence for an Oscillation Superimposed on the Hubble Flow. *Astrophysics and Space Science*, **349**, 337-442. http://dx.doi.org/10.1007/s10509-013-1601-4

[17] Rohde, R.A. and Muller, R.A. (2005) Cycles in Fossil Diversity. *Nature*, **434**, 208-210.

[18] Melott, A.L. and Bambach, R.K. (2010) A Ubiquitous 62 Myr Periodic Fluctuation Superimposed on General Trends in Fossil Biodiversity: II, Evolutionary Dynamics Associated with Periodic Fluctuation in Marine Diversity. http://arxiv.org/abs/1011.4496

[19] Melott, A.L. and Bambach, R.K. (2011) A Ubiquitous 62 Myr Periodic Fluctuation Superimposed on General Trends in Fossil Biodiversity: Documentation. *Paleobiology*, **37**, 92-112. http://dx.doi.org/10.1666/09054.1

[20] Melott, A.L. and Bambach, R.K. (2011) A Ubiquitous 62 Myr Periodic Fluctuation Superimposed on General Trends in Fossil Biodiversity: II, Evolutionary Dynamics Associated with Periodic Fluctuation in Marine Diversity. *Paleobiology*, **37**, 383-408. http://dx.doi.org/10.1666/09055.1

[21] Feng, F. and Bailer-Jones, C.A.L. (2013) Assessing the Influence of the Solar Orbit on Terrestrial Biodiversity. *Astrophysical Journal*, **768**, 152-173. http://dx.doi.org/10.1088/0004-637X/768/2/152

[22] Buniy, R.V. and Hsu, S.D.H. (2012) Everything Is Entangled. *Physics Letters B*, **718**, 233-236. http://dx.doi.org/10.1016/j.physletb.2012.09.047

27

Five Dimensional String Universes in Lyra Manifold

Mahbubur Rahman Mollah[1], Kangujam Priyokumar Singh[2], Koijam Manihar Singh[3]

[1]Department of Mathematics, Commerce College Kokrajhar, Kokrajhar, BTC, Assam, India
[2]Department of Mathematical Sciences, Bodoland University, Kokrajhar, BTC, Assam, India
[3]Department of Mathematics Sciences, National Institute of Technology Manipur, Imphal, India
Email: mr.mollah123@gmail.com, pk_mathematics@yahoo.co.in, drmanihar@rediffmail.cim

Abstract

Considering five dimensional plane symmetric metric, we discuss a model universe with different situations, by solving the modified Einstein field equations within the framework of Lyra geometry. We obtain many interesting realistic solutions governing the present day model of the universe. Physical and kinematical properties of the models are discussed in detail.

Keywords

Cosmic Strings, Lyra Geometry, Dark Energy, Evolution, Clouds, Early Universe, Dark Matter

1. Introduction

A lot of remarkable knowledge of cosmology is made by various experimental and theoretical results which have been made still today. But still now it is difficult to explain exactly the physical situation of the formation of our universe at the very early stage. To describe the events at the early stages of the universe, we are required to develop and study the concept of string theory. It is believed that universe may have many phase transitions after big-bang.

Einstein formulation of General Relativity is the foundation of other geometric theories in order to explain the actual gravitational phenomena. A more general theory in which both gravitation and electromagnetism were described geometrically was proposed by [1]. Later, [2] suggested a modification of Riemannian geometry by introducing a gauge function which removed the non-integrability condition of the length of a vector under parallel transport, which was known as Lyra's geometry. In Lyra's geometry, the connection is metric preserving as Riemannian geometry, and length transfers as integrable in contrast to Weyl's geometry. He also introduced a gauge function into the structure-less manifold, as a result of which a displacement field arose naturally. This alternating theory is of interest since it produces effects similar to Einstein's theory.

Many authors have investigated cosmology in Lyra's geometry with both a constant displacement field and time dependent one. Also cosmological models in the frame work of Lyra's geometry in different contexts are investigated by [3]-[11]. Cosmological models based on Lyra's manifold with constant displacement field vector were also studied by Bhamra [12]-[18]. But with this condition it is found as one of conveniences and there is no priori reason for it. Recently, several authors like [19]-[24] have studied cosmological models in the frame work of Lyra's geometry in various contexts.

We know that the constant vector displacement field in Lyra's geometry plays the role of cosmological constant in the normal general relativistic study as suggested by [25]. Also, [26] shows that the scalar-tensor treatment based on Lyra's geometry predicts the same effects, within observational limits, as the Einstein theory.

As the necessity of study of higher-dimensional space-time in this field aiming to unify gravity with other interactions, the concept of extra dimension is relevant in cosmology, particularly for the early stage of universe and theoretically the present four dimensional stage of the universe may have been preceded by a multi-dimensional stage. So, in this paper we discussed about the five dimensional cosmological models in Lyra's geometry by considering plane symmetric metric with some conditions to find out some solutions which were realistic with the observational facts.

2. Field Equations and Their Solutions

Here we consider the five dimensional plane symmetric metric in the form

$$ds^2 = A^2\left(dx^2 - dt^2\right) + B^2\left(dy^2 + dz^2\right) + C^2 dm^2 \tag{1}$$

where A, B and C are functions of time "t" only.

Einstein's field equations based on Lyra's Geometry is

$$R_{ij} - \frac{1}{2}g_{ij}R + \frac{3}{2}\phi_i\phi_j - \frac{3}{4}g_{ij}\phi^k\phi_k = -T_{ij} \tag{2}$$

where we use the units in which $\dfrac{8\pi G}{c^4} = 1$ (Wesson 1992; Baysal *et al.* 2001; Bali and Dave 2002), and ϕ_i is the displacement vector defined by

$$\phi_i = \left(\beta(t),0,0,0,0\right) \tag{3}$$

The energy momentum tensor of cosmic strings is

$$T_{ij} = \rho U_i U_j - \lambda X_i X_j \tag{4}$$

where, $\rho = \rho_p + \lambda$, is the energy density of the cloud of string, ρ_p being the rest energy density of particles attached to the strings and λ is the string tension density. $U^i = \left(0,0,0,0,A^{-1}\right)$ is the five velocity vector for the cloud of particles and $X^i = \left(A^{-1},0,0,0,0\right)$ is the direction of strings. Moreover the directions of strings satisfies

$$U^i U_i = -X^i X_i = -1, \text{ and } U^i X_i = 0 \tag{5}$$

Using the commoving coordinate system and Equations (3), (4) and (5), the field equations (2) for the metric (1) yield

$$2\frac{\ddot{B}}{B} + \frac{\ddot{C}}{C} + \frac{\dot{B}^2}{B^2} + 2\frac{\dot{B}\dot{C}}{BC} - 2\frac{\dot{A}\dot{B}}{AB} - \frac{\dot{A}\dot{C}}{AC} + \frac{3}{4}\beta^2 = \lambda A^2 \tag{6}$$

$$\frac{\ddot{A}}{A} + \frac{\ddot{B}}{B} + \frac{\ddot{C}}{C} - \frac{\dot{A}^2}{A^2} + \frac{\dot{B}\dot{C}}{BC} - \frac{3}{4}\beta^2 = 0 \tag{7}$$

$$\frac{\ddot{A}}{A} + 2\frac{\ddot{B}}{B} - \frac{\dot{A}^2}{A^2} + \frac{\dot{B}^2}{B^2} - \frac{3}{4}\beta^2 = 0 \tag{8}$$

$$\frac{\dot{B}^2}{B^2} + 2\frac{\dot{A}\dot{B}}{AB} + 2\frac{\dot{B}\dot{C}}{BC} + \frac{\dot{A}\dot{C}}{AC} - \frac{3}{4}\beta^2 = \rho A^2 \tag{9}$$

Now, (7) and (8) give

$$\frac{\ddot{B}}{B} + \frac{\dot{B}^2}{B^2} - \frac{\ddot{C}}{C} - \frac{\dot{B}\dot{C}}{BC} = 0 \tag{10}$$

A solution of (10) is

$$B = e^{b_0 t + b_1} \tag{11}$$

$$C = e^{b_0 t - c_1} \tag{12}$$

Thus, (11) and (12) together with (7) and (8) give

$$A = \left(a_0 t + a_1\right)^{\frac{1}{2}} \tag{13}$$

And

$$\beta^2 = \frac{4}{3}\left[3b_0^2 - \frac{a_0^2}{2}\left(a_0 t + a_1\right)^{-2}\right] \tag{14}$$

Now from (9), we have,

$$\rho = \frac{3}{2}a_0 b_0 \left(a_0 t + a_1\right)^{-2} + \frac{a_0^2}{2}\left(a_0 t + a_1\right)^{-3} \tag{15}$$

And from (6), we have,

$$\lambda = 9b_0^2 \left(a_0 t + a_1\right)^{-1} - \frac{3}{2}a_0 b_0 \left(a_0 t + a_1\right)^{-2} - \frac{a_0^2}{2}\left(a_0 t + a_1\right)^{-3} \tag{16}$$

Therefore, from the relation $\rho_p = \rho - \lambda$ we have

$$\rho_p = a_0^2 \left(a_0 t + a_1\right)^{-3} + 3a_0 b_0 \left(a_0 t + a_1\right)^{-2} - 9b_0^2 \left(a_0 t + a_1\right)^{-1} \tag{17}$$

For the metric (1), the expansion factor Θ is obtained as

$$\Theta = a_0 \left(a_0 t + a_1\right)^{-1} + 3b_0 \tag{18}$$

and

$$\sigma = \frac{1}{\sqrt{6}}\left[2b_0 - a_0 \left(a_0 t + a_1\right)^{-1}\right] \tag{19}$$

Therefore, from Equations (18) and (19) we have

$$\frac{\sigma}{\Theta} = \frac{1}{\sqrt{6}}\frac{2b_0 \left(a_0 t + a_1\right) - a_0}{3b_0 \left(a_0 t + a_1\right) + a_0} \tag{20}$$

Here, the deceleration parameter q is given by

$$q = \frac{4}{3}\frac{\left[a_0^2 - 2b_0^2 \left(a_0 t + a_1\right)^2\right]\cdot\left[a_0^2 - 12b_0^2 \left(a_0 t + a_1\right)^2\right]}{\left[a_0^2 - 4b_0^2 \left(a_0 t + a_1\right)^2\right]^2}. \tag{21}$$

3. Physical Interpretations of the Solutions

In the universe we obtain here, it is seen that the energy density has a finite value at the beginning and then it gradually decreases until it shrinks almost to zero at infinite time. The string tension density is also found to be a decreasing function of time until it almost tends to zero as time tends to infinity. Here, with the advent of time, the density of the string decreases more rapidly than density of the particles attached to them. Thus, our universe ultimately becomes a universe dominated by particles, where strings are becoming invisible in course of time. Here, for our universe, we see that the special dimensions expand isotropically, implying the expansion of our universe which bears testimony to our universe being a realistic one.

Moreover, from the expressions of the expansion factor and deceleration parameter obtained here, it can be inferred that our universe is expanding, but the rate of expansion is decreasing slowly until at infinite time where it is expanding at a constant rate. Here, the gauge function β^2 is found to be constant at the initial epoch of time and gradually increases with time until it becomes a finite constant $4b_0^2$ at infinite time.

Interacting with the pressureless matter here, the displacement vector can play the same role as a cosmological constant (term). Thus, it will be nice to study further whether the displacement vector plays a role in disturbing the rate of expansion of the universe.

Though our model universe seems to be anisotropic in the beginning it will become gradually an isotropic one until it becomes perfectly isotropic at time given by $t = -a_1 \dfrac{1}{a_0}\left(\dfrac{a_0}{b_0}\right)$. It can be seen that even though an anisotropic parameter is produced in this universe, its anisotropy does not promote anisotropy in the expansion, and thus in course of time our universe becomes an isotropic one.

References

[1] Weyl, H. (1918) Sitzungsberichte Der Preussischen Akademie Der Wissenschaften. Academy Wiss, Berlin, 465.

[2] Lyra, G. (1951) Über-eine Modifikation der Riemannschen Geometrie. *Mathematische Zeitschrift*, **54**, 52-64. http://dx.doi.org/10.1007/BF01175135

[3] Pradhan, A. and Kumar, S.S. (2009) Plane Symmetric Inhomogeneous Perfect Fluid Universe with Electromagnetic Field in Lyra Geometry. *Astrophysics and Space Science*, **321**, 137-146. http://dx.doi.org/10.1007/s10509-009-0015-9

[4] Pradhan, A. and Mathur, P. (2009) Inhomogeneous Perfect Fluid Universe with Electromagnetic Field in Lyra Geometry. *Fizika B*, **18**, 243-264. (gr-qc/0806.4815)

[5] Pradhan, A. and Yadav, P. (2009) Accelerated Lyra's Cosmology Driven by Electromagnetic Field in Inhomogeneous Universe. *International Journal of Mathematics and Mathematical Sciences* (*IJMMS*), **2009**, Article ID: 471938, 20 p. http://dx.doi.org/10.1155/2009/471938

[6] Pradhan, A. (2009) Cylindrically Symmetric Viscous Fluid Universe in Lyra Geometry. *Journal of Mathematical Physics*, **50**, 022501-022513. http://dx.doi.org/10.1063/1.3075571

[7] Pradhan, A., Amirhashchi, H. and Zainuddin, H. (2011) A New Class of Inhomogeneous Cosmological Models with Electromagnetic Field in Normal Gauge for Lyra's Manifold. *International Journal of Theoretical Physics*, **50**, 56-69. http://dx.doi.org/10.1007/s10773-010-0493-0

[8] Pradhan, A. and Singh, A.K. (2011) Anisotropic Bianchi Type-I String Cosmological Models in Normal Gauge for Lyra's Manifold with Constant Deceleration Parameter. *International Journal of Theoretical Physics*, **50**, 916-933. http://dx.doi.org/10.1007/s10773-010-0636-3

[9] Yadav, A.K. (2010) Lyra's Cosmology of Inhomogeneous Universe with Electromagnetic Field. *Fizika B*, **19**, 53-80.

[10] Agarwal, S., Pandey, R and Pradhan, A. (2011) LRS Bianchi Type II Perfect Fluid Cosmological Models in Normal Gauge for Lyra's Manifold. *International Journal of Theoretical Physics*, **50**, 296-307. http://dx.doi.org/10.1007/s10773-010-0523-y

[11] Singh, R.S. and Singh, A. (2012) A New Class of Magnetized Inhomogeneous Cosmological Models of Perfect Fluid Distribution with Variable Magnetic Permiability in Lyra Geometry. *Electronic Journal of Theoretical Physics*, **9**, 265-282.

[12] Bhamra, K.S. (1974) A Cosmological Model of Class One in Lyra's Manifold. *Australian Journal of Physics*, **27**, 541-547. http://dx.doi.org/10.1071/PH740541

[13] Kalyanshetti, S.B. and Waghmode, B.B. (1982) A Static Cosmological Model in Einstein-Cartan Theory. *General Relativity and Gravitation*, **14**, 823-830.

[14] Soleng, H.H. (1987) Cosmologies Based on Lyra's Geometry. *General Relativity and Gravitation*, **19**, 1213-1216.

[15] Sen, D.K. and Vanstone, J.R. (1972) On Weyl and Lyra Manifolds. *Journal of Mathematical Physics*, **13**, 990-994. http://dx.doi.org/10.1063/1.1666099

[16] Karade, T.M. and Borikar, S.M. (1978) Thermodynamic Equilibrium of a Gravitating Sphere in Lyra's Geometry. *General Relativity and Gravitation*, **9**, 431-436.

[17] Reddy, D.R.K. and Innaiah, P. (1986) A Plane Symmetric Cosmological Model in Lyra Manifold. *Astrophysics and*

Space Science, **123**, 49-52. http://dx.doi.org/10.1007/BF00649122

[18] Reddy, D.R.K. and Venkateswarlu, R. (1987) Birkhoff-Type Theorem in the Scale Covariant Theory of Gravitation. *Astrophysics and Space Science*, **136**, 191-194.

[19] Asgar, A. and Ansary, M. (2014) Accelerating Bianchi Type-VI_0 Bulk Viscous Cosmological Models in Lyra Geometry. *Journal of Theoretical and Applied Physics*, **8**, 219-224. http://dx.doi.org/10.1007/s40094-014-0151-7

[20] Kumari, P., Singh, M.K. and Ram, S. (2013) Anisotropic Bianchi Type-III Bulk Viscous Fluid Universe in Lyra Geometry. *Advances in Mathematical Physics*, **2013**, Article ID: 416294. http://dx.doi.org/10.1155/2013/416294

[21] Asgar, A. and Ansary, M. (2014) Bianchi Type-V Universe with Anisotropic Dark Energy in Lyra's Geometry. *The African Review of Physics*, **9**, 145-151.

[22] Zia, R. and Singh, R.P. (2012) Bulk Viscous Inhomogeneous Cosmological Models with Electromagnetic Field in Lyra Geometry. *Romanian Journal of Physics*, **57**, 761-778.

[23] Asgar, A. and Ansary, M. (2014) Exact Solutions of Axially Symmetric Bianchi Type-I Cosmological Model in Lyra Geometry. *IOSR Journal of Applied Physics (IOSR-JAP)*, **5**, 1-5. www.iosrjournals.org

[24] Panigrahi, U.K. and Nayak, B. (2014) Five Dimensional Stiff Fluids with Variable Displacement Vector in Lyra Manifold. *International Journal of Mathematical Archive*, **5**, 123-128. www.ijma.info

[25] Halford, W.D. (1970) Cosmological Theory Based on Lyra's Geometry. *Australian Journal of Physics*, **23**, 863-869. http://dx.doi.org/10.1071/PH700863

[26] Halford, W.D. (1972) Scalar-Tensor Theory of Gravitation in a Lyra Manifold. *Journal of Mathematical Physics*, **13**, 1699-1703. http://dx.doi.org/10.1063/1.1665894

28

Distribution of Mass and Energy in Closed Model of the Universe

Fadel A. Bukhari

Department of Astronomy, Faculty of Science, King Abdulaziz University, Jeddah, Saudi Arabia
Email: fdbukhari@gmail.com

Abstract

The universe's horizon distance and volume are constructed in the closed cosmic model. The universe horizon distance distribution increases constantly for $t < t_{me}$ and decreases for $t > t_{me}$. However, the universe's horizon volume shows a sudden reduction in the range $t = 0.5$ Gyr $- t_{me}$ due to the change of the universe space from flat to curved then closed in the interval 15.1261 Gyr $\leq t \leq t_{me}$. On the other hand, this distribution exhibits an abrupt rise in the range $t = t_{me} - t_*$ due to the change of the universe space from closed then curved to flat in the interval $39.3822 \leq t \leq 40.7521$ Gyr. The mass of radiation, matter and dark energy within the horizon volume of the universe are also investigated. These distributions reveal similar noticeable changes as the universe's horizon volume distribution for the same reasons. The mass of radiation dominates up to $t = 53221.5$ yr, then the mass of matter becomes larger. Afterwards, both distributions of radiation and matter decrease while the distribution of dark energy rises until $t = 10.1007$ Gyr, where the mass of dark energy prevails up to $t = t_{me}$. Hence, the distribution of dark energy reduces until $t = 40.2892$ Gyr, where the mass of matter becomes prominent again. At $t = 53.6246$ Gyr the masses of both matter and radiation become appreciably high such that the intercluster space will vanish and clusters of galaxies interfere with each other. Furthermore, not only the intergalactic medium will disappear, but also galaxies will collide and merge with each other to form extremely dense and close cosmological bodies. These very dense bodies will undergo further successive collisions and mergers under the action of central gravity, where the interstellar medium will vanish and the universe would develop to big crunch at $t_{bc} = 53.6251$ Gyr. It is interesting to note that the horizon distance of the universe in the closed model at $t = t_{me}$ is in very good agreement with the maximum horizon distances in the five general cosmic models.

Keywords

Dark Energy, Radiation, Closed Cosmic Model

1. Introduction

The distribution of density parameters of radiation, matter and dark energy in the closed cosmic model were investigated in a previous study [1], where we discovered the main epochs of the universe history in this model. It is worthy now to study the distributions of equivalent mass of radiation, mass of matter and equivalent mass of dark energy within the horizon volume of the universe to get deeper sight of the universe evolution in the closed model.

The reason for considering the equivalent mass of radiation in this study is the significant value of the radiation density parameter in the early universe and before the big crunch as we have seen in [1].

Therefore, it is vital to develop the distributions of the horizon distance and horizon volume of the universe in the closed model at various time ranges depending on the bases presented in [2]. Description of methodology is illustrated in Section 2, while algorithm would be shown in Section 3. Results and discussion are displayed in Section 4. Conclusion is given in Section 5.

2. Methodology

It is obvious from [2] that the horizon distance and horizon volume of the universe in closed cosmic model at the present time are respectively

$$d_h(t_o) = \frac{c}{H_o} \int_0^1 \frac{1}{a} \left[1 - S^2 \Omega''_{\Lambda,t} \left(1 - a^2\right) + S^2 \Omega''_{m,t} \left(a^2 - a^3\right) + S^2 \Omega''_{r,t} \left(a^2 - a^4\right) \right]^{-1/2} da. \tag{1}$$

$$V_h(t_o) = \frac{8\pi}{3} d_h^3(t_o). \tag{2}$$

where $\Omega''_{\Lambda,t}$, $\Omega''_{m,t}$ and $\Omega''_{r,t}$ are given by

$$\Omega''_{\Lambda,t} = \frac{\rho'_{\Lambda,t}}{c^2 \rho''_{c,t}}. \tag{3}$$

$$\Omega''_{m,t} = \frac{\rho'_{m,t}}{\rho''_{c,t}}. \tag{4}$$

$$\Omega''_{r,t} = \frac{\rho'_{r,t}}{c^2 \rho''_{c,t}}. \tag{5}$$

$$\rho''_{c,t} = \frac{3H'^2(t)}{8\pi G}. \tag{6}$$

$$H'(t) = \frac{H_o}{a} \left[1 - s^2 \Omega'_{\Lambda,t} \left(1 - a^2\right) + s^2 \Omega'_{m,t} \left(a^2 - a^3\right) + s^2 \Omega'_{r,t} \left(a^2 - a^4\right) \right]^{\frac{1}{2}}.$$

$$\text{where} \quad s = \frac{H(t)}{H_o}, \quad S = \frac{H'(t)}{H_o} \tag{7}$$

$$H(t) = \frac{H_o}{a} \left[1 - \Omega_{\Lambda,o} \left(1 - a^2\right) + \Omega_{m,o} \left(\frac{1}{a} - 1\right) + \Omega_{r,o} \left(\frac{1}{a^2} - 1\right) \right]^{\frac{1}{2}}. \tag{8}$$

$$\Omega'_{\Lambda,t} = \frac{\rho'_{\Lambda,t}}{c^2 \rho_{c,t}}. \tag{9}$$

$$\Omega'_{m,t} = \frac{\rho'_{m,t}}{\rho_{c,t}}. \tag{10}$$

$$\Omega'_{r,t} = \frac{\rho'_{r,t}}{c^2 \rho_{c,t}}. \tag{11}$$

$$\frac{\rho'_{\Lambda,t}}{c^2} = \rho_{c,o}\Omega_{\Lambda,o} - \Delta\left(\frac{\rho_{\Lambda,t}}{c^2}\right). \tag{12}$$

$$\rho'_{m,t} = \rho_{c,o}\frac{\Omega_{m,o}}{a^3} + \frac{1}{2}\Delta\left(\frac{\rho_{\Lambda,t}}{c^2}\right). \tag{13}$$

$$\frac{\rho'_{r,t}}{c^2} = \rho_{c,o}\frac{\Omega_{r,o}}{a^4} + \frac{1}{2}\Delta\left(\frac{\rho_{\Lambda,t}}{c^2}\right). \tag{14}$$

$$\Delta\left(\frac{\rho_{\Lambda,t}}{c^2}\right) = 0.01\frac{\rho_{\Lambda,t}}{c^2}t. \tag{15}$$

where t is the cosmic time in Gyr.

$$\rho_{c,t} = \frac{3H^2}{8\pi G}. \tag{16}$$

The horizon distance of the universe in the closed cosmic model at any given time is given by

$$d_h(t) = \frac{c}{H_o}\int_0^a \frac{1}{a}\left[1 - S^2\Omega''_{\Lambda,t}\left(1-a^2\right) + S^2\Omega''_{m,t}\left(a^2-a^3\right) + S^2\Omega''_{r,t}\left(a^2-a^4\right)\right]^{-1/2} da. \tag{17-a}$$

Consequently, the change in the horizon distance of the universe in the time interval between two instants of scale factors a_1, a_2 is written as

$$\Delta d_h(t) = \frac{c}{H_o}\int_{a_1}^{a_2} \frac{1}{a}\left[1 - S^2\Omega''_{\Lambda,t}\left(1-a^2\right) + S^2\Omega''_{m,t}\left(a^2-a^3\right) + S^2\Omega''_{r,t}\left(a^2-a^4\right)\right]^{-1/2} da. \tag{17-b}$$

The horizon volume of the universe in the closed model at any given time is expressed as

$$V_h(t) = f\left(d_h(t), k\right). \tag{18}$$

Equation (18) indicates that the horizon volume of the universe at t is a function of $d_h(t)$ and the curvature of space k at t. Since this curvature could be flat, open and closed from the big bang to big crunch as evident from Table 3 in [1]. Thus, the low of $V_h(t)$ can be determined according to the value of k at t, as explained in the following cases:
(1) Flat space ($k = 0$)
We have seen in [2] that the horizon volume of the universe at time t in this case is given by

$$V_h(t) = \frac{8\pi}{3}d_h^3(t). \tag{19}$$

Therefore, it is obvious from Table 3 in [1] that Equation (19) is used in the time intervals $t \le 6.5321\,\text{Gyr}$, $10.0751 < t \le 15.1261\,\text{Gyr}$, $39.3822 < t \le 40.7521\,\text{Gyr}$, and $53.48 < t \le t_{bc}$.
(2) Closed space ($k = +1$)
We recall the equation of proper distance of extragalactic object

$$d_p(t) = R(t)f(r_o). \tag{20}$$

where

$$f(r_o) = \int_0^{r_o} \frac{dr}{\sqrt{1-kr^2}} = \begin{cases} \sin^{-1} r_o & k = +1 \\ r_o & k = 0 \\ \sinh^{-1} r_o & k = -1 \end{cases} \tag{21}$$

And the volume of space within $d_p(t)$ is expressed as

$$V_p(t) = 2R^3(t)\int_0^{2\pi}d\phi\int_0^{\pi}\sin\theta d\theta\int_0^{r_o}\frac{r^2}{\sqrt{1-kr^2}}dr. \tag{22}$$

where $R(t), r_o, r, \theta$ and ϕ are defined as in [3] [4]. For $k = +1$, Equations (21), (20) and (22) yield

$$d_p(t) = R(t)\sin^{-1} r_o \tag{23}$$

Hence Equation (22) gives

$$V_p(t) = 8\pi R^3(t) \int_0^{r_o} \frac{r^2 dr}{\sqrt{1-r^2}}. \tag{24}$$

Assume

$$r = \sin\alpha, \ dr = \cos\alpha d\alpha, \ \cos\alpha = \sqrt{1-r^2} \tag{25}$$

Substituting by (25) in (24) we get

$$V_p(t) = 4\pi R^3(t) \int_0^{\sin^{-1} r_o} (1 - \cos 2\alpha) d\alpha. \tag{26}$$

Let

$$\beta = 2\alpha, \ d\beta = 2d\alpha. \tag{27}$$

Substituting by (27) in (26) we have

$$V_p(t) = 4\pi R^3(t)\left(\sin^{-1} r_o\right)^3 \left[\frac{1}{\left(\sin^{-1} r_o\right)^2} - \frac{r_o\sqrt{1-r_o^2}}{\left(\sin^{-1} r_o\right)^3}\right]. \tag{28}$$

Substituting by (23) in (28) yields

$$V_p(t) = 4\pi d_p^3(t)\left[\frac{1}{\left(\sin^{-1} r_o\right)^2} - \frac{r_o\sqrt{1-r_o^2}}{\left(\sin^{-1} r_o\right)^3}\right]. \tag{29}$$

Suppose $r_o = 1$, hence $\alpha_o = \frac{\pi}{2}$ and Equation (29) becomes

$$V_p(t) = \frac{16}{\pi} d_p^3(t). \tag{30}$$

Thus, the horizon volume of the universe in the closed cosmic model at time t in this case is expressed as

$$V_h(t) = \frac{16}{\pi} d_h^3(t). \tag{31}$$

It is evident form Table 3 in [1] that Equation (23) is used in two time intervals extending through $15.1261 < t \le 39.3822$ Gyr.
(3) Open space ($k = -1$)
   Equations (21), (20) and (22) give

$$d_p(t) = R(t)\sinh^{-1} r_o \tag{32}$$

$$V_p(t) = 8\pi R^3(t) \int_0^{r_o} \frac{r^2 dr}{\sqrt{1+r^2}}. \tag{33}$$

Assume

$$r = \sinh\xi, \ dr = \cosh\xi d\xi, \ \cosh\xi = \sqrt{1+r^2} \tag{34}$$

Substituting by (34) in (33) we have

$$V_p(t) = 4\pi R^3(t) \int_0^{\sinh^{-1} r_o} (\cosh 2\xi - 1) d\xi. \tag{35}$$

Let

$$\eta = 2\xi, \ \mathrm{d}\eta = 2\mathrm{d}\xi. \tag{36}$$

Substituting by (36) in (35) we get

$$V_p(t) = 4\pi R^3(t)\left(\sinh^{-1} r_o\right)^3 \left[\frac{r_o \cosh\left(\sinh^{-1} r_o\right)}{\left(\sinh^{-1} r_o\right)^3} - \frac{1}{\left(\sinh^{-1} r_o\right)^2}\right]. \tag{37}$$

Substituting by (32) in (37) yields

$$V_p(t) = 4\pi d_p^3(t)\left[\frac{r_o \cosh\left(\sinh^{-1} r_o\right)}{\left(\sinh^{-1} r_o\right)^3} - \frac{1}{\left(\sinh^{-1} r_o\right)^2}\right]. \tag{38}$$

When $r_o = 1$, Equation (38) reduces to

$$V_p(t) = 3.112983\pi d_p^3(t). \tag{39}$$

Therefore, the horizon volume of the universe in closed cosmic model at time t in this case is written as

$$V_h(t) = 3.112983\pi d_h^3(t). \tag{40}$$

It is clear form Table 3 in [1] that Equation (40) is used in the time intervals $6.5321 < t \le 10.0751\,\text{Gyr}$, $40.7521 < t \le 53.48\,\text{Gyr}$.

The total density of the universe in the closed cosmic model at time t is

$$\rho'(t) = \rho'_{m,t} + \frac{\rho'_{r,t}}{c^2} + \frac{\rho'_{\Lambda,t}}{c^2}. \tag{41}$$

Substituting by (3)-(5) in (41) we get

$$\rho'(t) = \rho''_{c,t}\Omega''_{m,t} + \rho''_{c,t}\Omega''_{r,t} + \rho''_{c,t}\Omega''_{\Lambda,t}$$

$$\rho'(t) = \rho''_{c,t}\Omega''(t). \tag{42}$$

where

$$\Omega''(t) = \Omega''_{m,t} + \Omega''_{r,t} + \Omega''_{\Lambda,t}. \tag{43}$$

From Equation (19), (30), (40) and (42) the total mass of the universe within the horizon volume in closed cosmic model at time t is

$$M_h(t) = V_h(t)\rho'(t). \tag{44}$$

The masses of matter, radiation and dark energy within the horizon volume of the universe in closed cosmic model at time t are respectively

$$M_{m,t} = M_h(t)\frac{\Omega''_{m,t}}{\Omega''(t)}. \tag{45}$$

$$M_{r,t} = M_h(t)\frac{\Omega''_{r,t}}{\Omega''(t)}. \tag{46}$$

$$M_{\Lambda,t} = M_h(t)\frac{\Omega''_{\Lambda,t}}{\Omega''(t)}. \tag{47}$$

The time interval between two instants with scale factors a_1, a_2 during the universe expansion is given by Equation (16) in [4] [5] as

$$\Delta t = \frac{1}{H_o}\int_{a_1}^{a_2}\left[1 - \Omega_{\Lambda,o}\left(1-a^2\right) + \Omega_{m,o}\left(\frac{1}{a}-1\right) + \Omega_{r,o}\left(\frac{1}{a^2}-1\right)\right]^{-\frac{1}{2}} da. \tag{48}$$

However, during the universe contraction if $a_2 < a_1$ then modulus of the right hand side of (50) should be taken.

3. Algorithm

In determination of the distributions of $d_h(t), V_h(t), M_h(t), M_{m,t}, M_{r,t}$ and $M_{\Lambda,t}$ we use the following steps:

(1) Stage of the universe expansion.

a) Set $t = 0, d_h = 0$ and insert the value of $a_{max} = 0.093188, J = 1000$ for $t \le 0.5$ Gyr, $a_{max} = 2.3755873$, $J = 1600$ for $0.5 < t \le t_{me}$.

b) Calculate $DA = \dfrac{a_{max}}{DBLE(J)}$.

c) Start general DO loop $I = 2$, J which includes the following sub steps:

d) $a_1 = DA(I-1)$, $a_2 = DA\, I$.

e) Compute new value of cosmic time t numerically using (48), where $t = t + \Delta t$.

f) Obtain new value of the universe horizon distance d_h numerically using (17-b), where $d_h = d_h + \Delta d_h$.

g) Determine the corresponding values of V_h using (19), (31) and (40), in addition to the values of

$H'(t), \rho''_{c,t}, \Omega''_{m,t}, \Omega''_{r,t}, \Omega''_{\Lambda,t}, \Omega''(t), \rho'(t), M_h(t), M_{m,t}, M_{r,t}$ and $M_{\Lambda,t}$ using (7), (6), (4), (5), (3), (43), (42), (44), (45), (46) and (47) respectively.

h) Continue the general DO loop.

(2) Stage of the universe contraction

a) Set $t = t_{me} = 26.8125327$ Gyr, $d_h = d_h(t_{me}) = 18.619224$ Gpc, and insert the values of

$a_{max} = 2.3755873, J1 = 1600$ for $t_{me} < t \le t_*, t_* = t_{bc} - 0.5$ Gyr, $a_{max} = 0.093539, J1 = 90$ for $t > t_*$.

b) Evaluate $DA = \dfrac{a_{max}}{DBLE(J1)}$.

c) Start general DO loop $I = 1$, $J1$ which includes the following sub steps:

d) Set $J2 = J1 - I + 1$, $a_1 = DA(J2-1)$, $a_2 = DA\, J2$.

e) Obtain new value of cosmic time t numerically using (48), where $t = t + \Delta t$.

f) Compute new value of the universe horizon distance d_h numerically using (17-b), where $d_h = d_h - \Delta d_h$.

Sub steps (g) and (h) are similar to (g), (h) mentioned above in the stage of the universe expansion.

4. Results and Discussion

The distribution of the universe horizon distance in the closed cosmic model until $t = 0.5$ Gyr is shown in **Figure 1(a)**. The distribution increases quite slowly up to $t = 5.8780$ Myr, then the distribution starts raising rapidly. However, the distribution of the universe horizon distance in the range $t = 0.5$ Gyr - t_{me} increases very fast until about $t = 5.7237$ Gyr. Afterwards, it raises gradually as indicated in **Figure 1(b)**. Furthermore, the distribution of the universe horizon distance in the range $t = t_{me} - t_*$, where $t_* = t_{bc} - 0.5$ Gyr, decreases quite slowly up to $t = 46.5790$ Gyr, hence it decreases relatively fast as presented in **Figure 1(c)**. Nevertheless, the distribution of the universe horizon distance in the range $t = t_* - t_n$ decreases slowly until $t = 53.4733$ Gyr, $t_n = t_{bc} - 0.0005$ Gyr, then it starts reduction sharply towards $t = t_{bc}$ as displayed in **Figure 1(d)**.

The distribution of the universe horizon volume in the closed cosmic model up to $t = 0.5$ Gyr is shown in **Figure 2(a)**. The distribution increases very slowly until $t = 18.4785$ Myr. Afterwards, the distribution starts raising appreciably. During this cosmic time range the space of universe is flat. The distribution of the universe horizon volume continues raising up to $t = 14.8$ Gyr, hence the distribution suddenly decreases until $t = 16.6667$ Gyr, then it increases gradually up to $t = t_{me}$ as seen is **Figure 2(b)**. The sharp decrease of the distribution in the range $14.8 < t < 16.6667$ Gyr because the space of the universe changes from flat to curved then closed in the interval 15.1261 Gyr $\le t \le t_{me}$.

The distribution of the universe horizon volume in the range $t = t_{me} - t_*$ is disclosed in **Figure 2(c)**. The distribution decreases until $t = 37.6974$ Gyr, hence it shows abrupt raising up to $t = 40.6579$ Gyr, hence it reduces again until $t = t_*$. The abrupt increase of the distribution in the range $37.6974 < t < 40.6579$ Gyr is due to

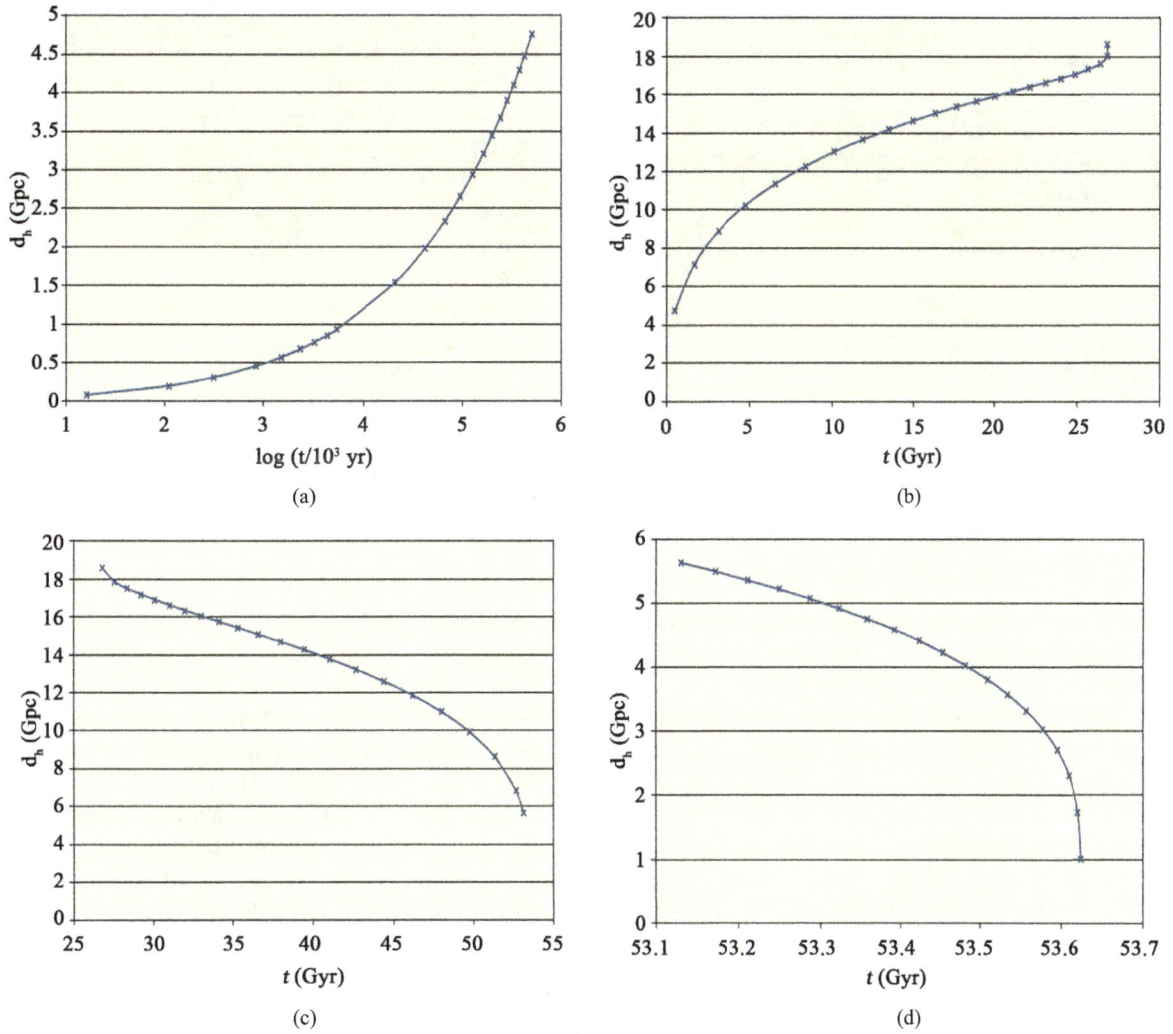

Figure 1. (a) The distribution of the universe horizon distance in the closed cosmic model up to $t = 0.5$ Gyr; (b) The distribution of the universe horizon distance in the closed cosmic model in the range $t = 0.5$ Gyr - t_{me}; (c) The distribution of the universe horizon distance in the closed cosmic model in the range $t = t_{me}$ - t_*; (d) The distribution of the universe horizon distance in the closed cosmic model in the range $t = t_*$ - t_n.

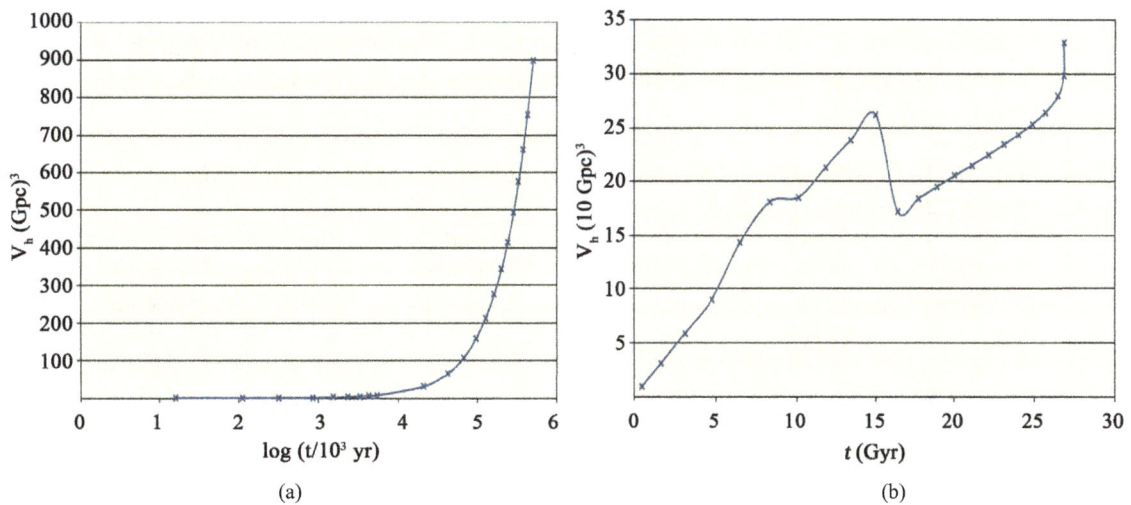

(c)

(d)

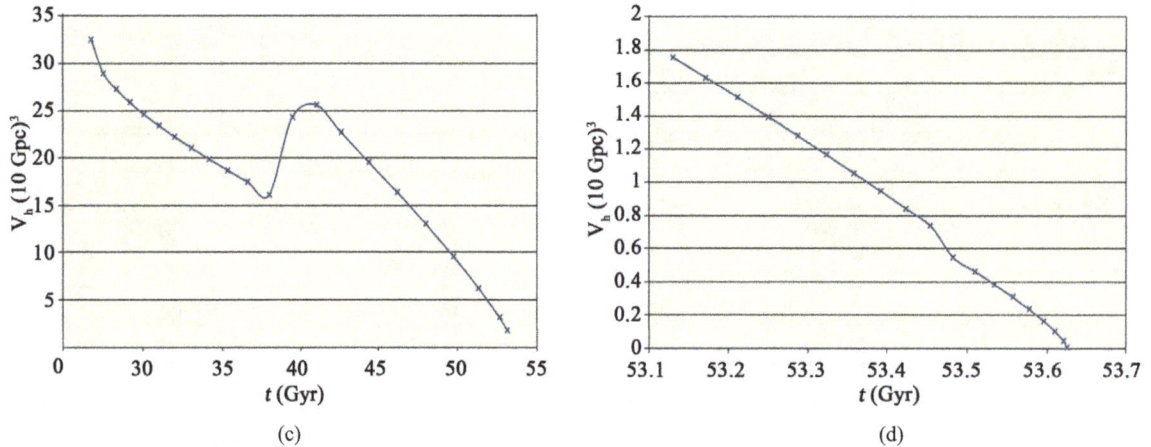

Figure 2. (a) The distribution of the universe horizon volume in the closed cosmic model up to $t = 0.5$ Gyr; (b) The distribution of the universe horizon volume in the closed cosmic model in the range $t = 0.5$ Gyr - t_{me}; (c) The distribution of the universe horizon volume in the closed cosmic model in the range $t = t_{me}$ - t_*; (d) The distribution of the universe horizon volume in the closed cosmic model in the range $t = t_*$ - t_n.

the fact that the universe space changes from closed then curved to flat in the interval $39.3822 \leq t \leq 40.7521$ Gyr. The distribution of the universe horizon volume in the range $t = t_*$ - t_n is presented in **Figure 2(d)**, which is reducing gradually towards $t = t_{bc}$.

The distribution of mass and energy within the horizon volume of the universe in the closed cosmic model until $t = 0.5$ Gyr is shown in **Figure 3(a)**. The distributions of radiation and total mass decrease very slowly up to $t = 134596$ yr and $t = 162061$ yr respectively, then they reduce more rapidly. However, the distribution of matter increases so slowly until $t = 181161$ yr and it decreases rapidly afterwards. This distribution intersects with the distribution of radiation at $t = 53221.5$ yr and coincides with the distribution of total mass at $t = 3.2819$ Myr. The distribution of dark energy increases continuously towards $t = 0.5$ Gyr. The distribution of mass and energy within the horizon volume of the universe in the closed model in the range $t = 0.5$ Gyr - t_{me} is illustrated in **Figure 3(b)**. It is noticeable that the distribution of radiation intersects the distribution of dark energy at $t = 0.6618$ Gyr, hence it reduces very fast up to $t = 3.0147$ Gyr. Afterwards, it raises gradually until $t = 14.9265$ Gyr, then it decreases abruptly up to $t = 16.3971$ Gyr, where it starts increasing again towards t_{me}. The distribution of matter decreases up to $t = 22.2794$ Gyr, hence it raises very slowly towards t_{me}. This distribution coincides with the distribution of total mass up to $t = 1.9853$ Gyr, and intersects with the dark energy distribution at $t = 10.1007$ Gyr. The matter distribution indicates obvious decrease in the range $t = 14.9265$ - 16.3971 Gyr. The distribution of dark matter increases until $t = 14.9265$ Gyr, hence it also shows marked decrease up to $t = 16.3971$ Gyr. Afterwards, it raises slightly towards t_{me}. The total mass distribution reduces up to $t = 14.9265$ Gyr, then it decreases abruptly until $t = 16.3971$ Gyr. Hence, this distribution increases gradually towards t_{me}. The noticeable decrease of all four distributions in the range about $14.9225 < t < 16.3971$ Gyr is due to the change of the universe space from flat to curved then closed in the range $15.1261 \leq t \leq t_{me}$.

The distribution of mass and energy within the horizon volume of the universe in the closed model in the range $t = t_{me}$ - t_* is illustrated in **Figure 3(c)**. The distribution of radiation reduces very slowly until $t = 37.9839$ Gyr, then it raises suddenly up to $t = 40.4032$ Gyr. Hence, this distribution decreases again until $t = 52.5807$ Gyr where it intersects with the distribution of dark energy, then it starts raising towards t_*. The matter distribution increases so slowly up to $t = 37.9839$ Gyr, hence it exhibits marked raise until $t = 40.4032$ Gyr and intersects with the dark energy distribution at $t = 40.2892$ Gyr. Afterwards, the distribution of matter shows substantial increase towards t_* and coincides with the total mass distribution at $t = 51.2903$ Gyr hence forth. The dark energy distribution reduces gradually until $t = 37.9839$ Gyr where it displays outstanding raise up to $t = 40.4032$ Gyr, hence it decreases again towards t_*. The total mass distribution reduces also gradually until $t = 37.9839$ Gyr, then it exhibits obvious increase up to $t = 40.4032$ Gyr. Afterwards, it raises very fast as the matter distribution. The prominent increase of the four distributions in the interval about $37.9839 < t < 40.4032$ Gyr is owing to the change of the universe space from closed then curved

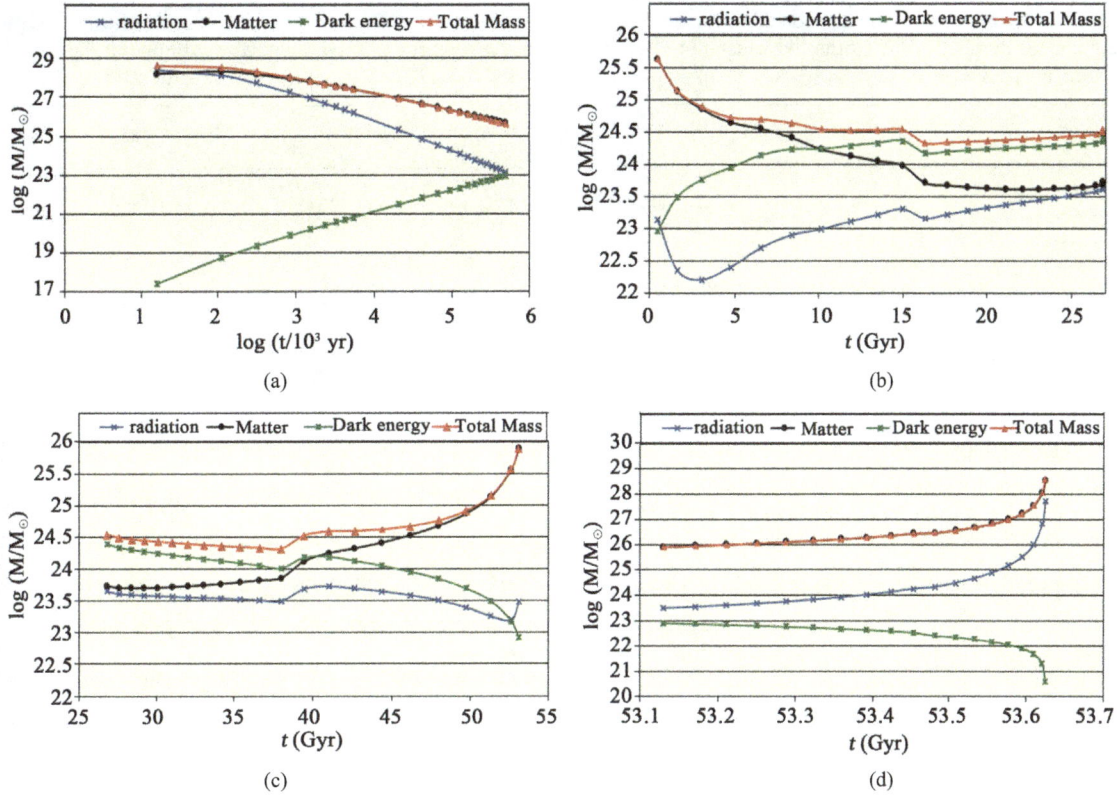

Figure 3. (a) The distribution of mass and energy in the closed cosmic model up to $t = 0.5$ Gyr; (b) The distribution of mass and energy in the closed cosmic model in the range $t = 0.5$ Gyr - t_{me}; (c) The distribution of mass and energy in the closed cosmic model in the range $t = t_{me}$ - t_*; (d) The distribution of the mass and energy in the closed cosmic model in the range $t = t_* - t_n$.

to flat in the range $39.3822 \leq t \leq 40.7521\,\text{Gyr}$. The distribution of mass and energy within the horizon volume of the universe in the closed model in the range $t = t_* - t_n$ is displayed in **Figure 3(d)**. The radiation distribution increases quite slowly until $t = 53.5032\,\text{Gyr}$, hence it starts raising appreciably fast.

The distribution of both matter and total mass coincide on each other and lie over the radiation distribution. The two distributions increase gradually up to $t = 53.5274\,\text{Gyr}$, then they raise up. However, the dark energy distribution decreases so slowly until $t = 53.5742\,\text{Gyr}$, afterwards it reduces substantially.

Estimations of $d_h(t), V_h(t), M_r(t), M_m(t), M_\wedge(t)$ and the equivalent number of the Coma-like clusters to the mass of matter within the universe horizon volume $N_{COMA}(t)$ in the closed cosmic model at special times are presented in **Table 1**. It is interesting to note that at $t = t_n = t_{bc} - 0.0005\,\text{Gyr}$, the horizon volume of the universe $V_h(t_n) = 0.0003(10\,\text{Gpc})^3$.

Since the radius of the Coma cluster is $r_{COMA} = 3.6\,\text{Mpc}$ [5], then $V_h(t_n) = 1.5738 \times 10^6 V_{COMA}$, where V_{COMA} is the Coma cluster volume. However, the mass of matter within the horizon volume of the universe at $t = t_n$ is $M_m(t_n) = 1.7455 \times 10^{13} M_{COMA}$. This indicates very clearly that the intercluster medium will disappear at $t = t_n$ and galaxy clusters will interfere with each other. Furthermore, the radius of the Milky Way galaxy is $r_{Mw} = 50\,\text{Kpc}$ [6]. Thus, $V_h(t_n) = 5.8742 \times 10^{11} V_{MW}$, where V_{MW} is the Milky Way galaxy volume. Nevertheless, it is found that $M_m(t_n) = 4.3638 \times 10^{16} M_{MW}$. Therefore, not only the intergalactic spaces will vanish at $t = t_n$, but also galaxies will collide and merge with each other to form extremely dense and close cosmological bodies. These very dense bodies will undergo further successive collisions and mergers under the action of central gravity, where the interstellar medium will vanish and the universe would develop to big crunch at $t_{bc} = 53.6251\,\text{Gyr}$.

It is also interesting to note from **Table 1** that the horizon distance of the universe at maximum expansion is $d_h(t_{me}) = 18.6192\,\text{Gpc}$. This horizon distance is in very good agreement with the maximum value of the universe horizon distance in the observed general cosmic model A, $d_h(t'_n) = 19.0103\,\text{Gpc}$, where $t'_n = 124\,\text{Gyr}$.

Table 1. Estimations of the horizon distance, horizon volume, mass of radiation, mass of matter, mass of dark energy and the equivalent number of the Coma-like clusters to the mass of matter within the universe horizon volume in the closed cosmic model at special times.

t	$d_h(t)$ Gpc	$V_h(t)$	$\log(M_r/M_\odot)$	$\log(M_m/M_\odot)$	$\log(M_\wedge/M_\odot)$	$N_{COMA}(t)$
t_{rm} (53221.5) yr	0.1356	$0.0209u_1$	28.4321	28.4321	18.3292	$1.352u_3$
$t_{m\wedge1}$ (10.1007) Gyr	12.9717	$18.2854u_2$	22.9777	24.2252	24.2252	$8.3970u_4$
t_o (13.7) Gyr	14.4420	$25.2348u_2$	23.2298	24.0515	24.3294	$5.6291u_4$
t_{me} (26.812.53) Gyr	18.6192	$32.8742u_2$	23.6535	23.7347	24.3906	$2.7141u_4$
$t_{m\wedge2}$ (40.2892) Gyr	14.0058	$23.0168u_2$	23.6758	24.1474	24.1474	$7.0198u_4$
t_n (53.6246) Gyr	0.9972	$0.0003u_2$	27.7134	28.5430	20.5951	$1.7455u_3$

where $t_n = t_{bc} - 0.0005$ Gyr , $u_1 = (\text{Gpc})^3$, $u_2 = (10 \text{ Gpc})^3$, $u_3 = 10^{13}$ and $u_4 = 10^8$.

The value of $d_h(t_{me})$ is also in high agreement with the values of $d_h(t'_n)$ in the other four general cosmic models as shown from **Table 1** in [7].

5. Conclusions

In this paper we have investigated the distributions of the universe horizon distance and the universe horizon volume in the closed cosmic model. It is found that the universe horizon distance distribution increases constantly for $t < t_{me}$ and decreases for $t > t_{me}$. However, the universe horizon volume distribution shows sudden reduction in the range $t = 0.5 \text{ Gyr} - t_{me}$ due to the change of the universe space from flat to curved then closed in the interval $15.1261 \text{ Gyr} \le t \le t_{me}$. On the other hand, this distribution exhibits abrupt raise in the range $t = t_{me} - t_*$ because of the change of the universe space from closed then curved to flat in the interval $39.3822 \le t \le 40.7521 \text{ Gyr}$.

Distributions of mass of radiation, matter and dark energy within the horizon volume of the universe were also investigated in the closed cosmic model. These distributions reveal similar noticeable changes as the universe horizon volume distribution for the same reasons. The mass of radiation dominates up to $t = 53221.5 \text{ yr}$, then the mass of matter becomes larger. Afterwards, both distributions of radiation and matter decrease while the distribution of dark energy rises until $t = 10.1007 \text{ Gyr}$, where the mass of dark energy prevails up to $t = t_{me}$. Hence, the distribution of dark energy reduces until $t = 40.2892 \text{ Gyr}$ where the mass of matter becomes prominent again. At $t = 53.6246 \text{ Gyr}$ the masses of both matter and radiation become appreciably high such that the intercluster space will vanish and clusters of galaxies will interfere with each other. Furthermore, not only the intergalactic medium will disappear, but also galaxies will collide and merge with each other to form extremely dense and close cosmological bodies. These very dense bodies will undergo further successive collisions and mergers under the action of central gravity, where the interstellar medium will vanish and the universe would develop to big crunch at $t_{bc} = 53.6251 \text{ Gyr}$.

References

[1]   Bukhari, F.A. (2013) A Closed Model of the Universe. *International Journal of Astronomy and Astrophysics*, **3**, 189-198. http://dx.doi.org/10.4236/ijaa.2013.32022

[2]   Bukhari, F.A. (2013) Cosmological Distances in Closed Model of the Universe. *International Journal of Astronomy and Astrophysics*, **3**, 199-203. http://dx.doi.org/10.4236/ijaa.2013.32023

[3]   Bukhari, F.A. (2013) Cosmological Distances in Five General Cosmic Models. *International Journal of Astronomy*

and Astrophysics, **3**, 183-188. http://dx.doi.org/10.4236/ijaa.2013.32021

[4]   Bukhari, F.A. (2013) Five General Cosmic Models. *Journal of King Abdulaziz University: Science*, **25**.

[5]   Ryden, B. (2003) Introduction to Cosmology. Addison & Wesley, Boston.

[6]   Schneider, P. (2010) Extragalactic Astronomy and Cosmology. Springer, New York.

[7]   Bukhari, F.A. (2015) Distribution of Mass and Energy in Five General Cosmic Models. *International Journal of Astronomy and Astrophysics*, **5**, 20-27. http://dx.doi.org/10.4236/ijaa.2015.51004

Permissions

The contributors of this book come from diverse backgrounds, making this book a truly international effort. This book will bring forth new frontiers with its revolutionizing research information and detailed analysis of the nascent developments around the world.

We would like to thank all the contributing authors for lending their expertise to make the book truly unique. They have played a crucial role in the development of this book. Without their invaluable contributions this book wouldn't have been possible. They have made vital efforts to compile up to date information on the varied aspects of this subject to make this book a valuable addition to the collection of many professionals and students.

This book was conceptualized with the vision of imparting up-to-date information and advanced data in this field. To ensure the same, a matchless editorial board was set up. Every individual on the board went through rigorous rounds of assessment to prove their worth. After which they invested a large part of their time researching and compiling the most relevant data for our readers.

The editorial board has been involved in producing this book since its inception. They have spent rigorous hours researching and exploring the diverse topics which have resulted in the successful publishing of this book. They have passed on their knowledge of decades through this book. To expedite this challenging task, the publisher supported the team at every step. A small team of assistant editors was also appointed to further simplify the editing procedure and attain best results for the readers.

Apart from the editorial board, the designing team has also invested a significant amount of their time in understanding the subject and creating the most relevant covers. They scrutinized every image to scout for the most suitable representation of the subject and create an appropriate cover for the book.

The publishing team has been an ardent support to the editorial, designing and production team. Their endless efforts to recruit the best for this project, has resulted in the accomplishment of this book. They are a veteran in the field of academics and their pool of knowledge is as vast as their experience in printing. Their expertise and guidance has proved useful at every step. Their uncompromising quality standards have made this book an exceptional effort. Their encouragement from time to time has been an inspiration for everyone.

The publisher and the editorial board hope that this book will prove to be a valuable piece of knowledge for researchers, students, practitioners and scholars across the globe.

List of Contributors

Francisco Frutos-Alfaro, Paulo Montero-Camacho and Miguel Araya
Space Research Center and School of Physics, University of Costa Rica, San José, Costa Rica

Javier Bonatti-González
Nuclear Research Center and School of Physics, University of Costa Rica, San José, Costa Rica

Trivedi Rajesh
Caterpillar Electric Pvt Limited, Delhi, India

Mohammed Adel Sharaf
Department of Astronomy, Faculty of Science, King Abdulaziz University, Jeddah, KSA

Abdel-naby Saad Saad
Department of Astronomy, National Research Institute of Astronomy and Geophysics, Cairo, Egypt
Department of Mathematics, Preparatory Year, Qassim University, Buraidah, KSA

Nihad Saad Abd El Motelp
Department of Astronomy, National Research Institute of Astronomy and Geophysics, Cairo, Egypt
Department of Mathematics, Preparatory Year for Girls Branch, Hail University, Hail, KSA

Fadel A. Bukhari
Department of Astronomy, Faculty of Science, King Abdulaziz University, Jeddah, KSA

A. F. Pugach
Main Astronomical Observatory of the NASU, Kiev, Ukraine

Bing Wang and Tao Wang
Institute of Fluid Physics, China Academy of Engineering Physics, Mianyang, China

Jing-Song Bai
Institute of Fluid Physics, China Academy of Engineering Physics, Mianyang, China
National Key Laboratory of Shock Wave and Detonation Physics, Institute of Fluid Physics, China Academy of Engineering Physics, Mianyang, China

R. C. Sahu
Department of Mathematics, K.S.U.B. College, Bhanjanagar, India

S. P. Misra
Department of Mathematics, Sri Jagannath Mahavidyalaya, Rambha, India

B. Behera
Department of Mathematics, U.N. College, Soro, India

L. C. Garcia de Andrade
Department of Theoretical Physics, State University of Rio de Janeiro (UERJ), Rio de Janeiro, Brazil

Antony Soosaleon and Blesson Jose
SPAP, Mahatma Gandhi University, Kottayam, India

Martin Tamm
Department of Mathematics, University of Stockholm, Stockholm, Sweden

M. Zafar Iqbal
Department of Physics, COMSATS Institute of Information Technology, Islamabad, Pakistan

Mamta Jain
Department of Mathematics, Shri Venkateshwara University, Gajraula, India

Rajiv Aggarwal
Department of Mathematics, Sri Aurobindo College, University of Delhi, Delhi, India

Beena R. Gupta
Department of Mathematics, Lakshmibai College, University of Delhi, New Delhi, India

Vinay Kumar
Department of Mathematics, Zakir Husain Delhi College, University of Delhi, New Delhi, India

Eugene Terry Tatum
760 Campbell Ln. Ste. 106 #161, Bowling Green, USA

U. V. S. Seshavatharam
Honorary Faculty, I-SERVE, Alakapuri, Hyderabad-35, Telangana, India

S. Lakshminarayana
Department of Nuclear Physics, Andhra University, Visakhapatnam, India

Baha T. Chiad and Abdhreda S. Hassani
Department of Physics, College of Science, University of Baghdad, Baghdad, Iraq

Lana T. Ali
Department of Astronomy and Space, College of Science, University of Baghdad, Baghdad, Iraq

Raul G. E. Morales
Centre for Environmental Sciences and Department of Chemistry Faculty of Sciences, Universidad de Chile, Santiago, Chile

Carlos Hernández
Department of Chemistry, Faculty of Basic Sciences
Universidad Metropolitana de Ciencias de la
Educación, Santiago, Chile

Ahmed Mostafa
Department of Mathematics, Faculty of Science, Ain
Shams University, Cairo, Egypt

Jagadish Singh
Department of Mathematics, Faculty of Science, Ahmadu
Bello University, Zaria, Nigeria

Joel John Taura
Department of Mathematics and Computer Science,
Federal University, Kashere, Nigeria

Morley B. Bell
National Research Council of Canada, Ottawa, Canada

Isao Satō
Nihon University, Sakata, Yamagata, Japan

Hiromi Hamanowa
Hamanowa Astronomical Observatory, Motomiya,
Fukushima, Japan

Hiroyuki Tomioka
Hitachi, Ibaraki, Japan

Sadaharu Uehara
KEK, Tsukuba, Ibaraki, Japan

Shubha S. Kotambkar
Department of Applied Mathematics, Laxminarayan
Institute of Technology, Rashtrasant Tukadoji Maharaj
Nagpur University, Nagpur, India

Gyan Prakash Singh
Department of Mathematics, Visvesvaraya National
Institute of Technology, Nagpur, India

Rupali R. Kelkar
Department of Applied Mathematics, S. B. Jain Institute
of Technology, Management and Research,
Nagpur, India

Mohamed S. El Naschie
Department of Physics, University of Alexandria,
Alexandria, Egypt

Sherif Mohamed Khalil
Physics Department, Faculty of Science, Princess Nora
Bent Abdurrahman University, Riyadh, Saudi Arabia
On Leave from Plasma Physics & Nuclear Fusion
Department, N.R.C., Atomic Energy Authority, Cairo,
Egypt

Weeam Saleh Albaltan
Physics Department, Faculty of Science, Princess Nora
Bent Abdurrahman University, Riyadh, Saudi Arabia

Jehangir A. Dar
Plasma Waves and Particle Acceleration Laboratory,
Department of Physics, Indian Institute of Technology
Delhi, New Delhi, India

Pawan Kumar Singh
Department of Physics, ARSD College, University of
Delhi (South Campus), New Delhi, India

Ram Swaroop
North Bengal Science Centre, National Council of Science
Museums, Ministry of Culture, Government of India,
Siliguri, India

Yikdem Mengesha Gebrehiwot
Astronomy and Astrophysics Research Division, Entoto
Observatory and Research Center (EORC),
Addis Ababa, Ethiopia

Solomon Belay Tessema and Oleg Malkov
Institute of Astronomy, Russian Academy Sciences,
Moscow, Russia

Mahbubur Rahman Mollah
Department of Mathematics, Commerce College
Kokrajhar, Kokrajhar, BTC, Assam, India

Kangujam Priyokumar Singh
Department of Mathematical Sciences, Bodoland
University, Kokrajhar, BTC, Assam, India

Koijam Manihar Singh
Department of Mathematics Sciences, National Institute
of Technology Manipur, Imphal, India

Fadel A. Bukhari
Department of Astronomy, Faculty of Science, King
Abdulaziz University, Jeddah, KSA